Emerging Trends in Mechatronics

Series Editor

Aydin Azizi, Oxford, UK

This book series helps understand how Mechatronics will impact the practice and research of developing advanced techniques to model, control and optimize complex systems. It presents recent state of advances in mechatronics and related technologies including: Automatic Control, Robotics, Agent Based systems, Smart Manufacturing and Industry 4.0. The selected topics give an overview of the state-of-the-art and present new research results and prospects of the future development in the interdisciplinary field of mechatronic systems. The book series will provide up-to-date and useful knowledge for researchers and engineers involved in mechatronics and related fields.

Within the above dimensions, the scope of the book series welcomes high quality original research and review articles that cover a broad range of topics related to theoretical and applied aspects of modern mechatronics. Potential topics include, but are not limited to the following:

Controller design

Many mechatronic systems have challenging control difficulties due to the high nonlinear structures and time-varying dynamic behaviors. In addition to these, there are external disturbances, which cannot be predicted and change according to uncontrolled working environments. Conventional controllers are insufficient to solve the problems and to compensate the environmental disturbances, therefore designing efficient controllers for complex systems has a great importance in the current book series.

Sensor Design & Data Collection Approaches

Essential part of a mechatronics system is the measurement system that senses the variations in the physical parameters, such as temperature, pressure, displacement, and so on, and converts it to voltage or current. In mechatronic systems, accurate measurement of the dynamic variables plays a vital role for the actuators to function properly, therefore on of the aims of this book series, to focus on understanding what should be done to design accurate sensors and efficient data collection methods to improve the performance of mechatronic systems.

Optimization Techniques

Designing a complex mechatronic system involves multiple variables, objectives, constraints, and evaluation criteria as well as their nonlinearly coupled relationships. One of the challenging research areas of this book series is utilizing optimization techniques to design an optimal mechatronic system.

Artificial Intelligence

From the point of view of dynamics, mechatronic systems can be characterized by model uncertainties, high nonlinearities, complicated coupling, and stringent performance requirements, among others. To deal with these facts, this book series would like to focus on investigating and utilizing Artificial Intelligence techniques such as Neural Networks, Fuzzy logic, Neuro-fuzzy, evolutionary techniques etc. to model, design and optimize complex mechatronic systems.

Robotics

In the field of Robotics, there is a need of focusing on kinematic, dynamic, control and optimization of innovative robotic systems. This book series can introduce new and novel robot designs such as hybrid (serial-parallel) redundant manipulators that have a high potential of application in special fields like space robots. State of the art researches on selection of optimum kinematic solutions and path planning based on such an optimization, dynamic analysis and new control methods developments and application of artificial intelligence in this field can be published in this book series.

Smart manufacturing

Application of networked sensors and machinery equipment using internet and intranet in order to monitor and automation of the manufacturing process and improve the performance efficiency and is a point of interest. Advanced robotics, multi input-multi output control approaches combined with the artificial intelligence and optimization techniques, big data processing approaches, multi-scale modelling and simulation of dynamic sub-systems etc. are the hot topics that will be focused in this book series.

AGVs

Automated Guided Vehicle (AGVs) are going to be very popular is industry logistics, surveillance, defense and social applications in order to make processes safer and more cost efficient. Innovative AGVs designs for various missions, Robot Operating System usage on AGV control, Indoor and outdoor navigation and control algorithms, intelligent path planning techniques, sensor fusion, simultaneous application of GPS, INS and offline maps are the topics that are planned to be published in this book series.

Aydin Azizi
Editor

Mechatronic Innovations in Sustainable Energy

Biofuel Systems and Thermal Storage Technologies

Springer

Editor
Aydin Azizi
School of Engineering, Computing
and Mathematics
Oxford Brookes University
Oxford, UK

ISSN 2731-4855 ISSN 2731-4863 (electronic)
Emerging Trends in Mechatronics
ISBN 978-981-95-5862-9 ISBN 978-981-95-5863-6 (eBook)
https://doi.org/10.1007/978-981-95-5863-6

This Springer imprint is published by the registered company Springer Nature Singapore Pte Ltd.
The registered company address is: 152 Beach Road, #21-01/04 Gateway East, Singapore 189721, Singapore

Preface

The proposed book, *Mechatronic Innovations in Sustainable Energy: Biofuel Systems and Thermal Storage Technologies*, brings together a diverse collection of pioneering studies focused on renewable energy generation and energy storage. The central aim of this work is to explore how mechatronic systems and advanced engineering principles can be applied to transform waste resources into valuable fuels and improve the performance of solar-based energy storage technologies. The volume serves as both a technical reference and an inspirational resource, showcasing practical applications, design methodologies, and experimental findings that highlight the role of engineering innovation in addressing global sustainability challenges. From biomass pyrolysis for biofuel production to the use of phase change materials for efficient solar energy storage, this book demonstrates how multidisciplinary approaches can lead to impactful and environmentally responsible energy solutions.

By drawing on research conducted at the German University of Technology in Oman and by experts in process and energy systems engineering, the book integrates theory with practical case studies that reflect real-world challenges and solutions. It emphasizes the importance of bridging mechanical, thermal, and control engineering in creating sustainable energy technologies that are not only innovative but also scalable and applicable to industry. The editor envisions this work as a resource that will inspire researchers, students, and practitioners to explore new pathways toward achieving energy efficiency, sustainability, and resilience.

Oxford, UK Aydin Azizi

About This Book

This book presents a focused exploration of sustainable energy technologies, with particular emphasis on biofuel production from agricultural and biomass waste through pyrolysis, and on the enhancement of thermal energy storage in solar systems using phase change materials (PCMs). The chapters collectively address key challenges and opportunities in converting waste resources into valuable energy carriers and improving the efficiency of renewable energy systems. Several chapters examine the thermochemical conversion of a wide range of feedstocks, including sugarcane waste, corn stover, rice straw, algae mixed with plastics, and almond hulls. Through detailed experimental investigations and analytical studies, these contributions highlight the versatility and potential of pyrolysis as a sustainable pathway for renewable fuel production. The book also covers advanced solar energy applications, with particular attention to the integration of PCMs for enhanced thermal energy storage. Topics include innovative uses of PCMs in solar collector systems and the design and analysis of thermal storage solutions for solar hot water applications. These studies demonstrate how advances in materials science, system design, and mechatronics can be combined to deliver efficient, reliable, and sustainable energy solutions. Overall, this volume provides a cohesive and interdisciplinary perspective on waste-to-energy technologies and thermal energy storage innovations, contributing to broader global sustainability goals. It is intended for researchers, engineers, graduate students, and industry professionals who seek to deepen their understanding and actively contribute to the development of sustainable energy systems.

Proposed Chapters

Chapter "Production of Biofuels from Pyrolysis of Sugarcane Waste": Production of Biofuels from pyrolysis of sugarcane waste
Chapter "Biofuel Production from Corn Stover by Pyrolysis": Biofuel production from corn stover by pyrolysis

Chapter "Thermochemical Conversion of Rice Straw to Produce Biofuel": Thermochemical conversion of rice straw to produce
Chapter "Production of Biofuel from Co-pyrolysis of Algae and Plastic Wastes": Production of biofuel from co-pyrolysis of algae and plastic wastes
Chapter "Production of Biofuels from Almond Hull Wastes": Production of Biofuels from Almond Hulls Wastes
Chapter "Applications of a Phase Change Materials (PCMs) in Solar Collector Systems": Applications of a Phase Change Materials (PCMs) in Solar Collector Systems
Chapter "Design of an Energy Storage System Using PCM for a Solar Hot Water System": Design of an energy storage system using PCM for a solar hot water system.

Contents

Production of Biofuels from Pyrolysis of Sugarcane Waste

Weam Al-Dughaishi and Seyed Mojtaba Sadrameli

Abstract The study focuses on the process of generating biofuel from rural waste that is made from sugarcane bagasse by pyrolysis. The objective is to enhance pyrolysis to generate advanced biofuels and enhance sustainable energy production for its global demand. An overview of biofuels includes characterization and significant effect in the field. The report also outlines the setup procedure, experimental setup, characteristics of materials, and worldwide production trends. Temperature, particle size, bed height, heating rate, and N_2 flow rate are adjusted to enhance pyrolysis efficiency. The study conducted three case studies focusing on certain conditions for a higher production of bio-oil yield. Therefore, the highest pyrolysis conditions are temperature 500 °C, particle size 0.3 m, bed height 16 cm, bio-oil 44.23 wt%. In the second case temperature 550 °C, 500 mm particle size, heating rate of 20 °C/min, with produced 43.32 wt%. Furthermore, in the third case study, at 500 °C, the bio-oil yield boosted to 45 wt% compared to the previous case, with a wide range of particle sizes, and heating rate of 30 °C/min. Also examining bio-oil extraction yields, biochar, and biogas. A maximum amount of bio-oil had been produced in case 2 that was 60.4 wt%, and in case three established the highest biogas and biochar that was 36 wt%. The following study focuses on environmentally friendly procedures and alternative resources that can be considered renewable by examining sugarcane bagasse, for example a source intended for biofuel production process. The outcomes associated with using biofuels instead of conventional energy sources might prevent the emission of greenhouse gas and minimize environmental degradation.

Keywords Sugarcane bagasse · Biofuel · Pyrolysis · Renewable · Biomass wastes

---

W. Al-Dughaishi · S. M. Sadrameli (✉)
Department of Engineering, German University of Technology in Oman, Muscat, Oman
e-mail: seyed.sadrameli@gutech.edu.om

A. Azizi (ed.), *Mechatronic Innovations in Sustainable Energy*, Emerging Trends in Mechatronics, https://doi.org/10.1007/978-981-95-5863-6_1

1 Introduction

1.1 Overview

Biofuel production from biomass waste has been observed to be an effective replacement for renewable energy sources particularly due to their economic plus environmental benefits [1]. We can convert different biomass sources from sewage sludge, agricultural, or food wastes into biofuel using different methods that include thermochemical treatment, and anaerobic digestion [2]. Consequentially, pretreatment techniques such as acid and alkaline treatments are fundamental to enhance the conversion efficiency of lignocellulosic biomass into biofuels. There exist several processes for producing biofuels such as pretreatments, hydrolysis, purification, and fermentation [3]. Promising outcomes have been observed in experimental case studies. Sugarcane bagasse with hemicellulose and cellulose is an exceptional fermentation environment for the manufacture of biofuel. The social, environmental, and economic properties have been examined in feasibility studies. To achieve effective biomass conversion of sugarcane into biofuels there is a necessity of availability of sugarcane, pretreatment, and manufacturing processes to develop biofuel production. Nevertheless, there are still hurdles that prevent biofuels from being economically competitive with fossil fuels. Future studies should focus on improving yields and production procedures to improve sustainability.

1.1.1 Annual Global Production

Sugarcane is a significant crop in agriculture that produces around 279,000,000 t containing leftover biomasses, such as bagasse and trees as well with an annual global production of 1.6 billion tons. Brazil leads the world in sugarcane production, other significant producers involve Indonesia, China, and Thailand [4].

1.1.2 Composition of Sugarcane Bagasse

This waste is a potential source of biofuel development. The composition of sugarcane bagasse involves a substance called cellulose (42–43%), cellulose derivative which is hemicellulose with (24–25%), and lignin with (20–22%) [5]. Sugars are extracted from sugarcane, so sugarcane waste is an important by-product of the sugar industry. Sugarcane bagasse is used to produce biofuel due to its high organic content, making it appealing raw source for lignocellulose and used to make biomass. Second generation biofuels made from lignocellulosic biomass can provide sustainability by eschewing the usage of food crops. Additionally, sugarcane bagasse is one of the major wastes from agricultural worldwide circulation, so making it simple to collect, therefore, the quantity of it is collected to be stored [6]. Agriculture waste is considered the most important renewable sources among lignocellulose; accordingly, sugarcane bagasse has been able to provide its benefits over other agricultural wastes including:

1. High quality of organic material.
2. Produced by the integrated mill, which makes collection and storage more efficient.
3. A small amount of water [7].

1.1.3 Pyrolysis of Sugarcane

Producing biofuels as an alternative to fossil fuel and as it was considered environmentally friendly. Biofuel has the possibility to improve the global economy while lowering pollution. Hence, pyrolysis of sugarcane bagasse is a strategy for manufacturing biofuel from agricultural leftovers [8]. The approach process is to produce biofuel through lignocellulosic bioconversion in the absences of oxygen. Pyrolysis of sugarcane bagasse can produce more than 45 wt% biofuel by pyrolysis of SCB at 500 °C, 0.5–1 mm size of particle, and 200 cm^3/min flow [9]. Figure 1 illustrates fast pyrolysis to generate bio-oil from sugarcane:

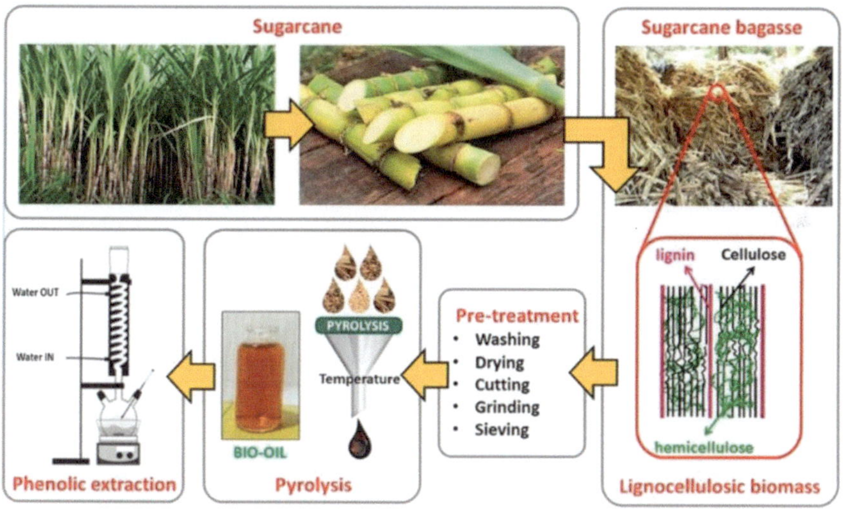

Fig. 1 By applying fast pyrolysis to produce bio-oil from sugarcane bagasse [10]

1.2 Motivation for Research

The motivation behind the research was that transforming biomass waste into biofuel would significantly decrease the release of greenhouse gases, minimize climate change, and lessen fossil fuel exhaustion.

1.3 Aim of the Research

This research project aims to explore the prospects and potential of converting biomass to biofuel as a substitute for fossil fuel. The purpose of the research is to emphasize the capability of sugarcane bagasse and its role as a renewable energy source.

1.4 Objective of the Research

The objective of the research is to provide the technology of converting agricultural waste into biofuel as a viable energy source. Its objective is to study the alternatives to traditional energy and fuel sources through the production of biofuel from biomass waste. Therefore, I will study producing biofuel from sugarcane bagasse since it among the biomasses that have been used most frequently in past several decades. Using biorefinery techniques, SCB can be transformed into a variety of biofuel products with additional potential. Several pretreatment techniques are capable of effectively breaking SCB to create biofuel and other useful compounds, through several procedures which involve:

- Pretreatment
- Hydrolysis

The research examines the energy efficiency of producing biofuel from pyrolysis without using oxygen from sugarcane material and refining the biofuel by hydrotreatment.

1.5 Limitation of Research

The limitation of the research is associated with the limited availability of seasonal biomass. Biofuel is limited to waste from farming due to the variety of raw materials, the need for higher productivity and sustainability, and the difficulties of large-scale manufacturing. However, sugarcane constructed biofuel production is not economically sustainable, green technology in business sector has been restricted to practical application.

1.6 Report Outline

The first chapter of the report is the introduction to the biofuel production using sugarcane waste mostly bagasse as it is an environmentally friendly substitute for fossil fuel. Objective of the research is to produce biofuels by effectively breaking sugarcane bagasse into beneficial products. In Chap. 2 the literature review focuses on the procedure of converting sugarcane bagasse industrial waste into biofuel. Chapter 3 studies an experimental case study related to the manufacture of biofuel from sugarcane bagasse. The case studies examine the conversion of SCB into biofuel in different countries. In Chap. 4, the comparison of the results from case studies examines the difficulties associated with creating biofuel from sugarcane bagasse in several countries and provides ways to improve the process's effectiveness. In conclusion, research paper reviews the current state, challenges, and future prediction of biofuel production from SCB considering technological and environmental concerns.

2 Literature Review

2.1 Overview

This chapter reviews several papers related to the production of biofuels from the pyrolysis of sugarcane bagasse, going into details for the different process methods of converting biomasses using pyrolysis techniques and enhancing pyrolysis for producing bio-oil. In addition to pretreatment used and catalyst engaged. Studying the chemical composition and heating properties of bio-oil formed by catalytic pyrolysis of sugarcane bagasse which is known as a renewable resource rich in carbon that can be further treated into bio-oil is the objective of this section.

2.2 Review of Experimental Studies

Miranda et al. [11] investigated that biomass pyrolysis can produce a variety of sustainable fuels such as biochar (solid), bio-oi (liquid), syngas (gas). It has concerned attention and its production of biofuels. Sugarcane bagasse is mainly manufactured in Thailand, Brazil, India, which represents a byproduct of the sugaralcohol industry, is mostly reflected popular good biomass for pyrolysis due to the quantity, present underutilization, and quality. The product characteristics, potential distribution, and distribution are impacted by individualities of sugarcane bagasse and by the pyrolysis employed in the future use of chemicals or fuels. In this research, more investigation is required for this issue, and a comprehensive summary of biomass and pyrolysis concepts are discussed. A method of using sugarcane bagasse is to enhance the process of manufacturing goods like gas, charcoal, and bio-oil. This report examines the connection between this biomass type and process operating conditions to determine how the characteristics and yield of the products are affected. By the given feedstocks and attributes, this study efforts to evaluate the concepts, product attributes of pyrolysis using sugarcane bagasse, and operational outcome. In several producer countries, Thailand, Brazil, India uses sugarcane bagasse as a significant raw material. These products yield, heating value, and composition can be determined by affecting the properties of the biomass and process parameters, bagasse pyrolysis can produce a variety of products that can be adapted into fuels and chemicals. This is a summary of how pyrolysis testing should be considered to achieve high production yield while with a lower amount of energy. Sugarcane bagasse pyrolysis can create different product types such as biochar up to 83%, gases up to 65%, and bio-oil up to 85%. Therefore, it's required to choose operating parameters depending on the process purpose. Figure 2 illustrates the potential production of sugarcane in the top 10 producing countries in 2019.

In this research paper Wibowo et al. [12] examine the various circular uses of trash and byproducts through sugar sector in a variety of purpose in the industry. Sugarcane molasses was decolored by activated carbon, which is composed of bagasse fly ash and sugarcane bagasse. Utilizing steam at 700°, 800°, and 900°. The investigation demonstrated that the adsorption capacity was drastically impacted by the adsorbent's

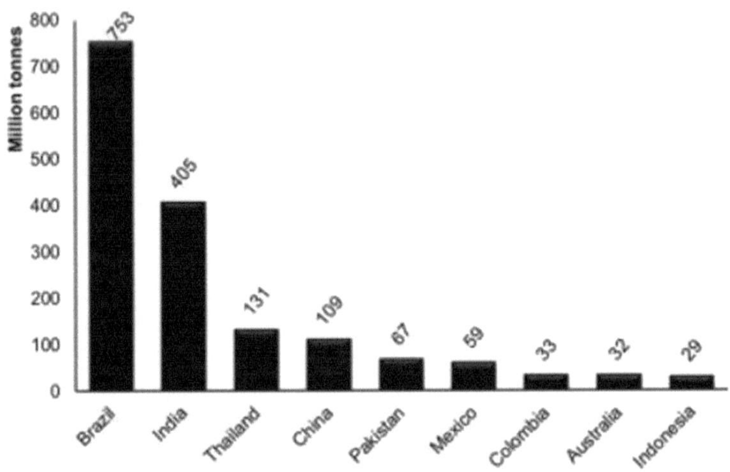

Fig. 2 Sugarcane production of the most 10 popular manufacturers in 2019 [11]

pore size. With an 87% reduction in efficiency and a 52% reduction in efficiency after thermal regeneration, initiated carbon developed from fly residue activated at 900 °C displayed the highest level of decolorization. The discarded adsorbents were evaluated for their regenerative characteristics and dying methods. Many absorbents were suitable for a regeneration cycle and a substantial amount of bulk and capacity.

Specified that the used adsorbent experienced a higher temperature of 17 MJ/kg, the energy recovery process appeared positive. Initially before storage as geopolymer to prevent contamination, a test regarding the energy recovery process fly ash the potential use in carbon dioxide capture. It is predicted that this review uses adsorbents based on bagasse fly ash for the decolorization of sugarcane syrup and their likely circular application. Figure 3 shows that for producing AC, bagasse from sugarcane alongside fly ash were used for energy recovery, syrup decolorization, production of geopolymers, and subsequent CO_2 adsorption. The results from this analysis are based upon the most efficient color removal efficiency from sugarcane syrup created by FAC900, which is 87%, then FAC800 by 75%, and FAC700 by 54%. Secondly, FAC900 demonstrated a reduction of 60% of the thermal regeneration of adsorbents spent, which may be recommended for a regeneration cycle. Thirdly, like wood pellets, spent adsorbents had an HVV of nearly 16.8–17.4 MJ/kg, according to the energy recovery by combustion. In addition, tests made using energy recovery processes for CO_2 adsorption presented a 0.2–0.4 mmol-CO_2/g-FA capacity and proved the process had no major effect by the steam activation process. Lastly, by mixing 10% fly ash with coal after twenty-eight days of drying, fly ash starting the energy recovery process formed a geopolymer besides a compressive strength of 25.6 MPa was created.

In an experiment conducted by Charusiri and Vitidsant [13] it was found that biofuels were developed by harvesting crops to obtain lignocellulosic biomass and sugarcane leaves as a means of pyrolysis feed through which making it easy to

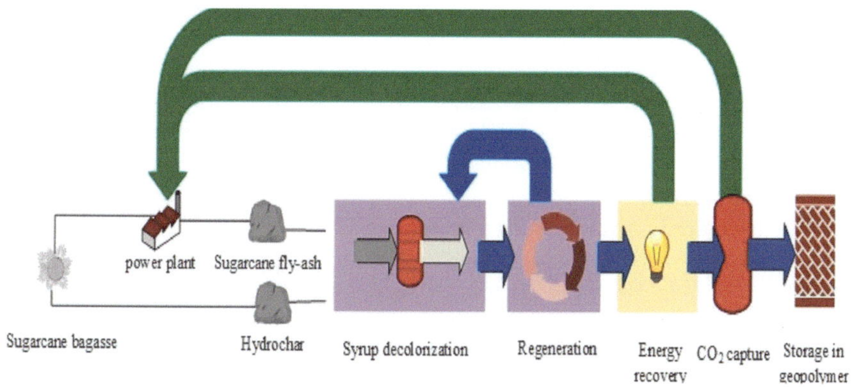

Fig. 3 Sugarcane bagasse fly-ash for syrup decolorization and their capability circular application [12]

manufacture and transport pyrolyzed oil. Considering yield and the pyrolyzed products which were separated into water and bio-oil fractions are studied in this report. Pyrolysis process was carried out using a screw-driven via pyrolysis reactor. The consequences of pyrolysis variables such as feedstock rate (0.3–1.8 kgh^{-1}), average size distribution (250, 500, 750 μm), nitrogen gas flow rate (80–240 cm^3/min), and temperature (400–650 °C) were thoroughly examined. The results from temperature and residence time as resolute by the nitrogen influence the yield of bio-oil and its characteristics, including its heating value, viscosity, chemical composition, and acidity. Achieving 500 μm is the average particle size of biomass, the maximum bio-oil yield at 40.16 wt%, and N_2 flow rate of 120 cm^3 min^{-1} when temperature is at 500 °C and 0.4 kg/h feed rate. Also included studies of gas chromatography with mass spectrometry by analyzing the chemical composition that classifies aromatics derivatives, oxygenized, molecules, and ketones that are appreciated chemical products. As result of this, it concludes that pyrolyzed oil needs to be processed before using it directly as a biofuel. Furthermore, oil needs to go into a co-catalytic method of pyrolysis that can be measured as important goods. It was performed in a required evaluation if the biochar could be utilized to generate activated carbon. Summarizing the report, it included pyrolysis parameters that are enhanced to create an appropriate distribution of SCL products. Chromatographic and physicochemical are used to identify factors that affect the manufacture of bio-oil. Bio-oil physicochemical features suggest that an approach for turning heavy solid into biomass interested in a moveable liquid for use as renewable energy source is pyrolysis of SCL. Findings indicated that combustion bio-oil HHV exceeded SCL raw material. Throughout the findings, fast pyrolysis oil was used to provide a range of chemical recovery options while it is possible to use slow pyrolysis as oil. However, it was discovered that corrosion could be caused by bio-oil generated through this pyrolysis outstanding by the presence of organic acids. Therefore, pretreatment is necessary for the pyrolysis oil to lower its acidity. Moreover, the solid chars can be tested as a source for activated carbon production and in fuel applications. Figure 4 displays the setup of the experiment for production of biofuel.

Pan et al. [7] observed the sugarcane bagasse is the most popular material after lignocellulosic material, which is used to produce biofuel due to its considerable amount of organic material, yield of biomass and ease of gathering and storing area. The chemical pathway for SCB-founded biofuel synresearch is an environmentally valuable approach. However, several technological challenges aimed at pretreatment of sugarcane hydrolysate, microbial resistance to inhibitors, purification, and biofuel transformation increased concerns about the long-term feasibility study about green technology. Through the development of economical pretreatment processes and engineering varieties that can permit inhibitors and efficiently transfer sugarcane bagasse obtained components, these challenges were overcome by researchers. The recent situation and issue surrounding the transformation of SCB across biofuels was discussed in this study. The feasibility of the primary environmental and economic development of sugarcane bagasse as fuel to produce biofuel is examined. In addition, focus is on the recent gaps of knowledge and the cost-effective solution to make biofuel. For producing biofuel SCB is a suitable feedstock for the manufacture of

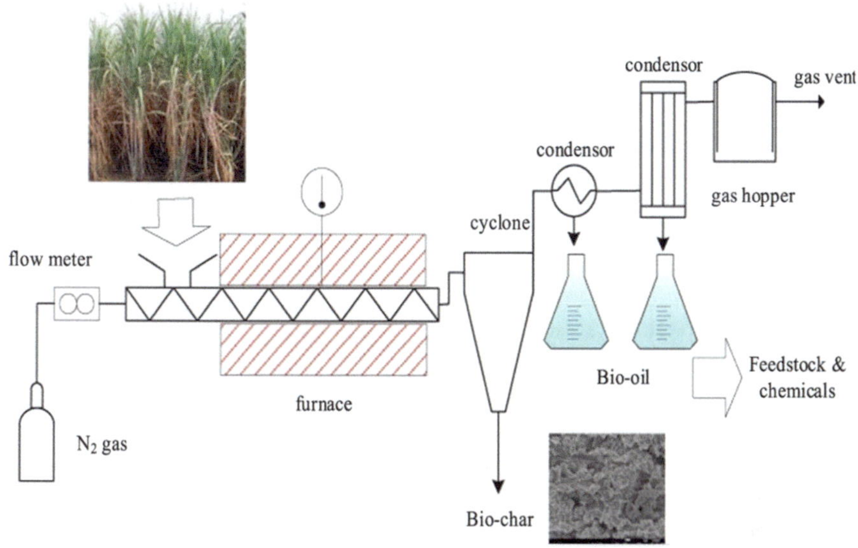

Fig. 4 Biofuel production through the pyrolysis of sugarcane and produced biochar for production of activated carbon [13]

value-added products. A review of various manufacturing possibilities of biofuels and its challenges with developments in process technologies are presented. The limited production of SCB must be improved to raise production level due to certain aspects which involves substrate compositions, toxic inhibition, inoculum, and types of selected conversion. Figure 5 shows the graphical abstract for production of biofuel from SCB.

Wani et al. [14] researched the challenges and ability for sugarcane bagasse that are both safe and sustainable. Sugar is extracted from the tall, perennial grass known as sugarcane that has the characteristics of fibrous stalks that are heavy in

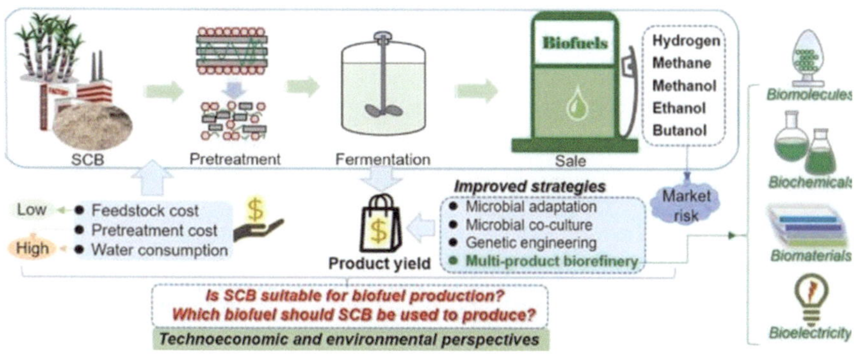

Fig. 5 Biofuel production from sugarcane bagasse [7]

sucrose. These sugarcanes are used to produce biofuels most commonly. Nevertheless, the waste from sugarcane processing and bagasse have a negative impact on the environment and cause safety hazards. It's crucial to take safety and sustainable processing methods for manufacturing biofuels. In addition to phytochemicals, sugarcane and its byproducts, phytosterols, flavonoids, phenolic acids, fatty acids, and terpenoids, phytochemicals, sugarcane and its byproducts flourish. Many potential microorganisms for industrial use comprise populations of bacteria and chemical supplementary among sugarcane living soil and surface. Microorganisms were identified and categorized using metagenomics and meta-omics methods to eliminate bagasse concerning primary source of lignin-based biomass. More precious sugarcane goods are black treacle that are flavored and bagasse, which is beneficial for producing paperboard, agricultural mulch, reconstituted panelboard, and kiln fuel. We can employ sugarcane in various applications such as animal feed additives, generation of electricity and biofuel production. The review highlights the need for effective and safe processing processes for producing bioactive compounds and biofuels from sugarcane and its bagasse. One of the biggest problems facing the sugarcane industry is the bioprospecting of sugarcane phytochemicals into products that are critical to industry. Due to the biological effectiveness of sugarcane, there has been a greater demand for it. Therefore, there was an increase in the intake of goods correlated to sugar, those caused the production of waste. Figure 6 illustrates the process of extracting biofuel, animal feed, and packing material from sugarcane waste, and the extraction of anticancer, antimicrobial, and nutraceutical from Saccharum Officinarum.

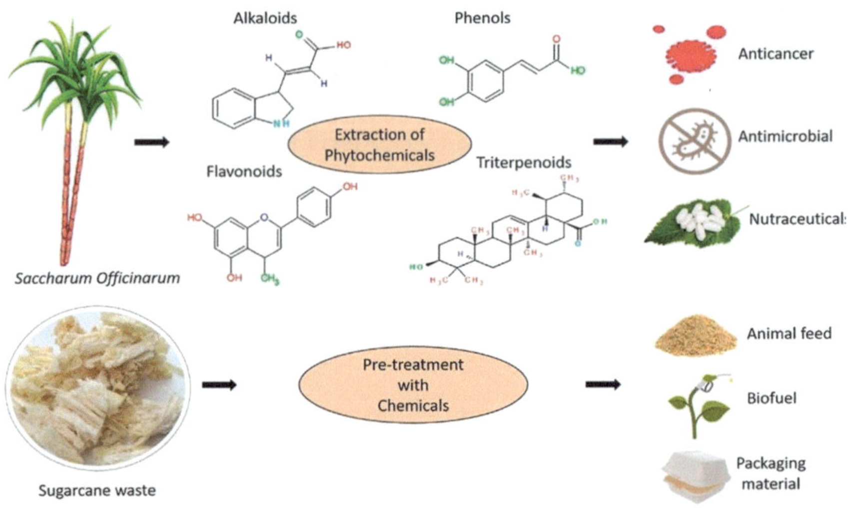

Fig. 6 Extraction of biofuel from sugarcane waste pretreatment and production of anticancer, antimicrobial, and nutraceutical from Saccharum [14]

Sitthikitpanya et al. [15] focused on the possibility of producing biofuels and additional value goods based on the economic and zero-waste principles from the microalgal biomass and sugarcane leaf co-digesting. In addition, a continuous stirred tank reactors HRT for continuous H_2 production is analyzed. A 12 Hrs HRT is ideal for both hydrogen generation and yield. The 43.3% streptococcus and 31.3% clostridium genera are among the numerous ecosystems uncovered by microbial research utilizing the Illumina platform for 16S rRNA amplicon sequencing. The procedure generates methane and hydrogen from leftover materials such as solid residue, acidified slurry, hydrogenic effluent, and yields a total energy of ~22.2 kJ/g VS. Vermicompost and densified waste-derived fuel are made from the solid residue excess after hydrogen synresearch, which is composting. Figure 7 shows that process reduces waste at every stage of the process by presenting the actual consumption of raw supplies in a zero-waste and circular economy environment. In this study, HRT was examined, which is considered one of several factors that regulate the increase in hydrogen production in a CSTR.

Continuous HRT has an impact on CSTR's hydrogen production. HRT produced hydrogen from hydrolyzed chlorella biomass for 12 h and sugarcane leaves.

The alkaline treatments impact on raw materials was inspected by Muñoz et al. [16] Sugarcane bagasse is a relatively distributed remaining pyrolysis derived from the sugar industry. The residues were treated with Na or KOH and rinsed to eliminate the hydroxides. Depending on the pyrolysis conditions employed, different fractions will have different spectra. The study explores the benefits that these methods can provide in bagasse valorization by contrasting the impact of fundamental pretreatment and the presence of alkaline ions catalytic action through pyrolysis on the compounds produced. According to the results, alkaline particles in the sample during the pyrolysis process significantly reduces methoxy groups, levoglucosane, and the

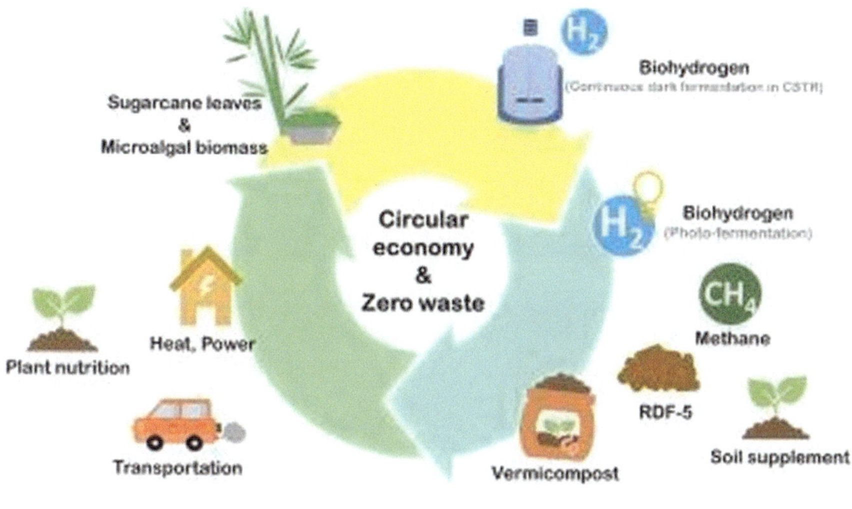

Fig. 7 Circular economy and zero-waste environment [15]

vinyl in volatiles acquired, whereas increasing the percentage of charm, cyclopentenone and its byproducts, and the quantity of CO_2 developed. In addition to alkaline saturation, the cleaning process until natural pH changes the range of pyrolytic products created, substantially improving the fraction of hydroxiacetaldehyde and levoglucosane while reducing phenol product. As an outcome of modifications to the bagasse structure and breakdown mechanism are made to increase carbon yield by producing activated C in a single stage, reducing O_2 subject matter in the subsequent natural oil to increase heating value, and narrowing its spectrum of pyrolytic compounds to improve the process selectivity. Observed pyrolysis is used to enhance sugarcane bagasse and create valuable products such as charcoal and volatiles. By slow pyrolysis, 18% biomass was produced from sugarcane bagasse. The number of catalysts in the model increases greatly. The percentage of biochar increases to 26 and 41% when sodium hydroxide and potassium hydroxide are used in a ratio of 1:1. To create activated carbon in a single step carbonization and chemical activation are combined, enhancing carbon yield below ideal operating conditions. Table 1 describes the various parameters that are used to characterize samples.

In this study Saif et al. [17] looked at the semi-batch reactors pyrolysis of sugarcane bagasse to ascertain an optimal condition for producing bio-oil. The pyrolysis processes parameters were all transformed such as temperature from 350 to 600 °C, sugarcane bagasse size 0.25–2 mm, and N_2 flow rate from 100 to 500 cm^3/min, correspondingly during the pyrolysis process. According to experiment tests various pyrolysis conditions formed char (24–36%), gas yield (23–37%), and oil (38–45%). The results of this investigation indicated that pyrolysis is a key component in product distribution. At 45 wt%, the highest pyrolysis bio-oil production was accomplished, with 500 °C 200 cm/min N_2 flowrate, and 0.5–1 mm particle size. Therefore, as pyrolysis temperature increases, gas yield will increase, but biochar yield will decrease. Consequently, bio-oil production is expected to reach 500 °C and

Table 1 Description of elemental composition of bagasse by applying a variety of parameters [16]

	S0	SK	SNa	SKW	SNaW
C (%)	47.6	21.9	26.5	44.2	44.5
H (%)	6.08	2.7	3.1	5.5	5.7
O (%)*	44.7	22.8	26.8	43.6	44.5
N (%)	0.21	0.1	0.12	0.25	0.23
Moisture (%)	3.8	8.4	8.3	7.0	7.4
Ash (%)	1.4	52.5	43.5	6.5	5.1
Volatile matter (%)	76.7	27.4	24.9	68.2	68.7
Fixed carbon (%)	18.1	11.7	23.2	18.3	18.8
$HHV_{dry\ basis}$ (MJ kg^{-1})	19.1	7.4	9.2	17.3	17.5
K	1997 ppm	31.0%	1115 ppm	4.2%	850 ppm
Na	10.4 ppm	4.8 ppm	24.5%	2.0 ppm	2.3%

then decline because of secondary cracking reactions of volatile. As well the gaseous yield increases with temperature. In relative to product yield a temperature of 500 °C, a particle size of $0.5 < Dp < 1$ mm, and a N_2 gas flow rate of 200 cm^3/min produced a maximum bio-oil output of 45 wt%. The amount of gas increases and the quantity of biochar decreases as pyrolysis temperature increases. The yield of biomass oil will continue to rise with temperature until it reaches ~500 °C, then decreases because of 2nd cracking reactions of volatiles, resulting in rises in the yield of gas. Below is a figure for different types of biomasses used with sugarcane bagasse. Table 2 summarizes different types of sugarcane bagasse used such as normal sugarcane, orange, olive, and grape.

A study done by Kumar et al. [18] used agricultural wastes as an elemental needed to create energy and goods with added value via thermal conversion and biochemical processes, which has significantly increased globally. Biomasses from agricultural waste, which are produced during the production of both grains and non-cereal crops, are regarded as sustainable and feasible substitutes for traditional fuels. The thermochemical properties inside a fixed-bed tubular reactor by batch pyrolysis, pyrolytic gases, biochar, and bio-oil were created. In this study, several effects were observed, such as temperature (350–650 °C), particle size (0.18–0.71 mm), bed size (4–16 cm), and heating rate (15–30 °C/min). Given a 16 cm deep bed of particle with proportions varying from 0.18 to 0.3 mm and at 25 °C/min an average heating rate, results highest yield were (44.7%) of bio-oil, biochar (36.82%), in addition biogas (36.82%) have been achieved on temperatures of 550, 650, 350 °C. For analyzing the composition of a gas chromatography mass spectrometry, Fourier transform infrared spectroscopy and Proton nuclear magnetic were employed. The bio-oil consists of aromatics, phenolic, aliphatic, acidic compounds, and ketonic. Due to its highly porous structure and multiple micronutrients, biochar is considered a suitable soil conditioner. Biogas at its higher temperature dominated H_2 while at its intermediate

Table 2 Sugarcane bagasse pyrolysis with different types of biomasses such as sugarcane, orange, olive, and grape bagasse [17]

Type of biomass	Reactor type	Temperature (°C)	Particle size (mm)	N_2 flow rate (cm^3/min)	Bio-oil yield (wt%)	References
Sugarcane bagasse	Semi-batch	500	0.5–1	200	45	Present study
	Semi-batch	500	0.5–0.6	100	45.23	[11]
	Semi-batch	500	0.5–1	200	66.1	[12]
	Batch	560	0.5–0.85	116	53.83	[11]
Orange bagasse	Semi-batch	525	0.425	200	35.53	[9]
Olive bagasse	Semi-batch	500	0.45–0.6	150	34.4	[6]
Grape bagasse	Fixed bed	550	–	100	41.6	[8]

Fig. 8 Extracting bio-oil, non-condensable gases, and biochar from sugarcane crops [18]

temperatures it had more carbon monoxide and methane than H_2. In overall, this study used powdered biomass for thermochemical aspects and thermal degradation attitude of sugarcane departs and caps. A fixed bed tubular was employed in the batch pyrolysis research. The yield of bio-oil, biochar, and biogas associated with temperature, the warmth rate, size of particles, as well as bed height is observed. Additionally, the highest possible yield of bio-oil and biogas is 44.7% and 36.67% respectively. Figure 8 shows the process of extracting non-condensable gases, bio-oil, and biochar from sugarcane crops.

Fardhyanti et al. [19] described the process of producing bio-oil from sugarcane bagasse through rapid pyrolysis in. The procedure was conducted at 500, 700 °C, with a heating rate of 10 °C/s. Investigating estimates of viscosity, flash point, density, and heating worth during these temperatures was comparable to those of gasoline closer toward 700 °C than those of 500 °C. Phenolic is the main component of the biomass pyrolysis formed bio-oil. Bio-oil formed by pyrolyzing sugarcane bagasse at 500 °C consumed an entire phenolic compound concentration of around 48.43%, while those produced at 700 °C had 58.89%. Therefore, these phenolic compounds can reduce bio-oil value. Consequently, repeating tests more than one time for extracting bio-oil produced in a phenolic compound mixed up with the mixture of chloroform and methanol conducting it at 50 min with changing temperatures each time 25, 40, 50 °C while stirring speed of two to three hundred rpm. Also, the quantity of phenolic compounds can be identified using the Folin-Ciocalteu process. In the bio-oil, its physical properties, and chemical composition were all determined. Bio-oil was produced using the main chemical elements, furfural and phenol. Figured out the best extraction parameters for phenolic compounds are at forty degrees Celsius and two hundred and fifty revaluations. During fast pyrolysis, the largest phenol profit and distribution coefficient was obtained at 500 °C (58.89%, 1.504) than that obtained at 700 °C (48.43%, 1.528). Figure 9 shows graphical abstract for the pyrolysis process.

In Summary, studies of literature review the pyrolysis of sugarcane bagasse that provided valuable perspectives to increase the yield of bio-il and relying on it as an

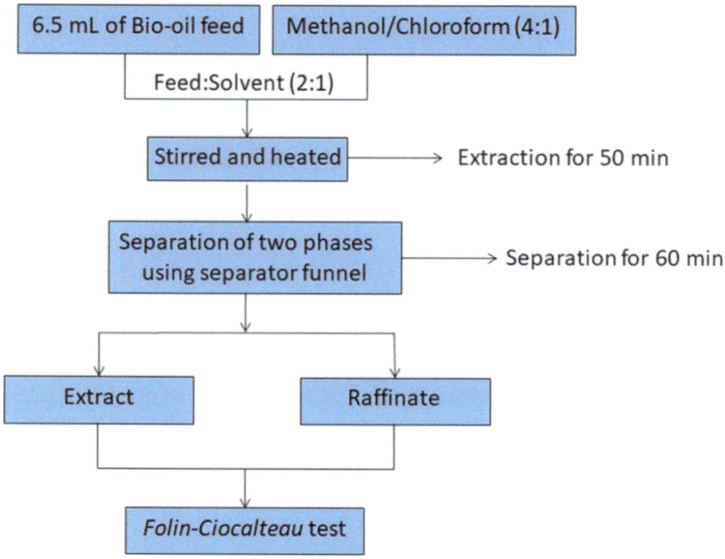

Fig. 9 Extraction method of bio-oil and methanol/Chloroform using FolinCiocalteau test [19]

environmentally friendly energy resource. Therefore, several methods are applied involving using catalysts and pretreatment. Studies are willing to make pyrolysis an applied process, so further optimization of the production of biofuels by pyrolysis is necessary to overcome the gaps for converting agricultural leftovers into biofuels.

3 Methodology: Case Studies and Results

3.1 Overview

This chapter examines the process of pyrolysis bagasse made from sugarcane with particular focus on the following case studies. These studies demonstrate the relationship between the final pyrolysis product and the impact of parameters on them, things like nitrogen gas circulation percentage, temperature, bed height, and dimensions of particles. Also, examine the experimental investigations setup, procedure, material, findings, and analysis of how parameters affect pyrolysis experiments.

3.2 *Case Study One*

Pyrolysis of Sugarcane Leaves and Characterization of Products [18]

Authors: Mohit Kumar, Siddh Nath Upadhyay, and P. K. Mishra.

Year of Publication: 2022.

3.2.1 The Aim of the Experiment

The experiment aim is to investigate how batch pyrolysis affects a fixed bed reactor's ambient temperature, its particle size, heating rate, and other operating parameters. The purpose is to classify thermochemical characterization from the leaves and tops of sugarcane that were collected, which are unused then dried and then converted into powder.

3.2.2 Material

Sugarcane bagasse, nitrogen.

3.2.3 Experimental Setup

First, the sugarcane biomass was washed and given three days to dry. The samples were kept in a hot air oven for 24 h, where the samples were split up into smaller pieces at 60 °C. Using a Wiley Mill to extract the dried sample and then filtered in a gyratory sieve shaker for 15 min using sieves of different sizes: >0.71, 0.710–0.425, 0.425–0.3, and 0.3–0.18 mm. The bulk densities of extracted samples were calculated using ASTM Standard Method-873, those with size range of 0.18–0.3 mm were used for the powdered biomass investigation. Also, the dimension of particles was found to influence the development of biochar, bio-oil, and biogas using different biomass sizes. In Fig. 10, an illustration diagram displays the setup for pyrolysis.

3.2.4 Chemicals Used

1. Dichloromethane
2. n-hexane
3. Deuterated dimethyl sulfoxide
4. Nitrogen
5. N-dimethyl formaldehyde
6. The distilled water created using quartz double in lab.

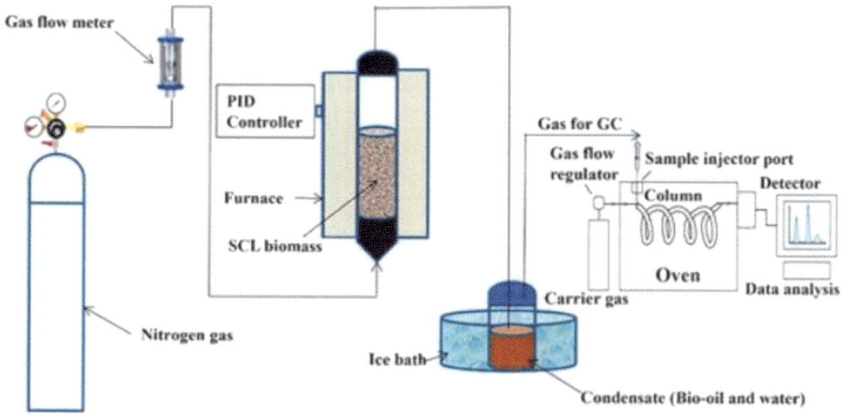

Fig. 10 A basic drawing of the pyrolysis process conducted on a laboratory-scale in a packed bed batch pyrolyzer unit [18]

3.2.5 SCL Biomass Thermochemical Characteristics

ASTM guidelines were used to examine the biomass sample concentrations of moisture, ash, and volatile matter. Conducting experiments using the following formula

$$100 - \{\mathrm{MC}(\%) + \mathrm{VM}(\%) + \text{ash content}(\%)\} \tag{1}$$

to determine fixed carbon by difference each experiment was repeated three times and lastly taking the average of each experiment. CHNS analyzers utilized the estimated carbon, nitrogen, hydrogen, and sulfur to obtain the estimated carbon. Therefore, the oxygen level was calculated using a difference.

3.2.6 Experimental Procedure of Batch Pyrolysis

This review uses packed bed batch pyrolyze unit for pyrolysis on the laboratory-scale system. The pyrolysis unit consists of a tubular reactor with a 44 mm internal diameter stainless tube that is 520 mm long that is rinsed into a vertically electrically heated tubular furnace that needs a controller with a PID. The impact that was made of bed height was observed with the addition of nearly 20 g of biomass powder sample in the tests. Additionally, the amount can be changed to achieve a different biomass bed height. A glass tube with the same internal diameter as the SS tube was filled with significant amount of powdered biomass to locate the bed height. Maintaining the nitrogen flow rate of 100 mL/min during the trials acts as a carrier gas in an inert atmosphere. A bottle that was immersed in an ice bath was used to extract condensable vapors. The furnace was switched on when the biomass was filled, and nitrogen began to flow. Immediately after the required temperature was reached, the warming up procedure existed to conduct an additional half hour to create biochar and accomplish the condensable aqueous bio-oil. By weighing the fluid fraction in the flask and the solid residue remaining in the reactor, the mass of the biochar residue and the aqueous biological oil fractions were established. The figure below shows the formulas used to estimate the product yield that include aqueous bio-oil, biochar, and condensed liquid. The following are the product yield:

$$\text{Bio - oil yield(wt\%)} = \frac{\text{bio - oil(g)}}{\text{biomass fed(g)}} \times 10 \tag{2}$$

$$\text{Bio - char yield(wt\%)} = \frac{\text{bio - char(g)}}{\text{biomass fed(g)}} \times 100 \tag{3}$$

The difference was implemented to calculate the gas yield

$$\text{Gas yield (wt\%)} = 100 - \{\text{bio - oil yield (wt\%)} + \text{bio - char yield(wt\%)}\} \tag{4}$$

And the conversion was calculated using the relation

$$\text{Conversion(\%)} = 100 - (\text{bio - char yield(wt\%)}) \tag{5}$$

3.2.7 Results and Discussion

Batch Pyrolysis Experiment

In the active pyrolysis zone, batch pyrolysis tests were carried out.

Effect of Process Temperature

The rate at which biomass was transformed into ended products increased with temperature. At 650 °C, the production of gaseous products rose from 26.27 to 36.2% at 350 °C, which is due to secondary degradation reactions that can affect biomass to thermally decompose quickly into more non-condensable gases at higher temperatures. Due to the ongoing breakdown of middle molecules of organic matter into gaseous forms, the yield of the bio-oil grown till it reached 550 °C at which point it declined. Char's secondary transformation into non-condensable gas and enhanced breakdown efficiency may be referred to.

Effect of Bed Height

As the reactor bed height decreases from 4 to 16 cm, the yield of gaseous products and biochar slightly decreased due to the growth of production of bio-oil. The longer the duration of products deprivation within the zone of pyrolysis the additional secondary fractures process may cause a tendency to occur.

Effect of Heating Rate

The bio-oil yield before dropping to 34.12% at 30 °C/min, it rose from 34.25% at 15 °C/min till 44.23% at 25 °C/min. Hence, pyrolysis of sugarcane bagasse biomass at low and medium temperatures lead to a reduction in biochar yield and a rise in the output of biofuel. As it was discovered that all heating rates except 30 °C/min, the liquid yield was not influenced by the gas and charcoal yield. The decline may result from either the subordinate breakdown of biochar that produces gaseous and condensable chemicals or the increased elementary decay of biomass.

Effect of Particle Size

Thermal degradation of biomass, thermal breakdown of oily results through pyrolysis, and volatilization to smaller molecular size elements in pyrolysis are all methods that are subjective by the size of biomass particles. The particle size division and biomass particle homogeneity contribute to the heat and mass transmission process. The pyrolysis of biomass occurs on the surface as the particle size increases. Since the internal temperature during the primary pyrolysis stage is inadequate to complete the thermal degradation process, the barrier to heat transfer affects the liquid products while the average size grows. Figure 11 shows the effect of parameters on the product yield.

4 Summary of Case Study One

This study was conducted in a fixed bed tubular reactor that was used for the batch pyrolysis experiments. The relationship between bio-oil yield, biochar, and biogas in temperature, particle measurements, and reheating is examined. The maximum bio-oil was obtained at 16 cm bed height and particles between 0.18 and 0.3 mm and a heating speed of 25 °C/min.

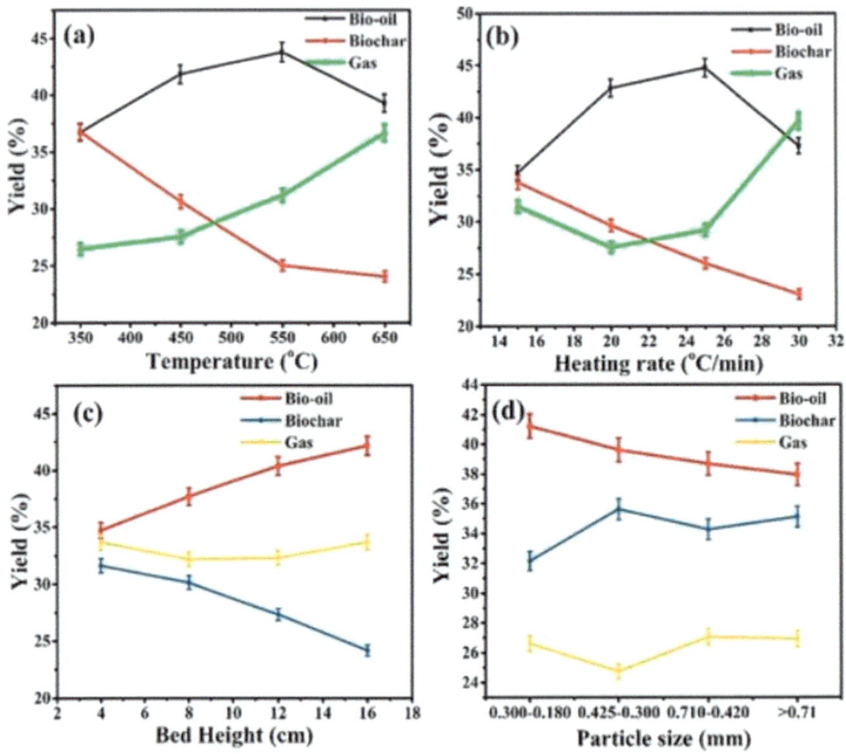

Fig. 11 Temperature, bed height, heating rate, and particle size affect the product yield [18]

4.1 Case Study Two

Biofuel Production via the Pyrolysis of Sugarcane Leaves [13]

Authors: Witchakorn Charusiri, Tharapong Vitidant.

Year of Publication: 2018.

4.1.1 The Aim of the Experiment

The experiment goal was to use a semi-batch reactor to pyrolyze sugarcane bagasse. The purpose of this research is to highlight the effects of various parameters on the characteristics of the produced bio-oil from pyrolysis.

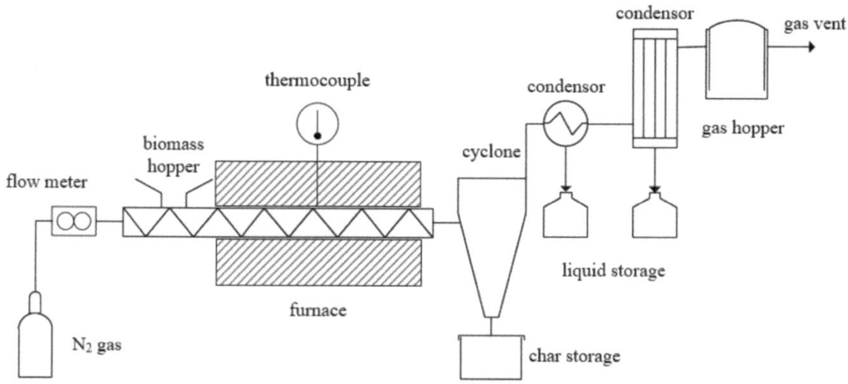

Fig. 12 Representation of experimental setup for the pyrolysis system [13]

4.1.2 Material

Sugarcane bagasse, nitrogen.

4.1.3 Experimental Setup

The pyrolysis system consists of pyrolysis reactor, biomass storage container, solid residue container, screw feeder for SCL, and cooling water condenser. An SS316 stainless steel pyrolysis reactor with dimensions of 1200 mm in length and 40 mm in diameter was used to perform the pyrolyze SCL. In addition, setting standard operating parameters for both temperature and electrical heating furnace is done. Hence, thermocouple of the K-type is used for calculating the system's temperature. The bio-oil and non-condensable gases were produced by condenser through cooling methods. Consequently, after cooling the biochar was drawn and weighed. Figure 12 displays the experimental setup of the pyrolysis process.

4.1.4 Experimental Procedure

1. Firstly, the pyrolysis system was heated with a proportional integral derivatives controller and electrical heating furnace.
2. Secondly, the temperature was raised from 400 to 650 °C for duration of 15 min at a rate of 20 °C min^{-1} to start isothermal process.
3. Thirdly, the feed flow rate to the screw feeder for the pyrolysis was changed between 0.3–1.8 kg/h.
4. Then, eliminating oxygen N_2 gas was incessantly introduced into the reactor for pyrolysis.

Table 3 The analysis of SCL characteristics, results of immediate and extreme investigates [13]

Proximate analysis (wt%)				Ultimate analysis (wt%)				HHV (kJ kg^{-1})
Moisture	Volatile matter	Fixed carbon	Ash	C	H	N	O	
5.44	72.13	13.13	10.26	42.70	7.40	0.45	33.32	19.12

5. Afterwards being loaded into biomass storage a 300 g of SCL samples were fed interested in the pyrolysis system employing an under-screw feeder that delivered SCL incessantly onto the unit.
6. The small rate of feeding was based on dwell time and the motor speed.
7. Using a water condenser to produce bio-oil through the transformation of volatile pyrolysis products and cooling.
8. The non-condensable gas was examined by micro-gas chromatography.
9. The resulting yield of solid residue and pyrolysis oil has been verified by weight.
10. The product distribution and conversion existed by applying the following formulas, the dry weights of SCL raw material expressed by $W_{\text{SCL,db}}$, and $W_{\text{SCL,}}$char:

$$\text{Conversion}(\%) = \left(\frac{W^{\text{SCL,db}} - W^{\text{char,db}}}{W_{\text{SCL,db}}} \right) \times 100 \tag{6}$$

$$\text{liquid yield}(\%\text{wt.}) = \frac{W_{\text{liquid}}}{W_{\text{SCL,db}}} \times 100 \tag{7}$$

$$\text{solid yield }(\%\text{wt.}) = \frac{W_{\text{char,db}}}{W_{\text{SCL,db}}} \times 100 \tag{8}$$

$$\text{gas yield }(\%\text{wt.}) = 100 - \text{liquid yield }(\%\text{wt.}) - \text{solid yield }(\%\text{wt.}) \tag{9}$$

4.1.5 Results and Discussion

Characteristics of Raw Materials

The SCL consists of 1.03% ash and 6.02% moisture content. The SCL composition considerably affected the efficiency of the pyrolysis process conversion. It comprised 0.45 weight percent N, 7.40 wt% H, 42.7 wt% C, and 33.32 wt% O. Table 3 demonstrates the analysis of the SCL characteristics:

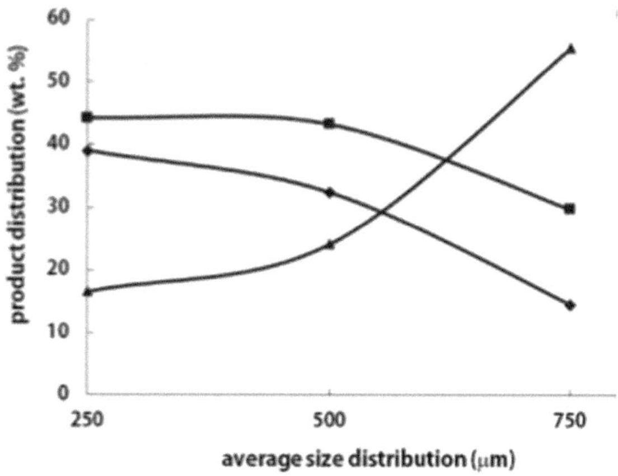

Fig. 13 Product distribution from the pyrolysis of SCL under average particle size [13]

4.2 *Effect on the Particle Supply of Pyrolyzed SCL*

4.2.1 Effect on the Particle Size

It affects tar thermal decomposition, devolatilization to volatile vapors, mass and heat transfer to feed stock, and biomass degradation caused by pyrolysis process. It was purposeful to discover it to determine major effects on the distribution of products, reducing heat transfer, and the characteristics of the pyrolysis oil produced. Figure 13 shows effect on average size on the product yield.

4.2.2 Effect on the Pyrolysis of Temperature

The non-condensable gas yield increased from 39.36 to 43.32 wt% due to the temperature growth from 400 to 500 °C.[1] This could be due to the primary failure of the low moisture content lignocellulosic material for the secondary breakdown of the char residue. Figure 14 displays impact on the temperature on the item yield.

4.2.3 Effect of N_2 Sweeping Gas Flow Rate

The effect of N_2 gas flow rate on pyrolysis at a target temperature of 550 °C has been investigated by varying the flow rate from 80 to 240 cm^3/min through a sequence of pyrolysis experiments. The pyrolysis oil yield fell as the moving rate of N_2 gas and gas yield rose, because of short duration, inadequate condensation by the cool water condenser that convert the flammable vapor into bio-oil products, and issue of the devolatilization vapors over the pyrolysis system before actual condensation

[1] C, the average particle size of 500 m, and 0.6 kg h^{-1} biomass feed rate are the effects.

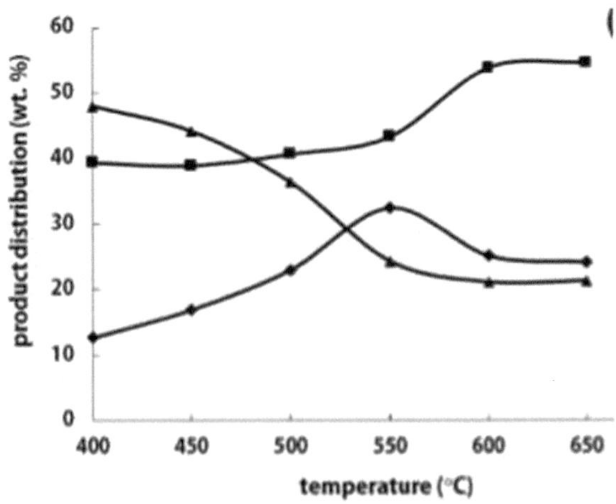

Fig. 14 Impact on the temperature on the item yield for different components on biochar [13]

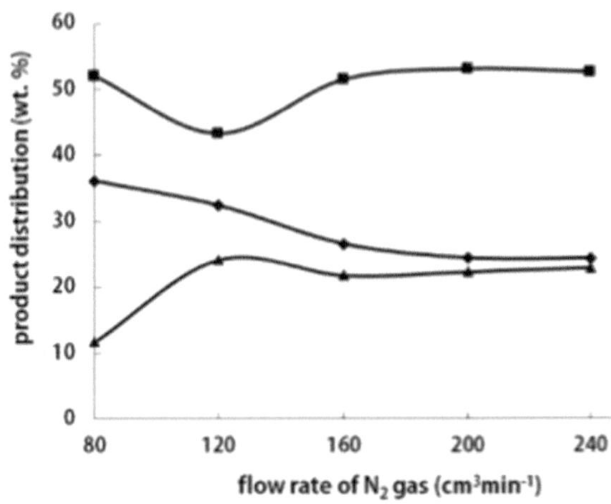

Fig. 15 Effect of flow rate of N_2 gas on the product yield for different component on bio-oil [13]

took place. Figure 15 shows the outcome of N_2 velocity of gas flow on the product yield.

4.2.4 Summary of Case Study 2

This case study focuses on the pyrolysis parameters that were optimized to provide a suitable distribution of SCL products. It also finds out the factors affecting the yield based production of bio-oil. The tendency of temperature rises the pyrolysis bio-oil output from 400 to 500 °C then resulting in a decrease when it reached 650 °C. The biomass feed rate is a factor in the conversion of bio-oil due to the heat transfer efficiency and residence duration between the SCL surface and the tubular furnace.

4.3 Case Study Three

Pyrolysis of Sugarcane Bagasse, the Effects of Process Parameters on the Product Yields [9]

Authors: Ahmed Gaber H. Saif, Seddik S. Wahid, Mohammed R. O. Ali.

Year of Publication: 2020

4.3.1 Aim of the Experiment

The case study aims to study the ideal conditions for increasing the production of bio-oil by examining the pyrolysis of sugarcane bagasse in a semi-batch reactor. The objective is to focus on how bio-oil yields, biogas, and biochar were influenced by process variables such as temperature, nitrogen gas flow rate, and sugarcane bagasse particle size.

4.3.2 Materials: Sugarcane Bagasse, Nitrogen

See Fig. 16.

4.3.3 Experimental Setup

This experiment utilizes the cylindrical semi-batch pyrolysis reactor to generate sugarcane bagasse pyrolysis as illustrated in Fig. 16. It is constructed of 304 stainless steel and its dimensions of 40 cm height and 10 cm diameter. Also, an electrical furnace was utilized with 2.5 kW of power to lower heat loss. Thermal insulation was employed. A K-type thermocouple was used to monitor the pyrolysis while a PID controller employed it to keep the reactor temperature constant. Moreover, pyrolysis vapors were transferred to the condensers and inside the reactor an inert condition was initiated using nitrogen gas. The condensers are two made of glass and one made of stainless steel. Subsequently after being accumulated in a flask, the condensed liquid weighed its yield. Using screw system, after each run, the char was collected and weighed after being taken from the reactor. In the end, obtaining the no condensable gas through subtracting. The following equations were used for calculations:

$$\text{Bio - oil yield } (\%) = \frac{\text{oil yield}}{\text{feeding weight}} \times 100 \tag{10}$$

$$\text{Bio - char yield } (\%) = \frac{\text{Char yield}}{\text{feeding weight}} \times 100 \tag{11}$$

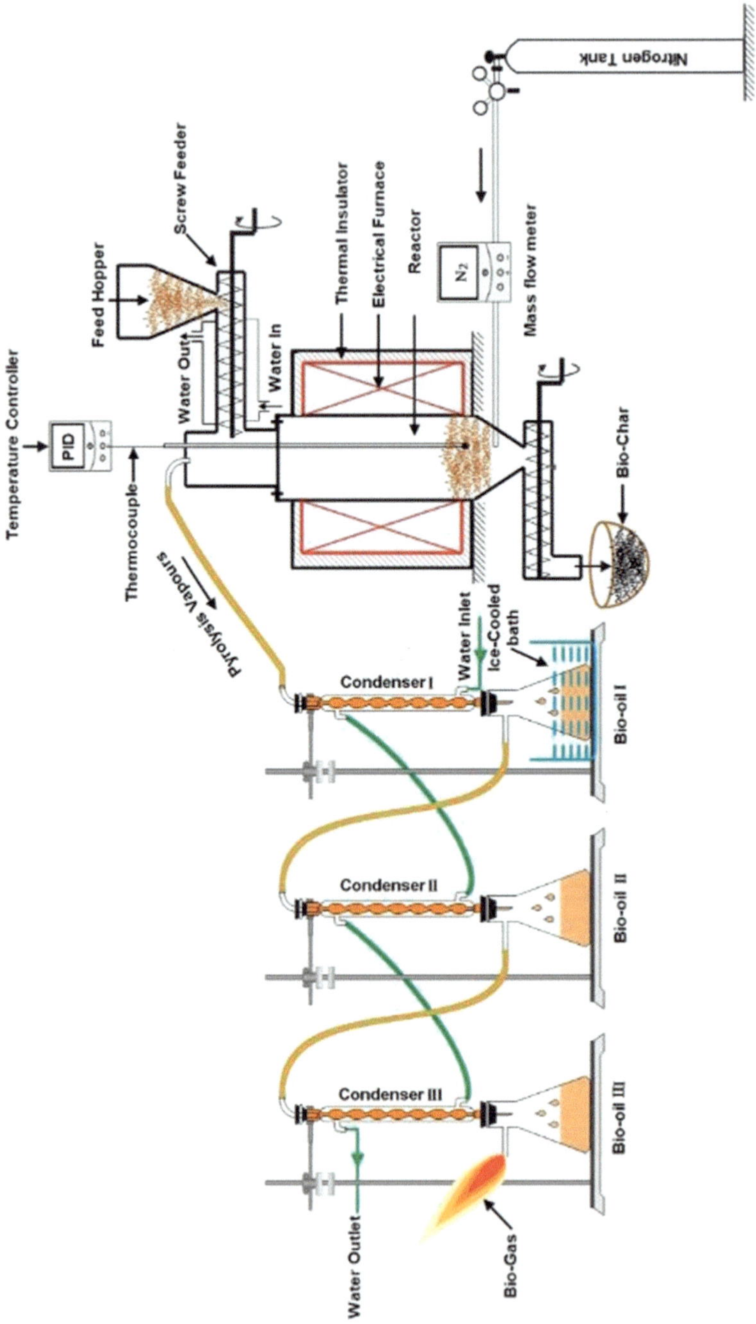

Fig. 16 Schematic of the experimental setup conducted in a test center cylindrical semi-batch pyrolysis reactor [9]

$$\text{Bio - gas } (\%) = 100 - (\text{oil yield} + \text{char yield}) \tag{12}$$

4.3.4 Experimental Procedure

To determine the optimal parameters for the maximum production of bio-oil, experiments were conducted into three groups:

1. First, the biomass feeding hooper filled with 100 g of feedstock sample for each run.
2. Secondly, using a screw feeder to add sample to the reactor after achieving desired temperature.
3. An amount of 20 cm^3/min of nitrogen gas is pumped into the reactor before each run.

For the three groups:

I. The first experimental group evaluated how temperature affected the sugarcane bagasse pyrolysis yield. Once the temperature reaches its final level a dried sugarcane bagasse sample is added that has a particle with $0.5 < Dp < 0.1$ mm. Then, considering a 200 cm^3/min nitrogen flow frequency the final temperature is achieved by heating the reactor. The conduct experiment took place until there was no discernible brownish vapor leak at the reactor outlet while considering predetermined temperature.

II. The second experimental group explored the impression of particle size for the range $0.25 < Dp < 0.5$, $0.5 < Dp < 1$, $1 < Dp < 2$, and $Dp < 2$ mm. The final measured 500 °C temperature and 200 cm^3/min nitrogen rate of flow in nitrogen.

III. Third experimental group studied the cause of nitrogen sweeping gas flow rate. Considering the results of the first and second group the testing was taken with nitrogen rates of (100, 200, 300, 400, and 500 cm^3/min), through a particle size of $0.5 < Dp < 0.1$ mm and at a regular temperature of 500 °C.

4.3.5 Results and Discussion

Effect of Pyrolysis Temperature

The temperature effects char yield to decrease from 36 to 26 wt% when it increases from 350 to 600 °C as gas also increases between 26 and 29 wt% as shown in Fig. 17. The amount of gas generated rises with increasing pyrolysis temperature, but the amount of biochar falls because of secondary split and char volatilization at better temperatures. It is determined that the distribution of product yield is impacted by the process temperature, gas yield, and the char yield decreases through temperature.

Therefore, the maximum bio-oil is produced at a pyrolysis of 500 °C.

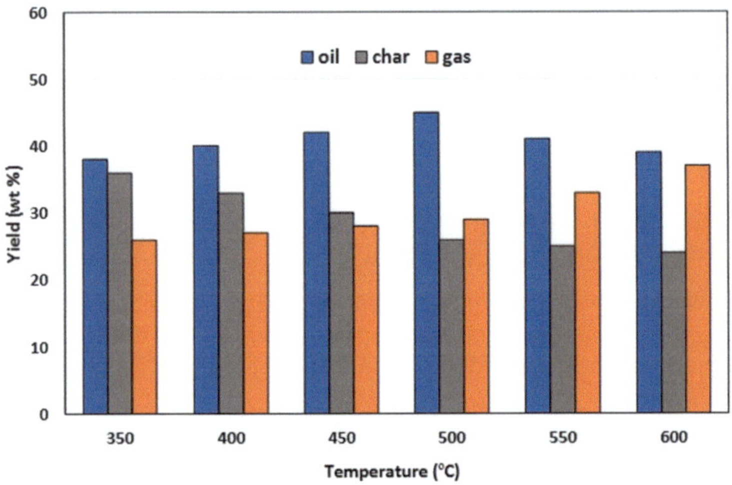

Fig. 17 Effect of Temperature on the product yield for different components [9]

Effect of Sugarcane Bagasse Particle Size

As the sugarcane bagasse particle size increases from $Dp < 0.5$ to $Dp > 2$ mm, the biochar yield increases from 27 to 36 wt%. In contrast, the biogas yield decreases from 35 to 23 wt%, as shown in Fig. 18. This behavior can be attributed to the development of larger internal temperature gradients in larger particles, causing the core temperature to remain lower than the surface temperature.

Effect of Nitrogen Gas Flow Rate

Creation of nitrogen gas has impact on the product at a fixed biomass particle size of $0.5 < Dp < 1$ mm and 500 °C presented in Fig. 19. Nitrogen flow rate is necessary for the procedure since it will intensify the bio-oil crop, which generates an inert state inside the pyrolysis while decreasing vapor residence time, potential for vapor thermal cracking.

4.3.6 Summary of Case Study 3

In this case, SCB pyrolysis is conducted in a semi-batch catalytic and therefore studies the effect of the ultimate temperature, nitrogen flow rate, and particle size. The experimental results indicate that the maximum bio-oil yield of 45 wt% was achieved at a temperature of 500 °C, a carrier gas flow rate of 200 cm^3 min^{-1}, and a particle size in the range of $0.5 < Dp < 1$ mm. An increase in temperature up to 500 °C enhances bio-oil production; however, further temperature increases lead to a reduction in biochar

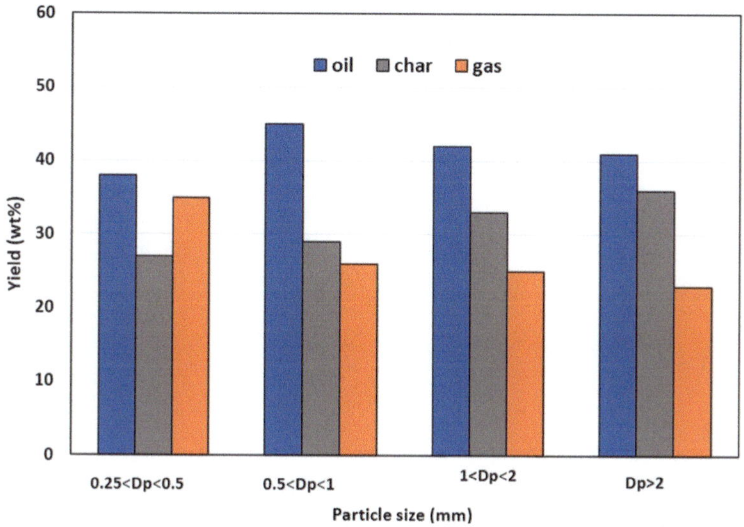

Fig. 18 Effect of Particle size on the product yield for different components on biochar and biogas [9]

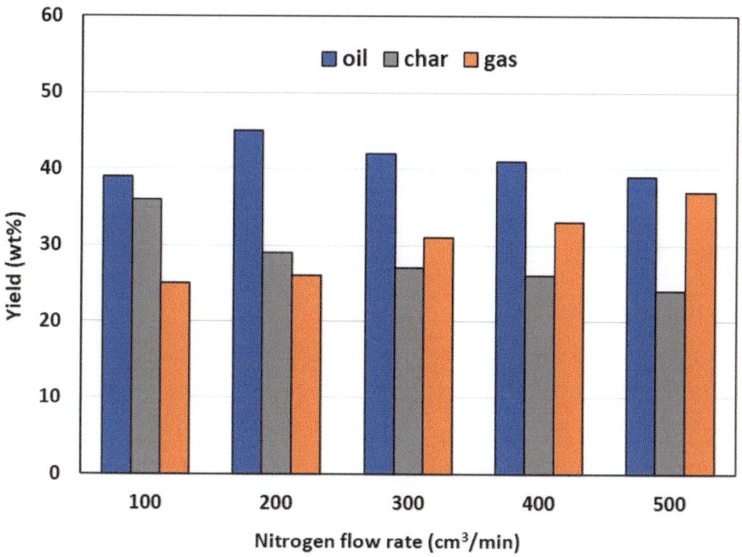

Fig. 19 Effect of Nitrogen flow rate on the product yield of different components [9]

Table 4 Comparison table of three case studies

Comparison	Case study no. 1	Case study no. 2	Case study no. 3
Reactor type	Fixed bed tubular reactor	Semi-batch reactor	Semi-batch reactor
Temperature (°C)	350–650	400–550	350–600
Particle size (mm)	0.18–0.3	500	$0.5 < Dp < 0.1, 0.25 < Dp < 0.5, 0.5 < Dp < 1, 1 < Dp < 2, Dp > 2$
Nitrogen gas flow rate (cm^3/min)	–	80–240	100, 200, 300, 400, 500
Bed height (cm)	16	–	–
Heating rate (°C/min)	25–30	20	–
Type of pyrolysis	Slow pyrolysis		
Material	Sugarcane bagasse		
Bio-oil yield (wt%)	34.12–44.23	39.36–43.32	26–36

yield. This trend is attributed to secondary cracking reactions of volatile compounds at higher temperatures, which favor the formation of gaseous products.

4.4 *Comparison Table for the Three Case Studies*

See Table 4.

5 Results and Discussion

5.1 *Overview*

In this chapter, details of the comparison table in Chap. 3 outline the results of the three case studies conducted. The factors that contribute to affecting the sugarcane bagasse pyrolysis process are temperature, particle size, nitrogen gas flow rate, bed height, and heating rate. In this report, conducted case studies to compare the various production of biofuel from sugarcane bagasse wastes using different reactors to examine the bio-oil yield percentage obtained.

5.2 Comparison Table Discussion

The table is composed of reactor types, temperature, pressure, N_2 stirring rate, particle size, warming up rate. The first case study conducted the process in a tubular reactor with a fixed bed, in a slow pyrolysis with a temperature of 350–650 °C, sizes among 0.18–0.3 mm, a height of 16 cm, and a flow of heat of 25–30 °C/min. The yield of bio-oil is between 34.12 and 44.23 wt%. Also, in the same reactor slow pyrolysis was conducted in the second case study with a pyrolysis temperature of 400–550 °C, particle size 500 mm, approximately 80–240 cm^3/min flow rates, and 20 °C/min heating that resulted in a bio-oil of least 39.36 wt% to highest of 43.32 wt%. In that order, in semi-batch reactor type case study three was similarly conducted at temperatures of 350–600 °C, with various sugarcane bagasse particle size ranges, nitrogen flow rate of ranges 100–500 cm^3/min, and as well as of slow pyrolysis type, and the bio-oil yield accomplished at 26–36 wt%. In comparison to the three-case study, it was observed that temperatures affected pyrolysis extensively meanwhile it increases yield increases as well at 500 °C due to the secondary thermal cracking to create non-condensable gases. Therefore, produced the highest bio-oil 44.23 wt% in case 1, where there is drop in temperature in case 2 to 450 °C lowered the bio-oil percent to 43.32 wt%, and again after achieving 500 °C in case 3 it enhanced the production of bio-oil to 45 wt%. Secondly, particle size would likely affect heat transfer, as it is observed in case 1 particle size is small between 0.18 and 0.3 mm, so will have the best bio-oil production. While increasing particle size to 500 m resulted in a reduction in bio-oil production in case 2 and dropped production to 43.32%. In case 3, a reduction was observed due to the impact of the ranges of particle size caused irregular pyrolysis. Thirdly, bed height was only considered in the first case that helped in reducing thermal cracking. Then, N_2 sweeping gas flow rate is used to reduce vapors. In addition, reducing heating rates at certain temperatures between 20 and 30 °C/min increased production. In summary, to strengthen temperature must be set to 500 °C.

5.3 Environmental and Economic Aspects

Making biofuel by pyrolyzing sugarcane bagasse is addressing both environmental and economic effects. The environmental impact was assessed using a Life cycle assessment method to produce biofuel for sugarcane bagasse. Therefore, reliance on renewable supplies biofuels is a prerequisite for fossil fuels to reduce greenhouse gas emissions since 80% of the globe's energy derives from fossil fuels. Therefore, it impacted 2.6 kg CO_2 on global warming, land ecotoxicity of 10.44, and 2.61 kg of human non-carcinogenic toxicity. This enables industry stakeholders and policy-makers to design more effective and sustainable energy policies [20]. The economic benefits of generating biofuel contributed to the development of interest from all over the world. For instance, employing sugarcane bagasse a product of sugar assembly reduces the cost of disposal. A study of technology conducted and simulated 84,000 tons of sugarcane, liquefaction, besides gasification popular Australia, generating 11.58 million liters annually. The total cost of the installation was expected to be USD 23 million, and an approximate amount of USD 52 million total capital cost. The majority of 73% of capital investment is based on site modified cost, worth USD 38 million. The local plant is likely not the same as the sugar mill that causes it to occur. Overall total annual operating costs contain feedstock for USD 2.04 million, catalysts replaced within less than USD 0.31 million, trade waste handling also to a lesser extent than USD 0.01 million, labor fee USD 1.17 million, preservation charges USD 0.87 million, electrical a smaller amount than USD 0.01 million, H_2 per USD 1.17, and water cost USD 0.60 million. Considering the larger cost of operating expenses, feedstock has 33% worth. Using sufficient biochar made by pyrolysis will reduce the use of electricity, because pyrolysis needs less heating than gasification and liquefaction, which makes it more economical [11].

6 Conclusion

In conclusion, the study examines the production of biofuel from sugarcane bagasse waste due to its potential to enhance the pyrolysis process to produce biofuel. Additionally, research conducted case studies that included pretreatment methods and operating variables like bed height, size of particles, nitrogen rate, and temperature. This study presents sustainable energy technology by providing that it is possible to transform agricultural waste into the best biofuel aids for the development of environmentally friendly energy resources. Such a study promotes the efficient use of biomasses and provides sustainability in the environment, and an affordable source of energy. Future research should focus on increasing the pyrolysis process and experimental industrial scale. Furthermore, included the environmental impact of producing biofuel and determined the economic aspects. Economic aspects offer both development of economic and sustainable energy generation. To promote sustainable and environmental energy resources, this report emphasizes the need to continue innovation and research in biofuels.

References

1. Ye Y et al (2024) Biofuel production for circular bioeconomy: present scenario and future scope. Sci Total Environ 935:172863. https://doi.org/10.1016/J.SCITOTENV.2024.172863
2. Sharmila VG, Shanmugavel SP, Banu JR (2024) A review on emerging technologies and machine learning approaches for sustainable production of biofuel from biomass waste. Biomass Bioenergy 180:106997. https://doi.org/10.1016/J.BIOMBIOE.2023.106997
3. Download citation of acid and alkaline pretreatment of lignocellulosic biomass to produce ethanol as biofuel. https://www.researchgate.net/publication/281199064_Acid_and_alkaline_pretreatment_of_lignocellulosic_biomass_to_produce_ethanol_as_biofuel/citation/download. Accessed 15 Feb 2025

4. Chandel AK, da Silva SS, Carvalho W, Singh OV (2012) Sugarcane bagasse and leaves: foreseeable biomass of biofuel and bio-products. J Chem Technol Biotechnol 87(1):11–20. https://doi.org/10.1002/JCTB.2742
5. Kim M, Day DF (2011) Composition of sugar cane, energy cane, and sweet sorghum suitable for ethanol production at Louisiana sugar mills. J Ind Microbiol Biotechnol 38(7):803–807. https://doi.org/10.1007/S10295-010-0812-8
6. Sikiru S, Abioye KJ, Adedayo HB, Adebukola SY, Soleimani H, Anar M (2024) Technology projection in biofuel production using agricultural waste materials as a source of energy sustainability: a comprehensive review. Renew Sustain Energy Rev 200:114535. https://doi.org/10.1016/J.RSER.2024.114535
7. Pan S, Zabed HM, Wei Y, Qi X (2022) Technoeconomic and environmental perspectives of biofuel production from sugarcane bagasse: current status, challenges and future outlook. Ind Crops Prod 188:115684. https://doi.org/10.1016/J.INDCROP.2022.115684
8. Fardhyanti DS, Megawati M, Prasetiawan H, Nabuasa N, Ardianta MA (2021) Chemical kinetics modeling on bio-oil production from pyrolysis of sugarcane bagasse. Mater Sci Forum 1034:199–205. https://doi.org/10.4028/www.scientific.net/MSF.1034.199
9. Saif AGH, Wahid SS, Ali MRO (2020) Pyrolysis of sugarcane bagasse: the effects of process parameters on the product yields. Mater Sci Forum 1008:159–167. https://doi.org/10.4028/www.scientific.net/MSF.1008.159
10. Fardhyanti DS et al (2024) Production of bio-oil from sugarcane bagasse by fast pyrolysis and removal of phenolic compounds. Biomass Convers Biorefin 14(1):217–227. https://doi.org/10.1007/S13399-022-025279/METRICS
11. Miranda NT, Motta IL, Filho RM, Maciel MRW (2021) Sugarcane bagasse pyrolysis: a review of operating conditions and products properties. Renew Sustain Energy Rev 149:111394. https://doi.org/10.1016/J.RSER.2021.111394
12. Wibowo H et al (2024) Application of sugarcane bagasse fly ash for syrup decolorization through activated carbon production and the circular utilization of its waste. Biomass Bioenergy 182:107066. https://doi.org/10.1016/J.BIOMBIOE.2024.107066
13. Charusiri W, Vitidsant T (2018) Biofuel production via the pyrolysis of sugarcane (*Saccharum officinarum* L.) leaves: characterization of the optimal conditions. Sustain Chem Pharm 10:71–78. https://doi.org/10.1016/J.SCP.2018.09.005
14. Wani AK, Rahayu F, Fauziah L, Suhara C (2023) Advances in safe processing of sugarcane and bagasse for the generation of biofuels and bioactive compounds. J Agric Food Res 12:100549. https://doi.org/10.1016/J.JAFR.2023.100549
15. Sitthikitpanya N, Khamtib S, Sittijunda S, Imai T, Reungsang A (2024) Valorization of sugarcane leaves and co-digestion with microalgal biomass to produce biofuels and value-added products under the circular economy and zero-waste concepts. Energy Convers Manag 299:117854. https://doi.org/10.1016/J.ENCONMAN.2023.117854
16. Muñoz M, Rosero M, García AN, Marcilla A (2024) Effect of alkaline catalysts on the valorization of sugarcane bagasse via pyrolysis. Ind Crops Prod 211:118225. https://doi.org/10.1016/J.INDCROP.2024.118225
17. Saif AGH, Ali MRO, Wahid SS (2020) Sugarcane bagasse pyrolysis: investigating the effect of process parameters on the product yields. https://doi.org/10.21203/RS.3.RS-21476/V1
18. Kumar M, Upadhyay SN, Mishra PK (2022) Pyrolysis of sugarcane (*Saccharum officinarum* L.) leaves and characterization of products. ACS Omega 7(32):28052–28064
19. Fardhyanti DS et al (2024) Production of bio-oil from sugarcane bagasse by fast pyrolysis and removal of phenolic compounds. Biomass Convers Biorefin 14(1):217–227. https://doi.org/10.1007/S13399-022-02527-9
20. Life cycle assessment of biofuel production from sugarcane bagasse|semantic scholar. https://www.semanticscholar.org/paper/Life-Cycle-Assessment-of-Biofuel-Production-from-Nasri-Hanafiah/89dfbb22c18ddf4d8b4cf81b547169682c60ba74. Accessed 23 April 2025

Biofuel Production from Corn Stover by Pyrolysis

Reem AL-Qassabi and Seyed Mojtaba Sadrameli

Abstract This comparative study investigates the production of biofuel from the pyrolysis of corn stover, where three different case studies were analyzed and compared to explore the potential of utilizing and optimizing the resulting biofuels as an alternative energy source. The effect of temperature on the product distribution and the biofuel properties was a major objective in this paper. Moreover, different reactors of different heating methods were evaluated along with other main variables such as particle sizes, the presence of catalysts as bed materials, acid and steam explosion pretreatments were also explored. The optimum temperature ranges for maximum bio-oil production are found to be 400 °C and 500 °C. Also, the highest bio-oil yield and HHV were obtained at 400 °C, which were about 43.3 wt% and 33.8 MJ/kg, respectively, from fast pyrolysis of corn stover in a catalytic fluidized bed reactor. While the highest biogas yield was about 73 wt% which was achieved by the microwave pyrolysis of corn stover, this reactor design also improved biogas properties but produced low bio-oil and biochar yields. Acid pretreatment had a positive effect on bio-oil yield using microwave pyrolysis, however reducing the particle size had an opposite effect. The products and feedstock properties such as pH level, ash content, volatile matter, and moisture were all analyzed by different standard analyses such as the GC–MS analysis and FTIR analysis. The elemental analysis was also performed to find the most significant compounds in feedstock and products in each case study.

Keywords Corn stover · Pyrolysis · Biomass · Biofuels · Bio-oil

---

R. AL-Qassabi · S. M. Sadrameli (✉)
Department of Engineering, German University of Technology in Oman, Muscat, Oman
e-mail: seyed.sadrameli@gutech.edu.om

A. Azizi (ed.), *Mechatronic Innovations in Sustainable Energy*, Emerging Trends in Mechatronics, https://doi.org/10.1007/978-981-95-5863-6_2

1 Introduction

1.1 The Importance of Renewable Energy

Although fossil fuel production of energy can contribute to approximately 80% of energy demand, it certainly causes significant environmental and health issues where using fossil fuels of all types: coal, oil, and gas have not been sufficiently addressed in the market pricing range, such as pollution of air and water, wildlife and the production of greenhouse emissions resulting in climate change and ocean acidification, which is all caused due to fossil fuels combustion [1]. On the other hand, the crucial need for sustainable renewable energy has not been met yet, where emission free electricity and transportation can be achieved through different renewable energy sources including solar, wind, geothermal, and biofuels [1]. Biofuels of all forms: solid, liquid, and gas that can be produced from various sustainable feedstocks can decrease the continuous demand for imported gasoline and diesel, significantly lower harmful contaminants and the release of greenhouse gas emissions, alongside relying on agricultural resources that encourage economic growth and development [2].

1.2 Biomass

Biomass refers to all those naturally renewable or recurrent organic materials like agricultural crops, trees planted for power generified wood residues, plants including aquatic plants and grasses, waste products, fiber waste, animal waste and other waste materials, fats, oils, and greases including recycled fats, oils, and greases [3]. Biomass is, therefore, an uninterrupted and continuous source of energy as compared to solar and wind intermittent renewable energy sources because it does not incur costs for battery storage, it is cheaper, and technologically much more efficient than other alternatives like hydrogen [4]. Thus, biomass would be an alternative way to fossil fuels, where it would be transformed into gaseous fuels, electricity, liquid fuels, and heat [4]. Decentralized applications are of great benefit, especially for modern energy-lacking access regions where they support development and poverty reduction [4]. Moreover, modern bioenergy technologies improve sustainability even further by replacing traditional cooking fuels with cleaner and renewable alternatives [4].

1.3 Different Sources of Biomass

The different biomass sources are shown in Fig. 1. The main biomass sources include wood waste, that is considered to be bioenergy source however, it need be managed sustainably in forestry, food waste which minimizes landfill waste can be converted by means of anaerobic digestion and fermentation, Microalgae are also another sources

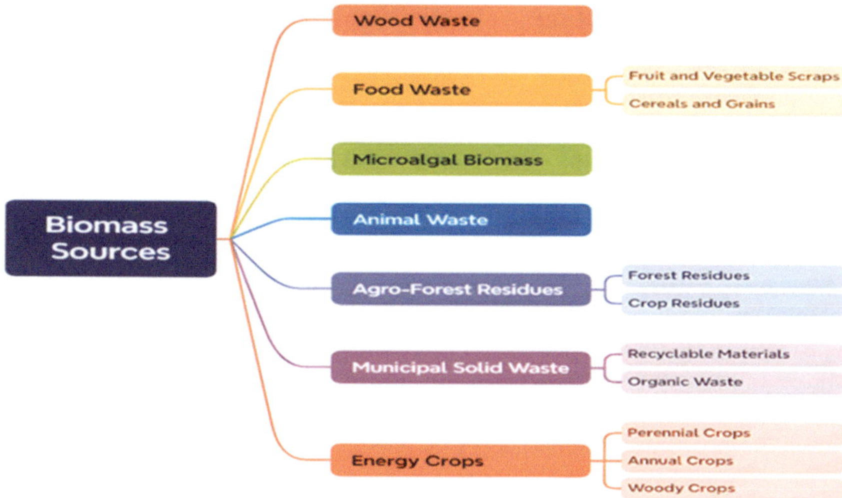

Fig. 1 Different biomass sources [5]

of biomass and although its harvesting cost is high, it has high productivity and carbon dioxide fixation, Animal waste can be converted into biogas when processed in a proper reactor design, Agricultural and forestry residues can provides huge amount of energy potential however it requires efficient treating technologies, and lastly Municipal solid waste (MSW) which is considered as an affordable source for biomass that include perennial and woody crops that are produced particularly for bioenergy [5].

1.4 Biofuel Classification

Biofuels can be classified into two main types: liquid and gaseous. However, they can be divided into four main categories based on feedstock source and production techniques [5]. Bioethanol and biodiesel are represented by the first-generation biofuels mainly generated from food crops that threaten agriculture [5]. Second-generation biofuels, including cellulosic ethanol and syngas, are produced from non-food biomass. While there might be some technical and economic complexity for their production, this type of biofuel can solve the problem issued with the first-generation biofuel by reducing competition in agriculture [5]. Moreover, third-generation biofuels such as biodiesel, biohydrogen, and biomethane are made from microalgae. This type of biofuel has a very high yield; however, it might not be as commercially valid in terms of scaling up. Lastly, fourth-generation biofuels, where microorganisms are genetically modified to enhance efficiency and CO_2 capture for the manufacturing of biofuel, but it has not been fully developed yet [5].

1.5 Corn Stover Biofuel

Corn stover has a significantly high potential as a feedstock for the production of biofuels due to its wide availability, renewable nature, and its ability to replace fossil fuels [6]. Corn stover is found mostly as a byproduct from corn production, and that is why it is available in large quantities in major corn-growing areas such as Iowa, Illinois, and Nebraska down to Minnesota [6]. Otherwise, this large quantity of corn stover would only decay in the field but can be converted into ethanol and enable the production of renewable fuel that does not compete with food supplies [6].

One of the major advantages of corn stover is the high cellulose content that can be converted into fermentable sugars for ethanol production through enzymatic hydrolysis. Moreover, according to different studies, it is estimated that if only 20–30% of corn stover is harvested, it will result in huge-scale ethanol production, and this will further reduce the dependency on imported fossil fuels [6].

1.6 Thermal Conversion Methods

There are many pathways of conversion that convert biomass into different forms of energy [7]. Different types of conversion methods can be influenced by several elements, such as the kind and characteristics of feedstock biomass and the kind of energy desired. Moreover, environmental regulation, policy, economic circumstances, and specific project considerations are other minor factors that influence the choice of conversion method [7]. Conversion processes include thermos chemical conversion processes such as combustion, pyrolysis, gasification, and liquefaction. These processes convert biomass into more valuable carriers using heat [7].

1.7 Pyrolysis Method

Pyrolysis is a type of thermochemical process in which biomass decomposes under severe temperatures without the presence of air or oxygen. Consequently, three main products are produced: bio-oil, liquid fuel; biochar, solid residue; and syngas that is a mixture of combustible gases. The conditions of pyrolysis, such as temperature, heating rate, and residence time, control the product distribution [7]. Furthermore, pyrolysis is performed mainly at a temperature of 500–800 °C, at which high complexity thermal decomposition mechanisms occur that are influenced by temperature and the rate of temperature elevation as well as pressure [8]. Biomass thermal decomposition starts from cellulose and involves two stages: initial decomposition at a temperature of less than 300 °C, followed by further degradation and depolymerization above that temperature [8]. In the first stage, cellulose may undergo decomposition and form biochar, CO_2, and H_2O or even tar that contains cellulose fragments [8]. While the temperature is increasing, the second stage includes additional decomposition, which results in the formation of liquid or gaseous products [8].

This research studies the production of biofuel from corn stover through pyrolysis. It investigates the detailed procedure of converting corn stover into different types of biofuels, analyzing the factors that influence efficiency, yield, and energy content of pyrolysis. While reviewing published studies on corn stover pyrolysis. Moreover, the research will evaluate the viability of pyrolysis as a conversion method for commercial biofuel production from different aspects and the potential environmental and economic benefits.

The research comprises four sections. Section 1 focuses on biomass-based biofuels, corn stover in particular as raw material, and the aims of the research. Previous research on corn stover pyrolysis, technology, how they were conducted, their main equipment, and advances is discussed in Sect. 2. Section 3 discusses case studies on the pyrolysis process and its key influencing factors. Finally, Sect. 4 presents the findings and suggestions based on these for optimizing the process for sustainable biofuel production.

2 Literature Review

2.1 Overview

In this section the latest and oldest research papers on the production of biofuel from corn stover via pyrolysis will be reviewed. It highlights process and product yield optimization through different effecting factors such as particle size, pressure, reactor design, and temperature.

2.2 Literature Review

In 2007, Yu et al. [9] explored the physical and chemical characteristics of the bio-oil obtained from microwave pyrolysis of corn stover in order to evaluate their ability to be utilized as gas turbine and home heating fuels. Various approaches were used to determine these properties like ASTM procedures, digital pH meters, and Inductive Coupled Plasma-Atomic Emission Spectrometers. Accordingly different bio-oil properties were identified, such as the water content, solid content, alkali metal content, dynamic viscosity, and gross high heating value for a typical bio-oil and they were found to be 15.2 wt%, 0.22 wt%, 12 ppm, 185 mPa · s, and 17.5 MJ/kg, respectively. Moreover, aging test indicated an increase in bio-oil product's viscosity over time as well as the water content during which phase separation was observed. Alcohols such as methanol and ethanol were tested and showed a significant enhancement in bio-oil stability by reducing viscosity and slowing water content increase during storage. Although the resulting bio-oil was low in ash and solids, it wasn't suitable for gas turbine fuel and applications due to its unstable pH, water content, and viscosity. Blending the bio-oil with alcohol, therefore, it was proven to be a simple, affordable, and effective method to improve bio-oil stability.

In 2010, Mullen et al. [10] experimented the production of bio-oil and biochar from both corn cobs and corn stover via fast pyrolysis in a pilot scale fluidized bed reactor at a temperature around 500 °C. Products were evaluated in terms of energy and soil amendments properties using different methods such as GC/MS and HPLC, to separate and analyze the products. Results showed that corn stover and corn cobs have different compositions, where the former is richer in cellulose and the latter is greater in potassium. A bio-oil yield of about 60% was achieved for both the corn cobs and corn stover, with high heating value of approximately 20 MJ/kg, which is 20–32 times higher than raw corn stover residue. Moreover, a yield of about 18.9% from corn cobs and 17.0% from corn stover was obtained for biochar. Biochar can be used as a soil amendment due to the minerals and carbon it contains, which can solve the environmental issues when removing and using crop leftovers for industrial and manufacturing purposes. Additionally, the resulting bio-oil has a feedstock energy that is 67% cobs and 74% stover. This high-density bio-oil lowers transportation costs in distributed refinery system.

In 2012, Shah et al. [11] examined fast pyrolysis of single-pass corn stover and corn cobs stored in different methods to determine the effect on the physiochemical properties of resulting pyrolysis product. The resulting properties are then compared to those of the multi-pass stover. The feedstock of both corn stover and corn cobs was harvested and stored for 5.5 months before the experiment. Different storage methods were used such as multi-pass corn stover bales stored inside a metal building, single-pass corn stover bales stored inside a metal building, single-pass corn stover bales stored outdoors anaerobically in a plastic tube-wrap (ensiled), and single-pass corn cobs stored anaerobically in a plastic bag (ensiled). Findings showed that the maximum yield of bio-oil obtained was for the ensiled single-pass corn stover, however the biochar yield was the lowest but the inside stored stover. Moreover, the most acidic bio-oil yield was obtained from ensiled single-pass cobs, and the lowest oxygen content of bio-oil was from inside-stored single-pass corn stover. Other bio-oil properties such as the ash contents, volatile matters, and fixed carbons were approximately the same for all four samples. The lowest ash content was achieved from the ensiled single pass stover, although high ash content was observed in the biochar produced from inside stored multi-pass stover.

In 2012, Capunitan et al. [12] investigated the biofuel production through the pyrolysis of corn stover at different temperatures which are 400, 500, and 600 °C using a pressurized batch reactor, to analyze and evaluate product yields, characteristics and conversion efficiencies. Results showed that the highest biochar yield as well as the highest liquid yield were obtained at 400 °C and were reported to be 7.3 wt% and 31.4 wt%, respectively. However, the maximum gas yield was 21.2 wt%

at 600 °C. Moreover, biochar was characterized as a possible alternative solid fuel due to its great amount of energy content and minimal oxygen content. Additionally, various bio-oil analyzing methods such as FTIR and GC–MS analyses indicated that it mainly contains phenolic compounds with high amounts of both aromatic and aliphatic compounds. However, methane, hydrogen, and carbon dioxide were the primary compounds of the gas product, while the gas product energy content was between 10.1 and 21.7 MJ/m^3. The analysis also stated that the bio-oil heating value decreased as the temperature increased where it was about 33.8 MJ kg^{-1} at 400 °C and 27.9 MJ kg^{-1} at 600 °C as reported.

In 2016, Rafiq et al. [13] explored effect of pyrolysis temperature on corn stover biochar both chemical and physical properties. Slow pyrolysis of corn stover was conducted at three different pyrolysis temperatures which are (300, 400, and 500 °C) at the same holding time and temperature rate. The main objective was to analyze carbon content, stability, and energy potential of corn stover biochar produced at these temperatures. The results indicated that the biochar yield decreased while the temperature increased with a maximum yield of 29.2% at 500 °C. However, the carbon content increased from 45.5% at 300 °C to 64.5% at 500 °C in which more volatile compounds were released as gas and bio-oil, indicating higher thermal conversion. Moreover, analysis showed a decrease in O/C and H/C ratios due to the decrease in the oxygen and hydrogen content at higher temperatures which further supports the pattern of improved biochar stability. The FTIR and NMR analysis confirmed that aromatic carbon concentration increased while the temperature increased. Energy potential of the biochar produced was evaluated and found the maximum energy output of pyrolysis to be 10.12 MJ/kg at 500 °C, validating its ability as a sustainable source. Additionally, porosity and surface area of biochar was improved at higher temperatures. The study also includes the environmental benefits of pyrolysis, in which it is considered to be carbon negative technology.

In 2017, Yang et al. [14] investigated the effect of pyrolysis temperature and different bed materials on bio-oil's product distribution, chemical composition, pH, kinetic viscosity and calorific values. The fast pyrolysis was conducted using dolomite as a catalytic bed material in a fluidized bed reactor at a temperature between 450 °C and 525 °C. Moreover, to avoid blocking the feeding system, two screws were fitted, a nozzle and a heat exchanger were connected to the condenser for rapid cooling of the cleaned hot gas into bio-oil. It was found that, when using dolomite as bed material, a maximum bio-oil yield of 43.3 wt% was recorded at a temperature of 500 °C. While the biochar yield decreased consequently at higher temperature, Fig. 2.6 indicates that bio-oil yield increases at higher pyrolysis temperature to a certain level before dropping while the temperature further increases. The GC/MS analysis indicated the bio-oil's pH, kinetic viscosity and calorific values to be in ranges of (3.18–3.99), (5.0 mPa · s–5.9 mPa · s), (7.63 MJ/kg–8.25 MJ/kg) respectively. This analysis showed that the bio-oil of corn stover has lower viscosity than other pyrolysis oils and hydrocarbon fuels. Additionally, the resulting bio-oil contains compounds such as acetol, butanone, acetic acid, propionic acid, ethylene glycol, and phenol. Furthermore, since pyrolysis oils are immiscible in diesel due to their nonpolar compounds, however a homogenous emulsion can be obtained when mixing different combinations of pyrolysis oil, emulsifier and diesel.

In 2018, Hung et al. [15] experimented with the effect of particle size, pretreatment (acid pretreatment and steam explosion, and aluminum oxide (Al_2O_3) catalysis on the microwave pyrolysis of corn stover. The results obtained show that the smaller particle size the greater the microwave heating efficiency which enhances gas production. In comparison with steam explosions, which have a low influence on product distribution, acid pretreatment had a significant impact on liquid yield of biofuel. Moreover, gas yield has increased by the addition of Al_2O_3 catalyst when using 10-mesh particles, however because of encapsulation by biomass particles this effect is lowered when using 50-mesh catalyst. In addition, the experiment found that while the yield of H_2, CH_4, and CO_2 was relatively low in the initial phase, about 50% of CO was produced during the first 5 min. First- and second-order models were used to analyze the reaction kinetics. Findings indicate that optimizing particle size, pretreatment methods, and catalyst selection can improve yield and efficiency of biofuel, proving that corn stover has a high potential as a feedstock to achieve sustainable bioenergy production.

In 2019, Polin et al. [16] examined conventional and autothermal pyrolysis of high ash corn stover using a pilot-scale fluidized bed reactor with the aim of enhancing the reactor stability and the overall bio-oil yield and that is to overcome challenges due to bed fouling which hindered steady state operation in the reactor. Results showed that stability was improved when using coarser bed material as it promotes biochar attrition, achieving high bio-oil yields for both conventional and autothermal pyrolysis. Autothermal pyrolysis required about 6.8% of oxygen to biomass ratio and in a reactor of 8.9 cm diameter the corn stover yield increased by 2.8 from 7.8 to 21.9 kg/h. Moreover, bio-oil yield remained the same with not much change while the biochar yield decreased to 18.5% and bio-oil light end yield decreased by 4.7%. Additionally, results from the carbon balance analysis showed that the non-condensable gas yield has increased by 10.2 wt%, indicating a 92% rise in CO_2 production.

In 2019, Ali et al. [17] examined the effect of various gas flow rates on the fast pyrolysis of corn stover biofuel yields and its composition using a pilot-scale fluidized bed reactor. The experiments were carried out at a temperature of 500 °C, with the use of silica sand as the bed material and at different nitrogen flow rates which are 45, 70, and 80 L/min. The compositions and pyrolysis products were evaluated by mass calculations and proximate and ultimate analysis. A maximum bio-oil yield of 52 wt% was observed when the fluidized bed reactor operated with a nitrogen flow rate of 70 L/min. The resulting bio-oil, when analyzed with GC–MS, showed a diverse composition with key components such as levoglucosan (a sugar derivative), various phenolic compounds, and organic acids. As the gas flow increased, the biochar yield decreased with higher ash content which may be better used for soil enhancement purposes. Moreover, the oxygen content in bio-oil was reduced through the replacement of nitrogen with the recycled pyrolysis gas in a tail gas reactive pyrolysis (TGRP) process, which helped improve its energy value.

In 2021, Hu et al. [18] investigated the infrared heated pyrolysis of corn stover in comparison with conventional electrically heated pyrolysis. The aim of this study is to improve biofuel yield and quality by optimizing heating methods and overcome challenges such as the thermal gradients and secondary reactions that take place during the traditional electrically heated pyrolysis process. The results showed that better heat transfer can be obtained through infrared heating, as it restrains the secondary cracking of volatiles. Thus, higher yield of bio-oil was achieved (34.34 wt% at 550 °C), while electric heating yield was (26.40 wt% at 500 °C). Moreover, the bio-oil produced contained higher phenolic compounds with a yield of 19.36%, which makes it more conventional for biofuel optimization. Additionally, syngas potential energy can be enhanced as lower CO_2 emissions along with higher yields of H_2 and CH_4 were observed from the (GC–MS) gas analysis performed. Both thermal degradation and pyrolysis kinetics were analyzed by the thermogravimetric method (TGA), and the activation energy was evaluated by FWO and KAS methods and were found to be 19.61–38.33 kJ/mol (FWO) and 11.40–29.51 kJ/mol (KAS). These values were quite lower than those obtained from previous research of pericarp pyrolysis.

In 2021, Meng et al. [19] experimented the effect of different pretreatment methods on biochar properties from pyrolysis of corn stover. Pretreatment methods such as drying, vacuum freezing drying (VFD), pelleting, and ultrasonic vibration-assisted pelleting (UV-AP) were examined to explore their influence on corn stover biochar physicochemical and morphological characteristics. Findings indicate that the internal and external surface and structure of biochar is highly influenced by pretreatment methods. Values of biochar highest and lowest specific area, micropores area, and total pore volume obtained from UV-AP and drying pretreatment methods were 25.84 m^2/g, 9.27 m^2/g, 0.53 cm^3/g, and 17.33 m^2/g, 4.39 m^2/g, 0.29 cm^3/g, respectively. In all the four pretreatment methods, drying had the highest carbon content followed by the VFD method, while pelleting had the lowest carbon content followed by UV-AP method. Furthermore, while pretreatment methods had no major effect on biochar pH level other parameters were significantly influenced such as biochar yield, enhancement factor, and energy yield.

In 2021, Fodah et al. [20] evaluated the quality of bio-oil and biochar made through microwave-assisted pyrolysis of corn stover using different types of adsorbents such as silicon carbide (SiC) and activated carbon (AC). Various adsorbent ratios (10, 20, and 30%) were examined to find out the effect on heating performance, product distribution, and energy recovery. In comparison with the absence of adsorbents, results showed that heating efficiency has improved while using adsorbents, increasing the maximum temperature and the heating rate by 170 °C and 36 °C/min respectively. While bio-oil yield reached about 45 wt% with the use of AC, SiC increased the gas yield to a maximum value of 48 wt%. using AC as an adsorbent produced bio-oil with improved physical properties, including 37% water content, a pH of 5.3, a viscosity of 2.9 mPa · s, and a density of 1 g/mL. Moreover, the content of carbon and hydrogen in bio-oil has increased with the use of AC, to calorific value of 17 MJ/kg. It also has increased hydrocarbons and phenols to 36% and 28% respectively while with relatively low oxygen content of 2%. Additionally, the biochar produced with the help of AC has improved characteristics such as cleaner pores, higher carbon content, and lower ash content with a calorific value of 28 MJ/kg, and a surface area of 105.2 m^2/g. A maximum value of 55% was achieved as the net energy recovery from corn stover when using AC. Further runs of the experiments at which both types of adsorbents were used with relatively low changes in heating behavior and product yield. Findings indicate that AC has the best results for improving bio-oi and biochar quality.

In 2022, Hu et al. [21] conducted an experiment using a new two stage rotary kiln with baffle to investigate the pyrolysis behavior of corn stover, highlighting its influence on yield, pyrolysis kinetics, and biochar properties. The activation energies were calculated by thermogravimetric method (TGA) and the Distributed Activation Energy Model (DAEM) and were in the range of 140.13–179.65 kJ/mol for the conversion of 0.2–0.8, respectively. The results showed that as the temperature rose, the biochar yield decreased while the bio-oil yield reached a maximum value of 23.02 wt% at 700 °C, however it decreased to 22.16 wt% at 800 °C. Additionally, the heat distribution was enhanced, due to the baffle that increased interaction between particles in the dead zone. The composition analysis of bio-oil indicated an increase in phenol content with the temperature, while the content of both ketones, aldehydes, and furfurans decreased. Moreover, as the temperature increased, higher content of H_2, CH_4, and C_2–C_3 was indicated, while the CO_2 content decreased. Increasing the temperature resulted in higher BET surface and total volume of biochar, however it decreased afterwards while the temperature increased. Findings also showed that biochar metal volatilization and pH increased at higher temperatures, however its high heating value decreased, which indicates energy loss at higher temperatures.

In 2022, Abatyough et al. [22] conducted an experiment with the aim of optimizing the reaction conditions of corn cobs pyrolysis to improve the resulting bio-oil yield. The response surface methodology (RSM) of central composite design matrix (CCD) was used to analyze and optimize the influence of temperature, catalyst load and retention time. Corn cobs were demineralized through three different which are nitric acid, acetic acid, and water to reduce inorganic content. Results showed that nitric acid was the most effective solvent in reducing alkali metals concentration. Optimal conditions to achieve maximum bio-oil yield of 20% were found to be at a temperature of 600 °C, catalyst load of 7 wt%, and retention time of 3 h. Moreover, bio-oil yield was predicted through a generated equation with an R^2 value of 0.9505, which indicate a great correlation between the predicted and experimental yields. Additionally, bio-oil yield is significantly influenced by the reaction parameters as the variance analysis (ANOVA) has proven the importance of the model with a p-value of 0.0317. The most notable parameter effect on the bio-oil yield was temperature then catalyst load followed by retention time. Furthermore, the interaction between the catalyst load and the retention time was high, however the interaction between temperature and catalyst load was not very high.

In 2023, Wang et al. [23] investigated the effect of pressure on the pyrolysis and gasification kinetics of corn stover, highlighting the influence of increasing pressure on activation energy, gas production and reaction efficiency. The results were obtained through an experimental thermogravimetric analysis at different levels of pressures: atmospheric pressure, high pressure of (0.8–1.0 MPa), and at supercritical water gasification. Reaction mechanisms were analyzed using several models of kinetic energy calculations. Findings showed that activation energy which is required for both pyrolysis and gasification has increased, affecting the product distribution. The activation energy of pyrolysis is 64.95 kJ/mol at atmospheric pressure and has increased to 71.38 kJ/mol at 0.8 MPa and 76.30 kJ/mol at 1.0 MPa. Results indicate that more complete pyrolysis and gasification can be achieved as high pressure not only prevents gas release but also improves molecular interaction. Eventually the study found that the optimal pressure was at 1.0 MPa, which promotes effects that overcome the process inhibitory effects.

In summary, improving the quality of bio-oil and biochar through the optimization of the process conditions was the main focus of the recent research on the pyrolysis of corn stover. Different elements have significantly influenced both the products' yields and properties such as particle size, pressure, reactor design, and temperature. In general, higher biochar yield and energy content is most likely produced at lower temperatures, however at higher temperatures produces phenolic -rich bio-oil and gas. While improvements in reactor technologies have sufficiently enhanced both the product recovery and thermal efficiency, some challenges to do with bio-oil quality, including high viscosity, acidity, and poor stability, remain unsolved. Although such challenges were addressed by the use of different additives and through catalytic upgrading to handle limitations and improve biofuel grade compatibility.

3 Case Studies Analysis

3.1 Overview

This section consists of three different case studies exploring the production of biofuel products such as bio-oil, biochar, and gas. Each experiment was discussed in detail including the procedure, the setup, and results. The first experiment investigated the optimal conditions for maximizing product yield of corn stover by pyrolysis in a pressurized bed, while the second case studied the impact of diverse pyrolysis temperatures and bed materials on the distribution of products and physicochemical attributes. The last experiment examined the microwave pyrolysis process of corn stover and evaluated the effect of particle size, pretreatment and catalyst on product yield.

3.2 *Case Study 1: Assessing the Potential for Biofuel Production of Corn Stover Pyrolysis Using a Pressurized Batch Reactor [12]*

Authors: Jewel A. Capunitan, Sergio C. Capareda

Year: 2012

3.2.1 Experimental Aim

The study aimed to find the optimal conditions for maximizing product yield of corn stover by pyrolysis in a pressurized bed. Both product yield and their properties were analyzed at different temperatures of 400, 500, and 600 °C, while the pressure was maintained at 100 psi. Conversion efficiencies were also assessed at each temperature to find out the possibility of utilizing products of corn stover pyrolysis as an alternative energy source highlighting the chemical composition and energy content of the resulting products.

3.2.2 Experimental Materials and Methods

1. **Collection and Preparation of Feedstock**

The particle sizes of ground corn stover samples used were passing through 3 mm-screens, and less than 10% of moisture content was maintained.

2. **Component Analysis**

ASTM standards were applied in the experiment to determine the Proximate analysis, ultimate analysis, and gross calorific value (GCV). The NREL Laboratory Analytical Procedures (LAPs) was also used to examine the structural composition of feedstock samples. The results are summarized in Tables 1 and 2, respectively.

3.2.3 Experiment Setup

The experiment was designed to be conducted in a pressurized batch reactor, using an automatic temperature controller along with a condenser as shown in Fig. 2. The volume of the generated gas was determined through the water displacement method. Another water jacketed tube condenser wrapped around the reactor, the bottom edge of this condenser was attached to a receiving vessel through which condensed vapor was gathered. A blend of ethylene glycol and water was cooled using a recirculating chiller to a temperature of −10 °C and got used as a refrigerant. The experiment examined only one factor which temperature at three different temperatures which

Table 1 Corn stover proximity and ultimate analysis [12]

Biomass characteristics	Experimental value
Proximate analysis (%w/w), dry basis	
Ash	6.62 ± 0.27
Volatile matter (VM)	78.7 ± 1.4
Fixed carbon (FC)	14.7 ± 1.7
Moisture content (%w/w), dry basis	
C	44.3 ± 0.4
H	6.28 ± 0.09
N	0.80 ± 0.06
S	0.23 ± 0.06
O (by difference)	41.8 ± 0.6

Table 2 Corn stover compositional analysis [12]

Biomass characteristics	Experimental value
Compositional analysis (% w/w), dry basis	
Lignin	13.1 ± 0.1
Extractives	18.1 ± 0.4
Glucan	31.9 ± 0.3
Xylan	19.6 ± 0.2
Galactan	1.0 ± 0.1
Arabinan	2.3 ± 0.1
Protein	5.8
Heating value (MJ kg^{-1})	17.0 ± 0.7

were 400, 500, and 600 °C, highlighting the fact that the order of these individual runs was random.

3.2.4 Experiment Procedure

1. A sample of around 100 g was fed to the reactor, the reactor was then closed tightly by inserting flanges and securing bolts around the head of the reactor.
2. For about 20 min nitrogen gas was purged into the reactor at less than 10 psi, to assure an oxygen free environment for the process before each run.
3. After turning on the temperature controller and specifying the temperature setpoint, the reactor is then heated to the desired temperature of each run.
4. The valve going to the condenser was closed until the pressure inside the reactor approached 100 psi, then this valve was lightly opened to sustain the pressure.
5. The pyrolysis process continues for approximately 20 min, once the desired temperature is reached.

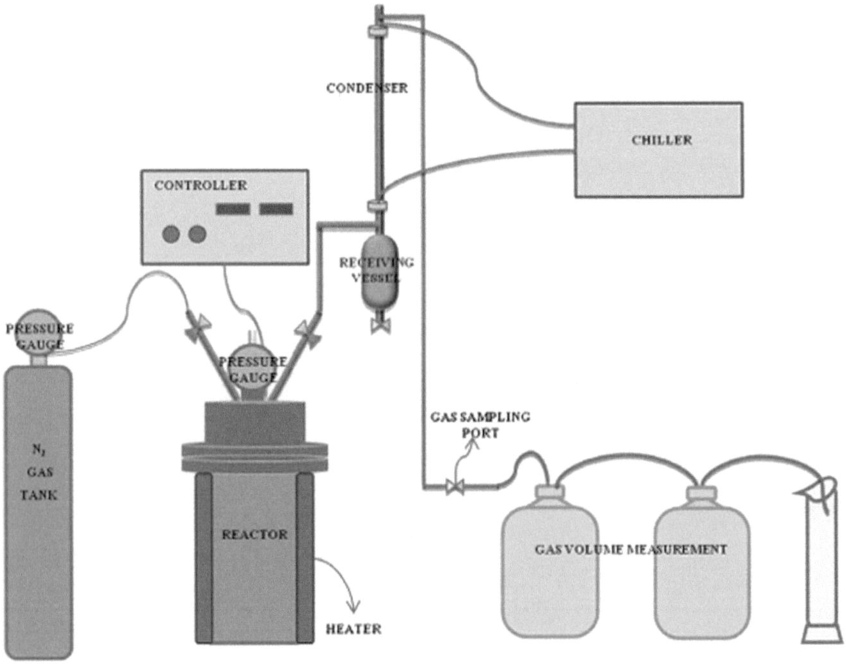

Fig. 2 Experiment setup of the pressurized batch reactor for pyrolysis corn stover at different temperatures [12]

6. Then gas composition analysis is carried out to the gas samples obtained from the reaction.
7. The reactor cools down after shutting off the heater.
8. Bio-oil and "aqueous product" were obtained from the condenser.
9. Biochar was obtained from the reactor after it cooled.
10. All products were weighted and analyzed.

3.2.5 Results and Discussion

1. **Product Analysis**

The corn stover pyrolysis was conducted at three different temperatures, to show the effect of temperature on product yields distribution. Figure 3 demonstrates the inverse relationship between the increasement in pyrolysis temperature and the decrease in both the bio-oil and biochar yields. The gas yield on the hand showed quite the opposite, as its product yield rose at higher temperatures. In comparison with all product yields, the biochar yield was the highest across all three temperatures which is (32–37 wt%), then the liquid products followed by gaseous product with a product yield of (28–31 wt%) and (15–21 wt%) respectively. Secondary cracking reactions were promoted due to the extended residence time of vapor in the reactor to reach a reaction

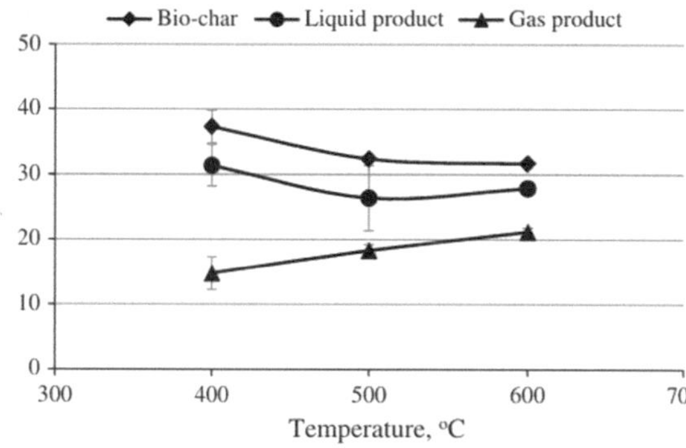

Fig. 3 Yields of different biofuel products at different corn stover pyrolysis temperatures in batch reactor [12]

pressure of 10 psi, which enhanced the formation of solid carbonaceous materials and resulted in the higher biochar yield. Pyrolysis temperatures and biochar production were evaluated using ANOVA analysis, which confirmed a statically significant effect of 95%, where increasing the temperature from 400 to 500 °C indicated an initial drop in the biochar yield, however further increasement in the temperature showed quite minor changes. This drop in biochar yield at higher temperatures was due to the release of large amounts of volatiles as well as minor moisture loss.

2. **Biochar Characteristics**

The results of the proximity analysis conducted on the biochar yield at various pyrolysis temperatures are demonstrated in Fig. 4. Results showed that the volatile matter (VM) of biochar samples decreased dramatically from 26.5 to 13.0%, as the temperature increased from 400 to 600 °C, which is much lower than the (VM) of the feedstock that was approximately 78.7%. Moreover, due to the reduction in the volatile matter at higher temperatures, the fixed carbon content (FC) increased from 56.5 to 67.0%, indicating higher thermal decomposition. The rise in (FC) content was more noticeable between 400 and 500 °C than between 500 and 600 °C and that is because FC in these biochar samples mainly originates from lignin which is more resistant to decomposition. Additionally, a minor increase in ash content was observed from 16.9 to 20.1% with increasing temperatures from 400 to 600 °C, implying that mineral residues become concentrated in the biochar.

Elemental analysis was also applied on the biochar samples as shown in Fig. 5. The analysis results indicate that carbon content increases with higher temperatures, while both of the hydrogen and oxygen contents decreased and that is due to the release of volatiles during the pyrolysis process. Consequently, lowered molar ratios of H/C and O/C were observed, especially at 600 °C where the biochar characteristics were more similar to the anthracite coal on the Van Krevelen diagram. And this trend of higher carbon content and lower contents of oxygen and hydrogen is very important as it increases the energy density. A drop in the gross calorific value (GCV) of the biochar produced from 27.9 MJ/kg at 400 °C to 22.9 MJ/kg at 500 °C was observed, however it increased slightly at 600 °C to 23.1 MJ/kg. Regardless of this, in comparison with raw corn stover, all of the biochar samples exhibited higher energy content which

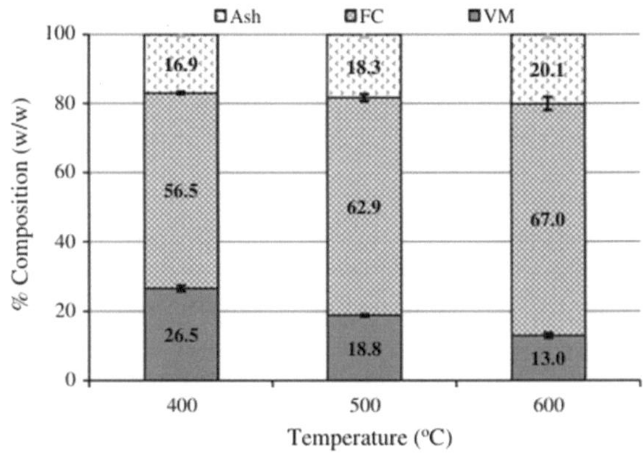

Fig. 4 Proximity analysis plot of biochar samples produced at various pyrolysis temperatures [12]

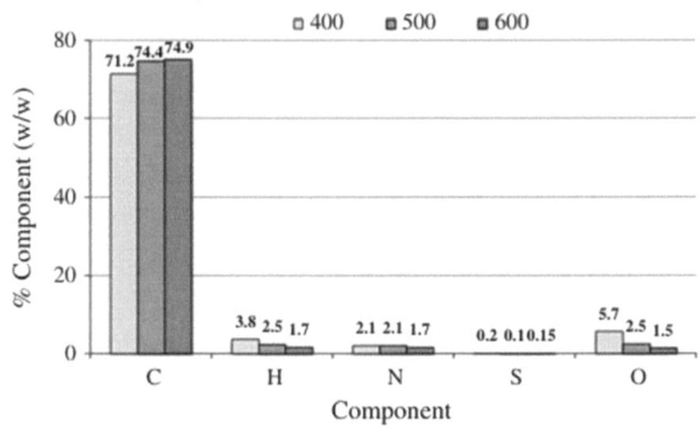

Fig. 5 Elemental composition of biochar produced at various pyrolysis temperatures [12]

proves its ability as solid fuel. Furthermore, the favorable balance of yield, energy value, and ash content was observed at 400 °C.

3. Bio-oil Properties

The resulting liquid product consisted of bio-oil, which is a very thick, dark organic phase, and a somewhat cloudy yellowish mixture that may be referred to as an aqueous product. The ultimate heating value of pyrolysis was about 33.8 MJ/kg at 400 °C, within a temperature range of (400, 500, and 600) °C, indicating negative results at higher temperatures. The ultimate analysis indicated the oil was low oxygen but high in carbon content which promotes a greater heating value. According to the H/C and O/C ratios, the oil required further treatment to be used as an alternative energy source. The analysis also found lower ash concentrations were found to be about 0.03–0.30% w/w which is quite low, and the moisture level was around 15–30% acceptable compared to past studies results. However, the acid number was not as much as resulted in other papers. Moreover, FTIR and GC–MS found different functional groups in the bio-oil sample using the FTIR spectroscopy approach that measures their infrared absorption. The results of this analysis indicated the existence of various functional groups as the bio-oil sample exhibited a higher absorption peak

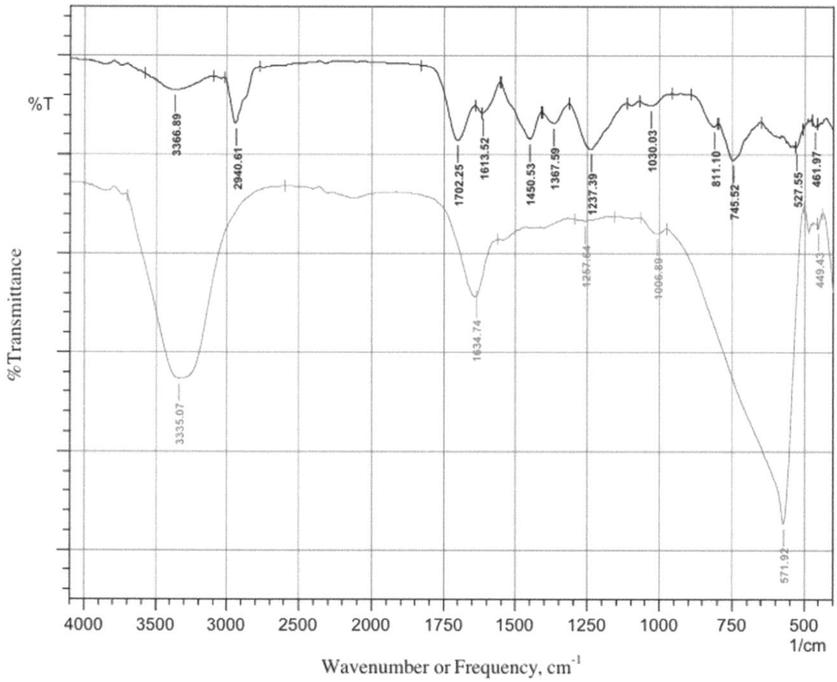

Fig. 6 FTIR spectra of the bio-oil produced from pyrolysis of corn stover (bio-oil at the top, and aqueous product at the bottom) [12]

than that of the aqueous product as it is shown in Fig. 6. Results showed that the oil was rich in phenolic compounds as it was about 26.7%, followed by 20.2% of benzene of different types. A very small number of aromatic compounds and ketones were also indicated.

4. Gas Product Compositions and Heating Values

Significant amounts of hydrogen, methane, carbon monoxide, and carbon dioxide were found in the resulting pyrolysis gas as it is shown in Fig. 7. The figure indicates a clear trend of increasement at higher temperatures for certain gases such as hydrogen and methane, while a slight decrement other gases like carbon monoxide, and carbon dioxide were observed. These findings are consistent with the previous studies which indicate that carboxyl groups release carbon dioxide at lower temperatures, the release of both hydrogen and methane requires secondary reactions at higher temperatures. Moreover, due to the increase formation of gases such as hydrogen and methane that are rich in energy, the resulting heating values of the gas product showed a significant increase at higher pyrolysis temperatures in which it grew from 10.1 MJ/m^3 (7.3 MJ/kg) to 21.7 MJ/m^3 (25.6 MJ/kg) at 400 °C and 600 °C respectively. The results obtained are consistent with or even exceed those found in other studies.

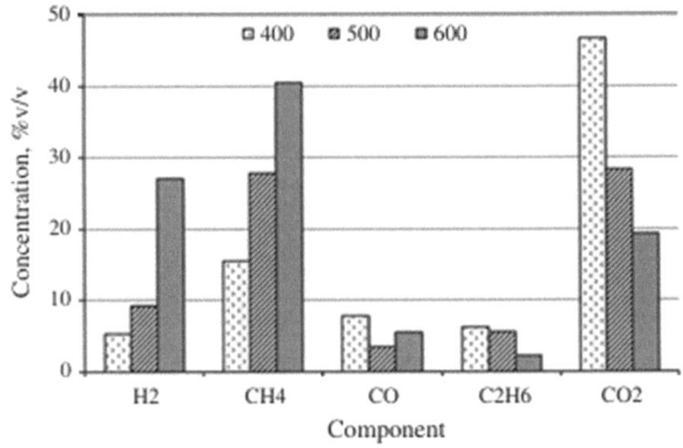

Fig. 7 Different gas components at various pyrolysis temperatures [12]

5. **Mass and Energy Conversion Analysis**

The efficiency of the pyrolysis process at different temperatures was evaluated using two equations: one for mass conversion and another for energy recovery. Mass conversion was calculated as the ratio of the product weight to the initial weight of corn stover, while energy recovery was determined by comparing the energy content of each product to that of the feedstock. Although the high conversion efficiency of 80% indicated for both mass and energy, significant mass losses were observed at higher pyrolysis temperature, as shown in Fig. 8. Moreover, the biochar conversion rate was found to be the most dominant among other products at 400 °C in terms of mass and energy, accounting for 37.3% of mass and 65.7% of energy. Additionally, the formation of combustible gases increased with temperature, resulting in higher energy recovery for the gas phase. Due to the sharp drop in bio-oil yield at higher pyrolysis temperatures, its energy recovery was the lowest among the gas and biochar products. In addition, a noticeable reduction in the biochar quantity was observed at the same elevated temperature which led to a lower contribution to the total energy recovery.

3.3 Case Study 2: Catalytic Fast Pyrolysis of Corn Stover in a Fluidized Bed Heated by Hot Flue Gas: Physicochemical Properties of Bio-oil and Its Application [14]

Authors: Yang Wen, Fu Peng, Yi Weiming

Year: 2017

Fig. 8 Mass conversion (**a**), and energy conversion (**b**) of different products at various pyrolysis temperatures [12]

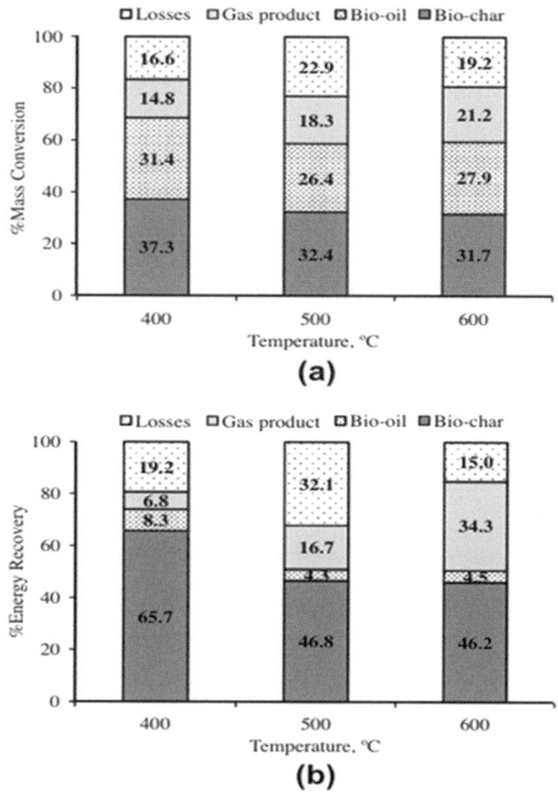

3.3.1 The Aim of the Experiment

The study aimed to explore the impact of diverse pyrolysis temperatures and bed materials on both the distribution of products derived from corn stover and the physicochemical attributes of the resulting bio-oil. A fluidized bed reactor was employed, maintained at temperatures between 450 and 525 °C, utilizing corn stover particles with dimensions less than 1.00 mm, a hot flue gas flow rate of 5.5 m^3/h, and dolomite as the catalytic bed constituent. To characterize the pyrolysis generated products, analytical techniques such as GC–MS were implemented to assess the bio-oil's chemical composition, while established testing protocols were used to ascertain its pH, kinetic viscosity, and calorific values.

Table 3 Proximity and ultimate analysis of feedstock [14]

Proximate analysis (wt%)	Value
Moisture	6.35 ± 0.12
Volatile matter	74.61 ± 0.34
Fixed carbon	13.45 ± 0.21
Ash	5.59 ± 0.17
Ultimate analysis (wt%)	
Carbon	48.62 ± 0.23
Hydrogen	5.86 ± 0.09
Nitrogen	0.85 ± 0.06
Oxygen (by difference)	44.67 ± 0.25
Calorific value (MJ kg)	17.3 ± 0.65

3.3.2 Material and Methods of the Experiment

1. **Feedstock Collection and Preparation**

The corn stover, which is the feedstock in this experiment, was crushed and then sieved, to a particle size was less than 1.0 mm.

2. **Feedstock Component and Elemental Analysis**

Proximate and ultimate analyses were performed to determine the main characteristics of the corn stover as shown in Table 3. The table includes moisture, volatile matter, the fixed carbon, ash, carbon, hydrogen, nitrogen, and the oxygen content as well as the calorific value of the corn stover was also measured.

3.3.3 Experiment Setup

The experiment was carried out in a fluidized bed reactor setup as shown in Fig. 9. The setup included different equipment such as a corn stover hopper, two identical screw feeders to avoid any sticking in the feed entry, a 2 m high reaction tube with a 70 mm diameter, an electrical heater to set the desired operating temperature, while the solid particles were isolated from the hot gas stream using two cyclones. Moreover, the nozzles and a heat exchanger were attached to the condenser which condenses the hot gas into liquid bio-oil. A pump was employed to recirculate the condensed oil and enhance the cooling process. Other monitoring devices were utilized such as thermocouples and pressure meters.

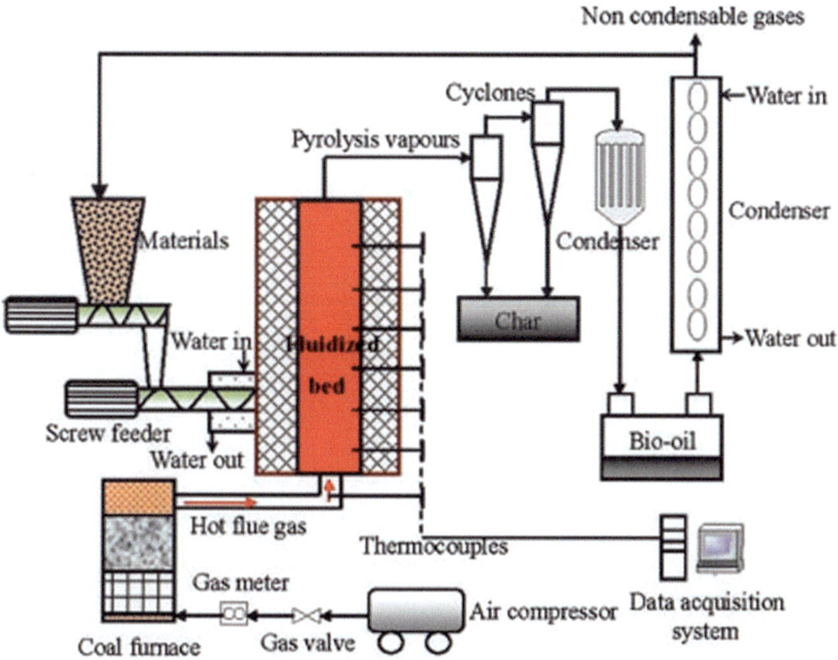

Fig. 9 Experimental setup of the corn stover fast pyrolysis process [14]

3.3.4 Experiment Procedure

1. The reactor was first loaded with a certain catalyst material such as dolomite, quartz sand, and aluminous soil.
2. The electric heater is turned on to raise the temperature to the desired operating temperature of the range between 450 and 525 °C.
3. Then ethanol was added to the condenser to assist in scrubbing the hot pyrolysis vapors.
4. With the help of a preheater, a flow of hot flue gas was kept at a rate of 5.5 m^3/h.
5. 3 kg of crushed corn stover particles were loaded inside the reactor.
6. The first screw feeder is enabled to regulate the amount of corn stover fed into the hopper.
7. Then the second screw feeder is enabled but at a higher speed, to avoid clogging and ensure smooth material transport.
8. The pyrolysis process is then allowed to run for more than 10 h, to ensure a complete conversion.
9. After that, products are collected using cyclones to isolate solid material such as charcoal and ash from the hot gas stream.

10. The hot gas stream is then directed to the condenser to cool and condensate it to bio-oil.
11. The leftover liquid is recirculated back to the condenser using an oil pump.
12. After the experiment, different tests were run on the products to analyze their properties.

3.3.5 Experiment Results

1. **Pyrolysis Product Distribution**

The pyrolysis temperature and the reactor bed material utilized have both affected the pyrolysis product distribution significantly. Figure 10 shows the product distribution when utilizing quartz sand, indicating a gradually increasing trend of bio-oil yield with temperature. As it reached a maximum value of 39.6 wt% at 500 °C, before it plateaued at further temperatures. However, the gas phase product showed an opposite trend of decreasing with temperature before it started to maximize at temperatures higher than 500 °C. This rise is due to the quicker secondary reactions at elevated temperatures, causing vapors to over crack. The gas yield can also be affected positively by the presence of inorganic materials such as K_2O, Na_2O, CaO, and Fe_2O_3. Moreover, biochar yield exhibited a constant decreasing trend with higher temperatures, which is likely because of the improved primary breakdown of feedstock at higher temperature, and the possible secondary cracking of the resulting biochar to gaseous compounds. Additionally, while using quartz sand as bed material, around 90% of the gas produced was composed primarily of CO_2, CO, and CH_4. The content of both CO and CH_4 exhibited an increasing trend with temperature in which their content reached a maximum value of 33.95 vol% and 8.82 vol% respectively at 525 °C. However, due to enhanced secondary reactions at higher temperatures, the content of CO_2 decreased with higher temperatures. As a result, a higher heating value of 9.11 MJ/kg was obtained for the resulting gas product supporting its validity as a possible energy source. In terms of bed material influence on the resulting bio-oil and heavy oil content, dolomite showed a great result achieving the highest yield of bio-oil of 43.3 wt% at 500 °C, as shown in Fig. 11. The figure demonstrates a fluctuating trend of heavy oil content with temperature. However, dolomite recorded the lowest heavy oil content among all other bed materials at temperatures exceeding 450 °C.

2. **Bio-oil Properties and Compositions**

A dark, brownish, thick, and heavy oil with a very acrid scent resulted from this experiment. Taking into consideration the effect of temperature and bed material variation, analysis conducted on the resulting light bio-oil showed a pH level of around 3.18–3.99, indicating its acidic nature. Compared with other pyrolysis oils, the bio-oil obtained from the process was less viscous with about 5.5 mPa · s at 18 °C, which is even lower than certain crude oils. The low viscosity of the light bio-oil was also identified to be between 5.0 and 5.9 mPa · s, while insignificant values

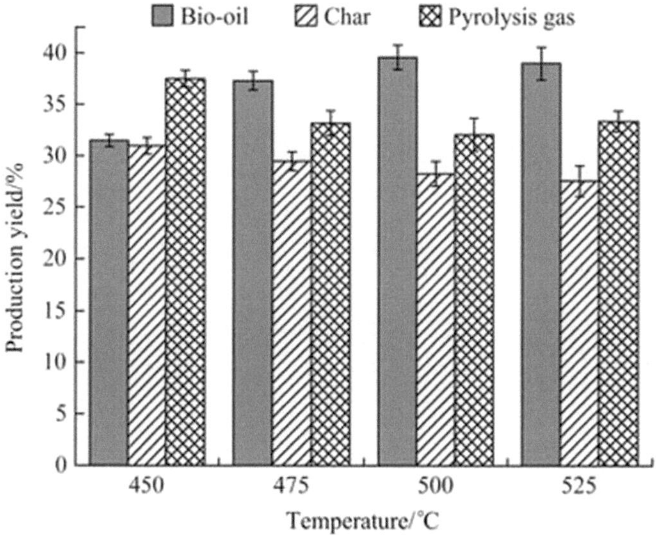

Fig. 10 Products yield at different pyrolysis temperatures while utilizing quartz sand as bed material [14]

of about 7.63–8.25 MJ/kg were observed for the calorific value, which was caused by the high amount of both oxygen and water. Moreover, during the experiment three different types of catalyst were used. However, the highest bio-oil yield was indicated at 500 °C using calcined dolomite. Despite having much larger energy density than the raw feedstock, the bio-oil's energy density was certainly lower than those of fossil fuels due to the high oxygen content. Additionally, over 100 peaks were indicated during the GC/MS analysis as demonstrated in Fig. 12, which are mainly oxygenated compounds such as acids, aldehydes, ketones, alcohols, phenols, and anhydrosugars. Such content of a nonpolar nature reflected on the bio-oil's ability to mix and react with other crude-based fuels as mentioned previously. However, emulsification processes could overcome such an issue.

3. **Pyrolysis Oil Stability Features**

Results indicated that the viscosity of pyrolysis oil increases over time during storage. Different storage conditions were tested, indicating that chemical processes such as polymerization and condensation reactions occur throughout the storage time at higher temperatures as shown in Fig. 13. These reactions cause oil molecules to bond and combine, which accelerates the increase in its viscosity. However, this behavior of increasing viscosity slows down after a period of time. Although this makes the resulting bio-oil somehow unstable, but it is still usable when certain types of Inhibitors such as hydroquinone, are added.

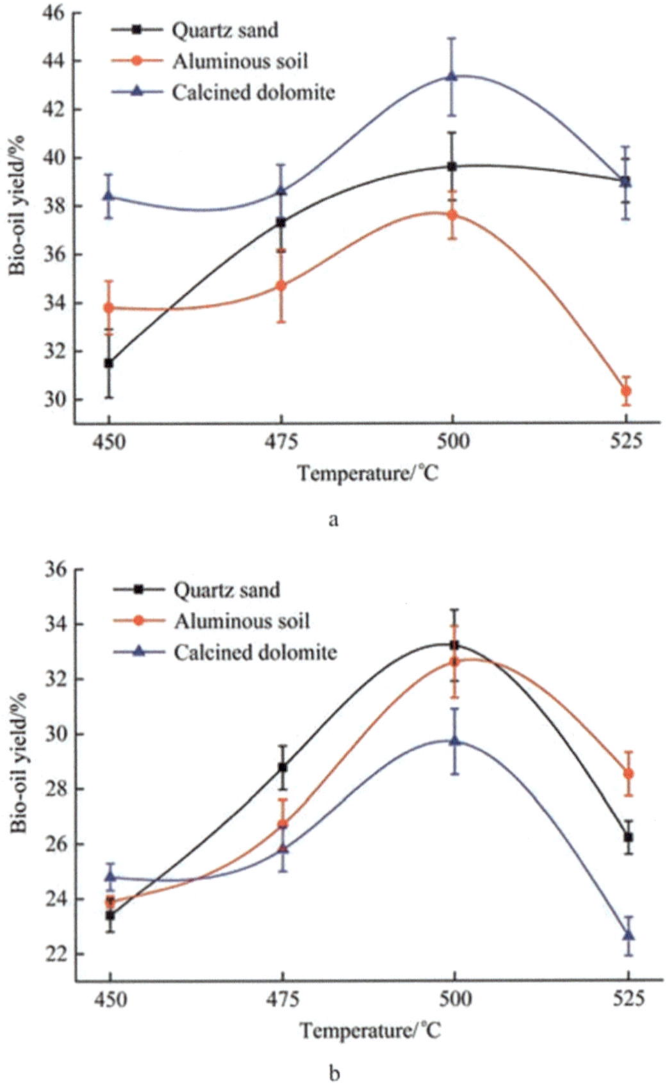

Fig. 11 **a** The influence of different bed materials bio-oil, **b** heavy oil content [14]

3.4 Case Study 3: Effects of Particle Size, Pretreatment, and Catalysis on Microwave Pyrolysis of Corn Stover [15]

Authors: Yu-Fong Huang, Wen-Hui Kuan, Chun-Yuan Chang

Year: 2018

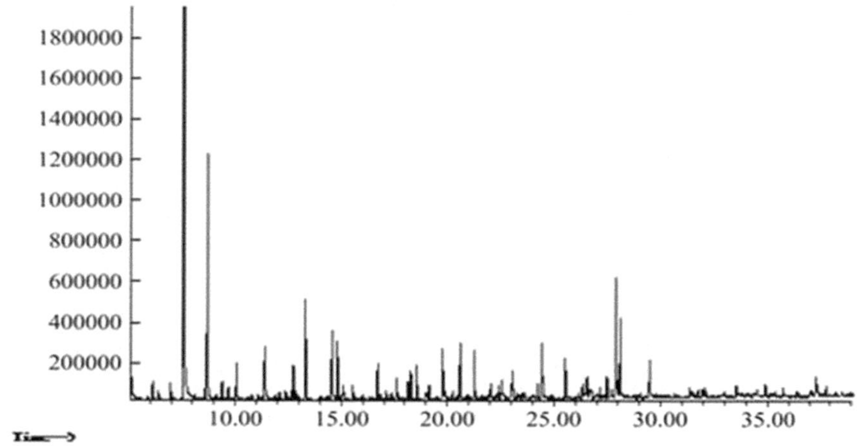

Fig. 12 An overall ion chromatogram of bio-oil obtained from corn stover fast pyrolysis [14]

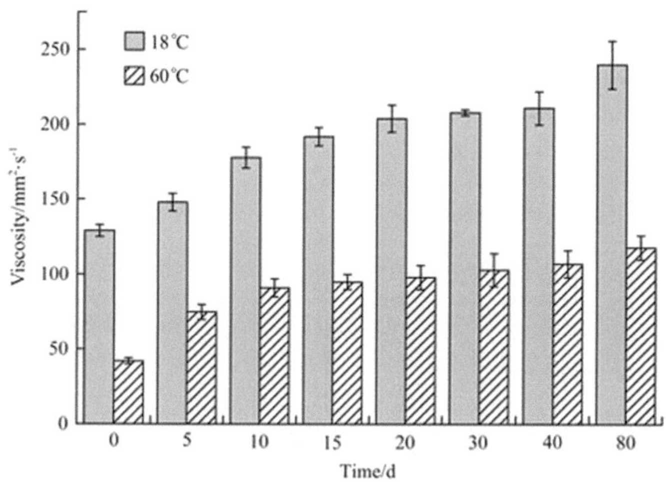

Fig. 13 Pyrolysis oil viscosity at different storing conditions over time [14]

3.4.1 Experiment Aim

This experiment examines three different parameters, which are particle sizes of less than 4.00 mm to 0.150–0.250 mm, pretreatment of two kinds (acid pretreatment and steam explosion), and the use of aluminum oxide as a catalyst. The aim was to find out how these parameters influence the distribution of products, yield, and the composition of the gaseous phase as well as improving the heating efficiency for the microwave pyrolysis process of corn stover.

3.4.2 Experiment Materials and Methods

1. **Feedstock Collection and Preparation**

Crude corn stover particles are gathered from corn farms in Taiwan. These corn grains were air dried for approximately two weeks and then crushed and screened through five different particle sizes of less than 4.00, 2.00–4.00, 0.425–2.00, 0.250–0.425, and 0.150–0.250 mm. The catalyst particles were also passed into screen of 10–50 mesh size.

2. **Pretreatments**

A couple of different types of pretreatments were conducted on feedstock throughout this experiment. The acid pretreatment involved mixing every 1 g of dry corn stover with about 1 ml of deionized water and 10 ml of phosphoric acid. This mixture was then warmed up at 323 k in a water bath and kept in there for 10 h. After that, the 50 ml of deionized water was added to the mixture and mixed heavily. After rinsing the mixture with water, NaOH solution was added to the mixture to naturalize its pH level. The other pretreatment method was steam explosion, in which used an autoclave at 160–260 °C and 6–34 bar for 1 min. After treating the corn stover, it was air-dried for approximately 14 days before crushing it to the desired size.

3. **Analytical Techniques**

The proximate analysis revealed that corn stover contained 4.97 wt% moisture along with 90.59 wt% volatile matter, 6.94 wt% fixed carbon, and 2.47 wt% ash. The higher heating value (HHV) was around 18.41 MJ/kg from calorimetry analysis, followed by gas chromatographic evaluation of gaseous products at a temperature-controlled 10 mL/min carrier gas flow rate.

3.4.3 Experiment Setup

The experiment was conducted using a solo approach microwave furnace preforming at 2.45 GHz, as shown in Fig. 14. The reaction was carried out in a 40 cm long quartz tube of a 50 cm outer diameter, the corn stover sample was held in a 3 cm long quartz container/crucible of a 4 cm outer diameter. The reaction temperature was monitored using a thermocouple tube of k-type positioned at the lower part of the container. To maintain an oxygen-deprived environment, nitrogen gas of 99.99% concentration was supplied through the system at 50 mL/min.

3.4.4 Experiment Steps

1. About 2 g of either pretreated or raw corn stover were mixed with 5 wt% of the catalyst (Al_2O_3).

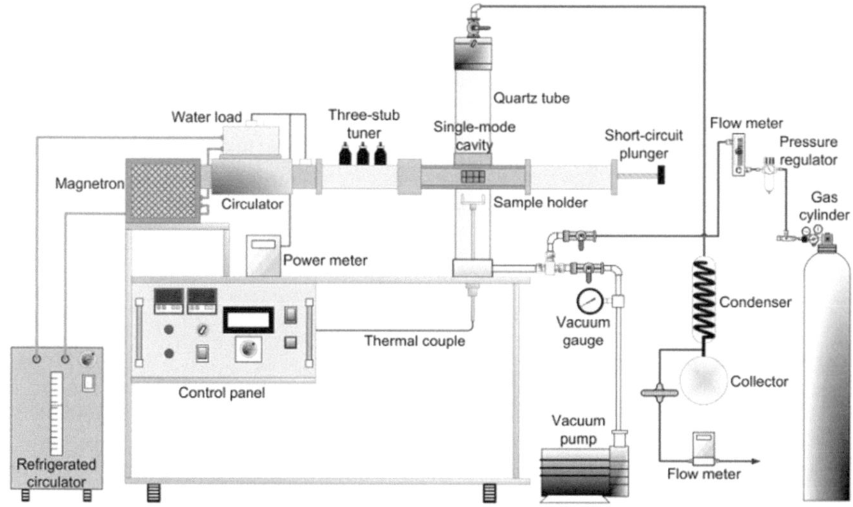

Fig. 14 Experimental setup of the microwave pyrolysis process of corn stover [24]

2. Then samples were placed in the quartz crucible which got inserted into the reaction tube.
3. The height of the quartz crucible was altered to fit the microwave diffusion zone.
4. Nitrogen gas was then purged into the system.
5. The microwave oven was set to 500 W.
6. After 20 min, both of the power and nitrogen flow were stopped.
7. The vapor phase products were entered into a 4 °C condenser, which allowed the condensable vapors to get separated and collected as a liquid product.
8. A 5 L Tedlar bag was used to collect the non-condensable vapors as the gaseous product, every 5 min.
9. The solid residue was collected after the system has cooled down.
10. Each experiment was conducted at least three times, and the average value of product yield was considered.

3.4.5 Experiment Results

1. **Temperature Profile**

The influence of particle size, pretreatment of two kinds, and the addition of a catalyst was found to be significant on the temperature profile of microwave pyrolysis. As can be seen in Fig. 15a, the temperature increases drastically in the first five minutes with the reduction in particle sizes before it stabilizes, marking a sharp increase from 475 °C at 5 mesh to 557 °C at 100 mesh. Figure 15b shows temperature variation over time influenced by both pretreatment and catalyst for a mesh size of 40, where a minor enhancement was noticed once applying Al_2O_3 as a catalyst without any pretreatment, but adding both the catalyst and any pretreatment, caused an excessive increasement in temperature profile. The highest temperature profile was recorded for the catalytic microwave pyrolysis when using the acid pretreatment specifically. Figure 16 shows that both maximum temperature and average heating rate in microwave pyrolysis were enhanced with decreasing particle size regardless of catalyst presence. Catalyst addition was moderately effective, while pretreatment (acid or steam explosion) considerably increased heating performance. Acid pretreatment in conjunction with catalysis produced the highest maximum temperature (654 °C) and heating rate (110 °C/min). Such improvements are credited to the changes in dielectric properties of the biomass.

2. **Product Distribution**

The resulting products of three different phases which are gas, liquid, and solid were analyzed in terms of yield at the influence of particle size, catalyst presence, and pretreatment, as shown in Fig. 17. Decreasing the particle size from 5 to 100 mesh caused an overall trend of increasement for the gaseous yield however, liquid

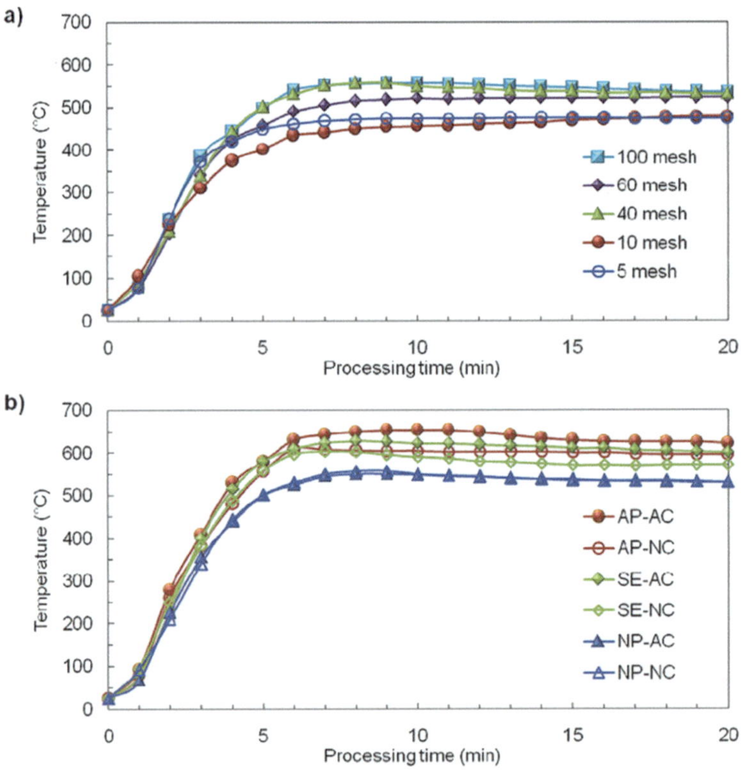

Fig. 15 Temperature variation over time at **a** different particle sizes and **b** catalytic pyrolysis with acid or steam explosion pretreatments [15]

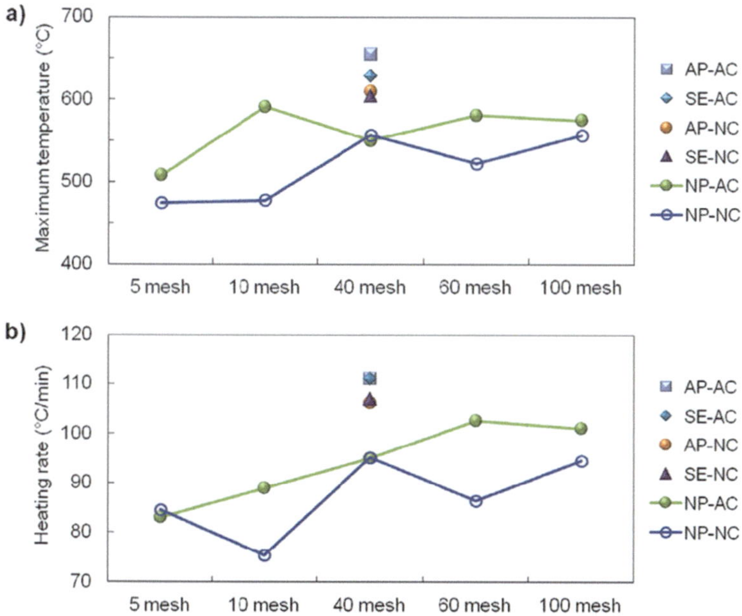

Fig. 16 Temperature (**a**), and heating rate (**b**) at (AP: acid pretreatment; SE: steam explosion; NP: no pretreatment; AC: adding catalyst; NC: no catalyst) [15]

and solid yields kept on decreasing. Regarding the influence of the catalyst, it was observed that the gaseous production has been improved by 4–9 wt%, when adding 10-mesh Al_2O_3. While using a smaller particle catalyst such as the 50-mesh Al_2O_3 caused a reduction of around 3–7 wt% in the yield of gaseous product, which is due to the fact that smaller particle sizes of a catalyst like the 50-mesh Al_2O_3 can be encapsulated by other surrounding particles, which limits their activity and results in lower gas production as if no catalyst was used. Therefore, using a catalyst can help converting vapors to gas and increases the gas yield however, choosing the right size is important. Moreover, acid pretreatment enhanced liquid yield with about 15 wt%, but decreased gas yield with around 14 wt%, because of the depolymerization of cellulose and hemicellulose, which inhibited the formation of volatile compounds that are converted to permanent gases. Steam explosion was less effective probably because of lower biomass solubilization. Catalyst addition increased gas yield by 3 wt% by favoring secondary vapor cracking. Furthermore, Solid yield was almost unchanged, which corresponds to the thermal conversion limit at 500 W. Additionally, it is important to highlight that the maximum gaseous yield of 73 wt% was observed with a 60-mesh corn stover without a catalyst, which resulted in improved devolatilization and lower vapor condensation. The highest liquid yield of 32 wt% was achieved with acid pretreatment with 10-mesh Al_2O_3, due to the hydrolysis of hemicellulose and cellulose caused by the acid pretreatment, which limited the formation of gas and kept the volatile intermediate in condensable nature, while the catalyst particle size hindered the secondary cracking. However, the solid phase was the highest at 10-mesh corn stover and 50-mesh Al_2O_3, which could be explained by catalyst particle encapsulation which decreased the catalytic activity and prevented complete thermal conversion.

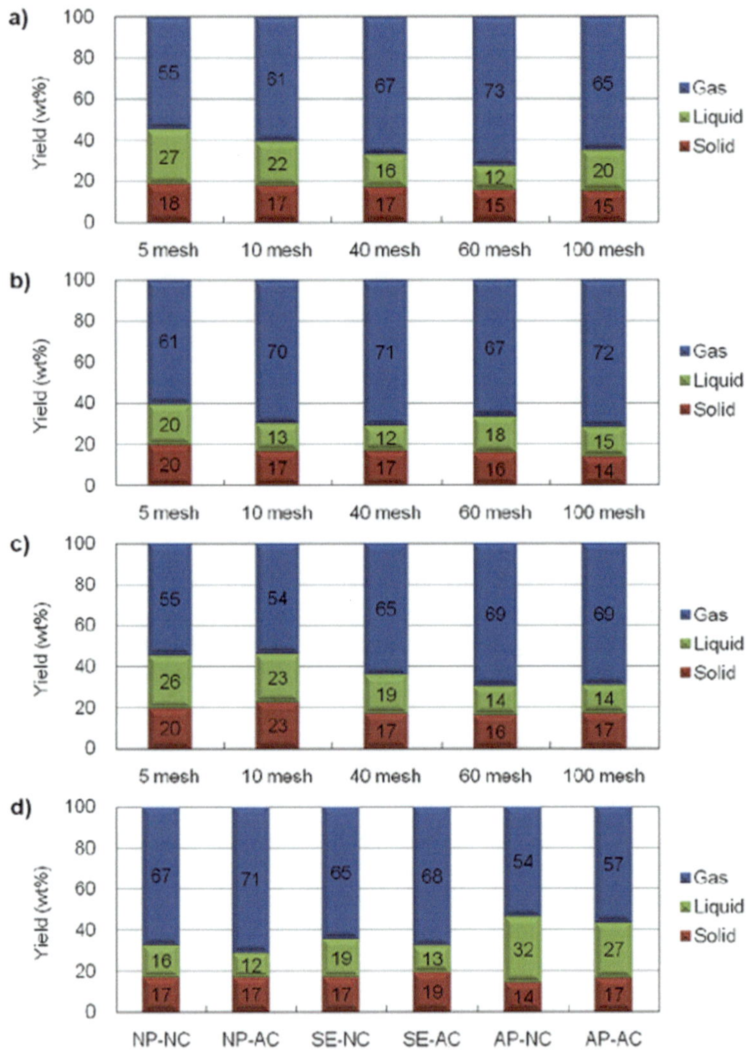

Fig. 17 Product yields from microwave pyrolysis at different conditions **a** without catalyst, **b** with 10-mesh Al_2O_3, **c** with 50-mesh Al_2O_3, and **d** using 40-mesh corn stover with and without pretreatment and catalyst [15]

3. Gas Product Composition Analysis

The resulting gaseous products from the microwave pyrolysis of corn stover were H_2, CH_4, CO_2, and CO. Different quantities of these gases were reported for a 40-mesh corn stover at various conditions as demonstrated in Fig. 18. Results indicated that CO had the highest recorded volume among all other gas products, of which it was about 0.91–1.15 L, varying based on pretreatment and catalyst use. On the other hand, the remaining gases were quite lower and typically between 0.001 and 0.05 L. It has been noticed that over the first 5 min, almost fifty percent of the CO was already released before it continues to form in a steady pattern throughout the reaction. While the formation of H_2, CH_4, and CO_2 occurred only at the first 5–20 min but then it stopped. The gas yields were mainly increased with the use of the Al_2O_3 catalyst however, acid pretreatment has excessively decreased them, particularly H_2 yield dropping from 0.04–0.05 to only 0.001–0.005 L. This trend can be explained by the removal of hemicellulose and cellulose due to the acid pretreatment, which are the main sources of these gases. Moreover, the production of gas is temperature dependent and is closely associated with the thermal degradation of certain biomass components as the formation of H_2, CH_4, and CO_2 was somewhat low at the first 5 min where temperature was not significant yet. Furthermore, it is important to mention that the results obtained were relatively different than those obtained in past papers due to the difference in catalyst type and reactivity.

4. Reaction Kinetics

The study determined the reaction kinetics of gas formation in microwave pyrolysis of corn stover using both first-order and second-order kinetic models. Even though the two models predicted approximately the same maximum gas quantities (1.34–2.30 L), the second-order model demonstrated higher accuracy, with higher R^2 values (0.95–0.99) and lower average absolute error. Reaction rate constants were 0.10–0.25 min^{-1} (first order) and 0.04–0.16 L^{-1} min^{-1} (second order). The maximum rate constants were observed for 5-mesh stover with and without 50-mesh Al_2O_3, but they produced less gas in total. On the other hand, 10-mesh Al_2O_3 produced greater gas yields, but lower rate constants, suggesting that the production of maximum gas is a more important factor in reaction than rate constant alone. Kinetics were slightly affected by acid pretreatment, while steam explosion with 10-mesh Al_2O_3 had the lowest rate constant. Using Arrhenius analysis of the second-order rate constants, the activation energy was ~ 21 kJ/mol and the pre-exponential factor was 0.02 s^{-1}, both in agreement with other microwave pyrolysis studies but lower than for conventional pyrolysis.

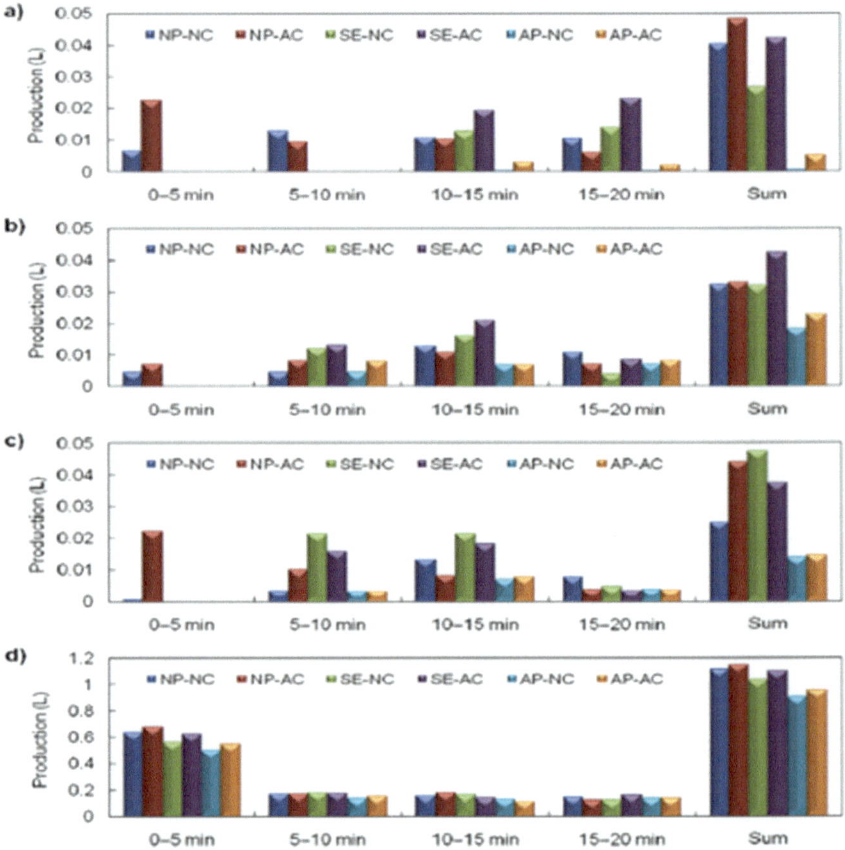

Fig. 18 Production of gaseous component from corn stover via microwave pyrolysis over time: **a** H_2, **b** CH_4, **c** CO_2, and **d** CO [15]

3.5 Comparison Table

See Table 4.

Table 4 Comparison table of the case studies

Parameter	Experiment 1	Experiment 2	Experiment 3
Title	Assessing the potential for biofuel production of corn stover pyrolysis using a pressurized batch reactor	Catalytic fast pyrolysis of corn stover in a fluidized bed heated by hot flue gas	Effects of particle size, pretreatment, and catalysis on microwave pyrolysis of corn stover
Reactor type	Pressurized batch reactor	Fluidized-bed reactor	Microwave reactor
Particles size	Passed through 3 mm-screen	< 1.0 mm	5–100 mesh (0.15–4 mm)
Temperature range	400, 500 and 600 °C	450–525 °C	500 °C, fixed
Carrier gas	Nitrogen	None	Nitrogen
Optimum temperature	400 °C	500 °C	–
Catalyst used	None	Dolomite	Al_2O_3 (10 & 50 mesh)
Pretreatment	None	None	Acid and steam explosion
Maximum bio-oil yield	31.4 wt%	43.3 wt%	32 wt% (with AP + AC)
Maximum biochar yield	37.3 wt%	–	23 wt% (10 mesh, 50-mesh catalyst)
Maximum bio-gas yield	21.2 wt%	–	73 wt% (60 mesh, no catalyst)
Maximum (HHV) of bio-oil	33.8 MJ/kg	17.2 MJ/kg	–

4 Results and Discussion

4.1 Overview

This section provides a comparison of the three pyrolysis case studies conducted on corn stover, analyzing the influence of various common parameters on the product yields which are bio-oil, biochar, and gas. Different types of reactors, the effect of particle size, catalyst and temperature ranges are the main focus of this section.

4.2 Table Discussion

Table 4 summarizes the three corn stover pyrolysis case studies, with the aim of biofuel production, highlighting the operating conditions of each case, their product distribution, and their product yields. The first case study was of the title "Assessing the potential for biofuel production of corn stover pyrolysis using a pressurized batch reactor." This study was conducted in a pressurized bed reactor at the following temperatures: 400 500, 600 °C, while other parameters were somewhat fixed. The influence on the product distribution and yield was evaluated and found out that the maximum yield of both bio-oil and biochar were obtained at 400 °C, which are 34.4 wt%, and 37.3 wt%, respectively. However, the maximum yield of the biogas was about 21.2 wt% at 600 °C. The results also indicate the maximum high heating value of bio-oil was about 33.8 MJ/kg. These findings suggest that lower temperatures favor the formation of solid phase products, however higher temperatures promote the production of gaseous products by leading to the breakdown of biochar through secondary cracking reactions and improving the thermal decomposition of biomass, which converts even more volatiles into gases instead of retaining them as liquid bio-oil. The second case study, "Catalytic fast pyrolysis of corn stover in a fluidized bed heated by hot flue gas: Physicochemical properties of bio-oil and its application," was carried out in a fluidized bed reactor using hot flue gas to heat up the reactor at a temperature range of 450–525 °C. The corn stover particle size was less than 0.1. The study investigated the effect of both temperature and catalytic bed materials on the properties of the resulting products. The optimum temperature for the highest bio-oil yield was at 500 °C using dolomite as bed material, which was approximately 43.3 wt%, while other products' yields were not reported in this study. It is important to note that this yield of bio-oil was the highest among the three studies, where using the right choice of bed material, such as Dolomite which had a dual role of a catalyst, and a bed material helped achieve such a yield. Both calcium carbonate and magnesium carbonate are the main compositions in dolomite and these two produce calcium oxide and magnesium oxide during calcination, where these oxides promote the cracking of tar and other heavy organic compounds into lighter fractions, which are favored in bio-oil. Moreover, the maximum high heating value of the resulting oil was about 17.2 MJ/kg. Results in this study indicate that specific temperature range, particularly around 500 °C, is essential to optimize the bio-oil production by enhancing the breakdown of biomass into liquid products while reducing solid char and gaseous by-products where excessive temperatures can cause over-cracking of bio-oil into gases, reducing overall liquid yield, which supports the claims observed in case study 1. In the last case study of the title "Effects of particle size, pretreatment, and catalysis on microwave pyrolysis of corn stover" was obtained in a microwave reactor to analyze the effect of (0.15–4 mm) particle sizes, acid and steam explosion pretreatment, as well as catalyst effect of aluminum oxide, while the temperature was fixed at 500 °C. The highest yield of all products in this study was about 73 wt% (60 mesh, no catalyst), while the bio-oil and the biochar yields were around 32 wt% (with AP + AC) 23 wt% (10 mesh, 50-mesh catalyst), respectively. These results indicate that as the particle size was reduced, the overall heating rate was increased and gas generation was improved, yet the formation of liquids and solids was reduced, and even though that is not the case for conventional pyrolysis were reducing the particle size increase the bio-oil yield but this is due to the different heating method, where smaller particles in microwave pyrolysis absorbed heat very quickly, leading to secondary cracking that converted vapors into gases, thus reducing bio-oil yield. Moreover, unlike the effect of catalyst in case 2, using aluminum oxide (Al_2O_3) showed that larger catalyst particles (10-mesh) improved gas formation by encouraging secondary cracking reactions, while smaller particles (50-mesh) had no significant impact as it got encapsulated by biomass, which hindered their catalytic activity. Finally, acid pretreatment caused a substantial 15 wt% rise in liquid yield and 14 wt% fall in gas yield which is due to the fact that hemicellulose and cellulose decomposition inhibited further gas forming reactions. In the other hand, Steam explosion pretreatment had a little effect on the overall composition of the products.

4.3 Economic and Environmental Aspects

The production of biofuel from the pyrolysis of corn stover may be a possible alternative to hydrocarbon based fuels, as they eventually will damage the environment causing environmental issues such as greenhouse gas emissions and air and water pollution [5]. From an economic perspective, biomass based fuels can lead to great advantages either in rural or central regions as it can offer job opportunities, energy surplus to back up the economic system. However, biomass based fuels face some challenges due to the high initial cost, infrastructure constraints, and the lack of interest from leading companies in the energy sector [5].

5 Conclusion

In conclusion, according to the results obtained from the various research papers conducted on the production of biofuel via the pyrolysis of corn stover, this kind of feedstock specifically offers a lot of potential in biofuel production and can contribute widely as an alternative clean energy source. As it was seen from the case studies that were analyzed deeply in Sect. 3, the pyrolysis process of corn stover can be influenced by a wide range of factors and parameters, and although the same type of feedstock was used in all of the three cases, and the same concept of pyrolysis means were used to convert these corn stover particles into biofuel, different pyrolysis methods, temperature ranges, catalysts, heating methods, pretreatments and even particle size affected the final results of product distribution and biofuel properties. As obtained from case studies, the highest bio-oil yield of about 43.3 wt% was obtained at 400 °C, in a catalytic bed reactor, which was enhanced by the use of dolomite catalyst. However, the highest HHV which is 33.8 MJ/kg, was reported at 400 °C, in a standard pyrolysis setup of fixed bed reactor and at the lowest pyrolysis temperature out of the three tested temperatures. This is because of the lower content of both oxygen and moisture, therefore making the resulting oil richer with hydrocarbons and reducing the energy lost in vaporizing water or oxygenates. Moreover, at higher pyrolysis temperatures second cracking reactions occur which produces higher gas and water resulting in lower HHV. From all three cases we can say that the ideal pyrolysis temperature was between 400 and 700 °C, when it comes to both oil yield and HHV. Although the lowest bio-oil yield was observed in the third case of the microwave pyrolysis reactor, the maximum biogas yield of 73 wt% for the same case. In this reactor heat transfer was carried out in a microwave reactor, which heats the material faster and uniformly reaches higher temperatures and leads to secondary reactions which reduces the bio-oil formation. Additionally, it is true that the results obtained from the fluidized bed reactor were optimal when it comes to bio-oil yield which most suitable product as an alternative fuel, but it is technically complex. Different outcomes were observed for the effect of catalyst on product yield, where dolomite has significantly increased the bio-oil yield (43.3 wt%) due to its dual role in cracking heavy compounds and stabilizing liquid product as in case 2, while it had the opposite effect in the last case study the use of aluminum oxide (Al_2O_3) increased gas formation instead, especially with larger catalyst particles, however smaller particles were ineffective because of encapsulation phenom. Furthermore, the effect of particle size was observed in the last case study of microwave pyrolysis, in which it caused reduction in the bio-oil yield when using smaller particles size. It is important to highlight that this result is quite different than what was observed for the effect of particle size on biomass yield in general, but that is due to the unique heating method of microwave pyrolysis as explained previously. The aim of this comparative analysis is to optimize the production of biofuel products of all kinds, especially bio-oil highlighting mainly the influence of temperature, heating method, particle size, pretreatments, and catalyst materials on improving the resulting products to be suitable alternatives as a clean energy source.

References

1. El-Araby R (2024) Biofuel production: exploring renewable energy solutions for a greener future. BioMed Central Ltd. https://doi.org/10.1186/s13068-024-02571-9
2. Suhara A et al (2024) Biodiesel sustainability: review of progress and challenges of biodiesel as sustainable biofuel. Clean Technol 6(3):886–906. https://doi.org/10.3390/cleantechnol6030045
3. Bracmort K (2012) CRS report for congress biomass: comparison of definitions in legislation through the 112th Congress Kelsi Bracmort specialist in agricultural conservation and natural resources policy [Online]. Available: www.crs.gov
4. De Castro JFM (2007) Biofuels-an overview final report
5. Chowdhury P, Mahi NA, Yeassin R, Chowdhury NUR, Farrok O (2025) Biomass to biofuel: impacts and mitigation of environmental, health, and socioeconomic challenges. Energy Convers Manage X 25. https://doi.org/10.1016/j.ecmx.2025.100889
6. Johnson J, Reicosky D, Sharratt B, Lindstrom M, Voorhees W, Corn stover as a biofuel
7. Adams P, Bridgwater T, Lea-Langton A, Ross A, Watson I (2017) Biomass conversion technologies. In: Greenhouse gas balances of bioenergy systems. Elsevier, pp 107–139. https://doi.org/10.1016/B978-0-08-101036-5.00008-2
8. Lewandowski WM, Ryms M, Kosakowski W (2020) Thermal biomass conversion: a review. Processes 8(5). https://doi.org/10.3390/PR8050516
9. Yu F et al (2007) Physical and chemical properties of bio-oils from microwave pyrolysis of corn stover

10. Mullen CA, Boateng AA, Goldberg NM, Lima IM, Laird DA, Hicks KB (2010) Bio-oil and biochar production from corn cobs and stover by fast pyrolysis. Biomass Bioenergy 34(1):67–74. https://doi.org/10.1016/j.biombioe.2009.09.012
11. Shah A, Darr MJ, Dalluge D, Medic D, Webster K, Brown RC (2012) Physicochemical properties of bio-oil and biochar produced by fast pyrolysis of stored single-pass corn stover and cobs. Bioresour Technol 125:348–352. https://doi.org/10.1016/j.biortech.2012.09.061
12. Capunitan JA, Capareda SC (2012) Assessing the potential for biofuel production of corn stover pyrolysis using a pressurized batch reactor. Fuel 95:563–572. https://doi.org/10.1016/j.fuel.2011.12.029
13. Rafiq MK, Bachmann RT, Rafiq MT, Shang Z, Joseph S, Long RL (2016) Influence of pyrolysis temperature on physico-chemical properties of corn stover (zea mays l.) biochar and feasibility for carbon capture and energy balance. PLoS One 11(6). https://doi.org/10.1371/journal.pone.0156894
14. Yang W, Fu P, Yi WM (2017) Catalytic fast pyrolysis of corn stover in a fluidized bed heated by hot flue gas: physicochemical properties of bio-oil and its application. Int J Agric Biol Eng 10(5):226–233. https://doi.org/10.25165/j.ijabe.20171005.2473
15. Huang YF, Kuan WH, Chang CY (2018) Effects of particle size, pretreatment, and catalysis on microwave pyrolysis of corn stover. Energy 143:696–703. https://doi.org/10.1016/j.energy.2017.11.022
16. Polin JP, Carr HD, Whitmer LE, Smith RG, Brown RC (2019) Conventional and autothermal pyrolysis of corn stover: overcoming the processing challenges of high-ash agricultural residues. J Anal Appl Pyrolysis 143. https://doi.org/10.1016/j.jaap.2019.104679
17. Ali N, Ashraf M, Shahzad K, Saleem M, Chughtai A (2019) Fluidized bed fast pyrolysis of corn stover: effects of fluidizing gas flow rate and composition. Energy Sources Part A Recov Util Environ Effects. https://doi.org/10.1080/15567036.2019.1694105
18. Hu E et al (2021) Infrared heated pyrolysis of corn stover: determination of kinetic and thermodynamic parameters. J Anal Appl Pyrolysis 158. https://doi.org/10.1016/j.jaap.2021.105273
19. Meng F, Wang D, Zhang M (2021) Effects of different pretreatment methods on biochar properties from pyrolysis of corn stover. J Energy Inst 98:294–302. https://doi.org/10.1016/j.joei.2021.07.008
20. Fodah AEM, Ghosal MK, Behera D (2021) Quality assessment of bio-oil and biochar from microwave-assisted pyrolysis of corn stover using different adsorbents. J Energy Inst 98:63–76. https://doi.org/10.1016/j.joei.2021.06.008
21. Hu E et al (2022) Pyrolysis behaviors of corn stover in new two-stage rotary kiln with baffle. J Anal Appl Pyrolysis 161. https://doi.org/10.1016/j.jaap.2021.105398
22. Abatyough MT, Ajibola VO, Agbaji EB, Yashim ZI (2022) Optimization of reaction conditions for bio-oil production from pyrolysis of corncobs
23. Wang Q et al (2023) Effect of pressure on the pyrolysis and gasification mechanism of corn stovers from kinetics. J Anal Appl Pyrolysis 176. https://doi.org/10.1016/j.jaap.2023.106267
24. Huang YF, Te Chiueh P, Kuan WH, Lo SL (2015) Effects of lignocellulosic composition and microwave power level on the gaseous product of microwave pyrolysis. Energy 89:974–981. https://doi.org/10.1016/j.energy.2015.06.035

Thermochemical Conversion of Rice Straw to Produce Biofuel

Nabaa Saif Al Khatri and Seyed Mojtaba Sadrameli

Abstract As more people look for sustainable sources of energy, there is now greater demand for developing renewable biofuels. Rice straw, which occurs in large amounts on farms, could be an excellent biofuel source by undergoing thermochemical processes like pyrolysis. This chapter looks into the use of rice straw as a source of biofuel, with a focus on the pyrolysis method that uses heat without oxygen. It looks at three important cases that study the use of pyrolysis processes to produce more biofuels. This process covers ethanol-based liquefaction, catalytic pyrolysis using metal salts and zeolites, as well as combining pyrolysis with low-temperature liquefaction. Study outcomes from these case studies demonstrate that the results and quality of bio-oil depend on things like temperature, the kind of catalyst, and the amount of time the material is processed. The results of this chapter have shown that catalytic pyrolysis using Y-zeolite leads to the highest production of bio-oil (55.2 wt%), and that the highest energy efficiency comes from the combined pyrolysis-liquefaction process (67.41% yield and a high heating value of 42.16 MJ/kg). Statistical modeling, such as response surface methodology, is often used to help improve how efficiently the process works. It is evident from the results that rice straw can be used for creating biofuels and has valuable uses in electing renewable energy and energy generated from waste. At the end of the research, I put forward suggestions for more research and ways to improve the process's scalability, feasibility in terms of costs, and sustainability.

Keywords Rice straw · Biofuels · Bio-oil · Pyrolysis · Renewable energy

N. S. Al Khatri · S. M. Sadrameli (✉)
Department of Engineering, German University of Technology in Oman, Muscat, Oman
e-mail: seyed.sadrameli@gutech.edu.om

A. Azizi (ed.), *Mechatronic Innovations in Sustainable Energy*, Emerging Trends in Mechatronics, https://doi.org/10.1007/978-981-95-5863-6_3

1 Introduction

1.1 Overview

Finding an alternative for efficiently generating energy is necessary given the emergence of multiple issues, including growing fuel prices, global warming, the depletion of fossil fuels, and increased fuel usage. Currently, biomass is one of the renewable resources used to produce biofuel and provides more than 45 ~ 10 EJ of the world's energy demands. In industrialized nations, biomass makes up less than 10% of the overall energy supply; in underdeveloped nations, it accounts for 20–30%. With energy demands predicted to increase by 1.3% annually until 2040, it is critical that all nations fortify their energy policies to guarantee global energy security. One such promising option for lowering greenhouse gas (GHG) emissions, preventing biodiversity loss, enhancing food production, and supporting the idea of sustainable development which entails efficient use of water, and several other resources is biofuels [1, 2]. A number of significant factors need to be taken into account when producing biofuels from biomass, such as land-use exploitation and environmental sensitivity. The generation of bioenergy from biomass, a renewable bioenergy source, has helped to solve several socioeconomic problems. Biomass has lower life cycle greenhouse gas (GHG) emissions than traditional fossil fuels, and there is less conflict over food production due to (1) crop residues, (2) mixed cropping patterns, (3) perennial plant growth on degraded lands that have been abandoned for agricultural use, (4) sustainable harvesting practices, and (5) municipal and industrial wastes [1].

1.2 Raw Materials and Processes

1.2.1 Rice Straw

One byproduct of rice after harvest is rice straw. Numerous elements, such as rice types, weather, nutrient management techniques, and soil management practices, affect the total biomass. The cutting height the height of the stubble that remains in the field determines how much rice straw is harvested overall from the field. After being gathered and burned, rice straw is allowed to integrate. After harvest, the uncut portion of rice is left over and either burned or mixed into the soil to prepare the following crop. Rice straw is primarily made up of cellulose, hemicellulose, lignin, and silica, as you can find in Fig. 1. Presence of these constituents greatly influences how rice straw behaves in thermochemical treatment processes such as pyrolysis and liquefaction [3]. One of the main crops grown in Europe, Asia, Africa, and the United States is rice, which produces between 730 and 1111 million tons annually [1].

S.No.	A component found in rice straw	Quantity (wt %)
1	Cellulose	43
2	Hemicellulose	25
3	Lignin	12
4	Ash and silica	34–44
5	Glucose	19–22
6	Arabinose	2–4
7	Mannose	1.8–2
8	Galactose	0.4
9	Non-structural components	16–20

Fig. 1 Rice straw compositions [3]

1.2.2 Biofuels

In contrast to the finite nature, geopolitical instability, and detrimental global repercussions of fossil fuel energy, it is becoming more and more evident that biofuels can be a sustainable source of renewable energy. All energy-enriched compounds produced directly by biological processes or obtained through chemical conversion from biomass of previous living creatures are collectively referred to as biofuels. The majority of biofuels are made from photosynthetic organisms, including vascular land plants, micro- and macro-algae, and photosynthetic bacteria. Biofuel's main byproducts might be solid, liquid, or gaseous. Thermochemical, physical, and biological processes can all be used to further transform these compounds. Primary and secondary biofuels are the two categories into which biofuels fall. Burning dry animal waste and woody or cellulosic plant material directly produces the main biofuels. There are three generations of secondary biofuels that are produced indirectly from plant and animal sources [4].

1.2.3 Pyrolysis Processes

Pyrolysis, which is simply described as the chemical changes that take place when heat is delivered to a material without oxygen, is the basic chemical reaction mechanism that precedes both the gasification and the combustion of solid fuels. Based on the process environment, parameters (temperature, heating rates, residence time), and products produced, pyrolysis can be broadly classified into eight categories: carbonization, conventional, fast, flash-liquid, flash-gas, ultra, vacuum, hydro pyrolysis, and methanopyrolysis. Fluidized beds (circulating, bubbling), spinning cones, augers, vacuum, and ablative pyrolysis reactors are the most crucial reactor configurations. Rotating cones and fluidized beds are simpler to scale and might be more economical. By carefully choosing the biomass feedstock, reactor types like fluidized bed, circulating fluidized bed, and rotating cone reactor may be able to provide a greater output of bio-oil (70–75%). The production of bio-H_2 using a variety of techniques is one strategy for the value addition of the pyrolysis bio-oil [5].

1.3 Challenges and Opportunity

The production of biofuels faces crucial barriers together with vital accepting circumstances. Numerous biofuel manufacturing systems produce small yields and operate inefficiently primarily because of processes that use both microalgae and bacterial organisms. High production costs and environmental control requirements for biomass cultivation act as major barriers toward large-scale production. Biofuel production enzymes such as hydrogenases function poorly when oxygen is present because they demonstrate sensitivities to this element. Industrial-scale implementation faces significant barriers when it comes to laboratory level scale-up. This field maintains substantial chances for advancement even though it faces certain obstacles. Several agricultural materials starting from rice straw can serve as sources for creating biofuels. Scientists who work with metabolic engineering and genetics have identified methods through which fuel production can be optimized along with procedures becoming more stable. When biofuels are incorporated into biorefinery systems which extract useful byproducts there emerges an improved economic potential. Biofuels function as an environmentally friendly power source that supports the reduction of climate change effects. Synthetic biology together with biofilm-based systems join emerging technologies which promote both innovation and cost reduction potentials in this sector [4].

1.4 Objective

The research determines the viability of producing biofuel from rice straw utilizing the pyrolysis technique. The research examines three key variables which determine both the production output and quality of biofuel derived from rice straw through pyrolysis processes. The investigation examines both physical and chemical attributes of bio-oil after production to evaluate its capabilities as clean renewable energy. The study seeks to enhance sustainable waste-to-energy applications by optimizing the operating conditions of the pyrolysis process for improved biofuel output quality suitable for renewable energy use.

1.5 Organization of the Research

The research splits into different sections which form its four main chapters. Chapter One establishes the expanding demand for renewable energy sources while showing how rice straw acts as a potential sustainable feedstock for biofuel manufacturing. A detailed literature review appears in Chapter Two because it reports and analyzes various case studies that investigate rice straw pyrolysis together with similar biomass. Selected research studies are assessed during Chapter Three through comparative analysis based on their research goals while examining experimental procedures and material types and pyrolysis conditions and main results. The research ends with Chapter Four by assessing the full feasibility of rice straw pyrolysis for biofuel production and presenting future research strategies to improve industrial commercialization along with environmental advantages.

2 Literature Review

2.1 Overview

As people's awareness of environmental issues and the depletion of fossil resources has grown, it is more important than ever to find sustainable and ecologically friendly energy alternatives. By turning waste into a valuable source of energy, the production of biofuels from waste materials provides a path ahead. In academia, where academics have been examining different types of biomass and conversion methods, the pyrolysis process in which organic material breaks down under intense heat without oxygen has garnered a lot of attention. Pyrolysis is an interesting process that has the potential to provide sustainable energy sources since it can transform a variety of waste materials into useful products including syngas, charcoal, and liquid biofuels. The agricultural waste known as rice straw functions as a practical source material for biofuel creation. The composition of lignocellulose makes rice straw appropriate for pyrolysis processing through its high content of cellulose and hemicellulose. The production of biofuel from rice straw leads to better environmental health as well as creating sustainable renewable energy. The analysis presented in this chapter focuses on how scientists study the utility of pyrolysis to create biofuel from rice straw. The production of bio-oil from rice straw depends heavily on specific elements studied by investigators such as pretreatment approaches and heating rates as well as retention time along with particle size and pyrolysis temperature and catalyst selection and rice straw composition and type.

2.2 Recent Literatures

Das et al. [1] conducted an exhaustive review which evaluated rice straw potential as a renewable biofuel feedstock. The research evaluated both environmental and energy-based advantages of rice straw usage in addition to the technological obstacles for converting it. The authors explained that physical, chemical, and biological pretreatments must be implemented to break down cellular barriers which protect hemicellulose and cellulose and lignin inside the lignocellulosic material. Biomethane and bioethanol generation through enhanced yields requires these pretreatments as a necessary step. Biochar application together with composting along with field and off-field management performed important roles in reducing greenhouse gas emissions according to the review. The research demonstrated that properly optimized bioenergy conversion practices combined with life cycle assessments show rice straw as a viable large-scale biomass material on condition that scientists solve economic challenges and manage silica content effectively.

Sharma et al. [3] performed a review extensive of rice straw biofuel utilization while analyzing its worldwide access and ecological consequences in their research. The authors developed a sustainable fossil fuel replacement methodology which converted the agricultural waste rice straw into biofuels through thermochemical and biochemical reaction pathways. The authors explained that rice straw contains cellulose and hemicellulose together with lignin while the combination of high silica levels and recalcitrant structure makes processability challenging. Different pretreatment methods were evaluated in the review to increase biodegradability potential and obtain better biofuel production yields. Studies confirmed that processed rice straw can generate biofuels through the production of bioethanol together with biodiesel and biogas and biochar. Research has shown that using sophisticated pretreatment in conjunction with conversion techniques effectively reduces greenhouse gas emissions while improving energy independence and encouraging growth that is ecologically conscious.

Uddin et al. [6] used a dual-catalyst bed system with cement (CEM) and ZSM-5 to study rice straw pyrolysis for enhancing fuel gas and value-added chemical production. Bio-oil and gas product quality improvement through the development of a volatile compound catalyzing dual catalytic system during pyrolysis was the main goal of this study. The investigators operated the fixed bed reactor at 600 °C under nitrogen atmosphere for thirty minutes. The dual-catalyst bed brought substantial changes to the pyrolysis reaction as it produced greater amounts of hydrogen, methane, and aromatic hydrocarbons alongside diminished oxygenated and water-containing compounds present in bio-oil. Valuable hydrocarbons emerged from the combination of cement's deoxygenation power with zeolite's acidity which improved the characteristics of resulting fuel products. The research findings demonstrated that combining two catalysts represents an efficient approach for generating high-quality refined biofuels from renewable energy sources while substituting fossil energy.

Zhou et al. [7] developed a two-step approach to create clean liquid biofuel from rice straw which combined pyrolysis pretreatment with low-temperature liquefaction. The objective involved overcoming difficulties associated with biomass conversion through technical solutions against high oxygen content and low energy density. The researchers followed two procedure steps to transform rice straw into biochar through pyrolysis before liquefying it through hydrogen-donor solvent treatment at different low temperatures. The best results from the tested conditions emerged through liquefaction performed at 150 °C because this generated bio-oil containing high heating value (42.16 MJ/kg) alongside 69.42% energy yield with both low sulfur and oxygen content. Research indicated that the commercial evaluation of the production process yielded a $171.29/bbl. cost with a zero-waste self-circulating preparation system. Rice straw-derived biochar emerges as a sustainable replacement for fossil fuels because researchers demonstrated its ability to generate high-quality eco-friendly biofuels through this method.

Seo et al. [8] investigated the production of modified biochar through rice straw co-pyrolysis with industrial waste materials such as spent tires and sulfur by-products for environmental remediation purposes. The research investigated if these waste streams could be processed in one pyrolysis system to create useful sorbents which remove lead, barium, chromate, selenate, and organic contamination. The authors achieved production of biochar through combined rice straw pyrolysis with different combinations of tire materials and sulfur waste substances under regulated conditions. The research showed that modified char materials demonstrated specific enhancing effects for pollutant sorption because spent tire–rice straw char improved organic pollutant uptake by increasing its carbon content and sulfur–rice straw char attracted heavy metals because of its elevated functional group density. Real-world industrial assessment requires additional research to establish the operational potential of waste co-pyrolysis as an effective disposal and environmental management method.

Moholkar and Singh [9] conducted research to assess the potential of rice straw hydrolysate as a different feedstock for acetone–butanol–ethanol (ABE) fermentation to produce biobutanol. This research evaluated if dilute acid hydrolyzed rice straw without treatment exhibited appropriate sugar recovery levels together with sufficient microorganism activity to produce solvents. Experimental work included hydrolyzing the material with acid before saccharification and a twelve-day fermentation of Clostridium species without optimized settings. The improved version of the research study introduced essential processing information about treatment concentrations as well as the quantities of generated sugars alongside fermentation time data. Fermentation started only after an extended adaptation period caused by the highly acidic pH value close to 1 which strongly decelerated metabolic operations. Although the study required a substantial adaptation phase before fermentation began it proved the ability to produce both biobutanol and additional solvents in sufficient amounts. The authors demonstrate that rice straw serves as a promising raw material for biobutanol production, but production yields are affected by fermentation pH along with pretreatment methods and process optimization.

Izzatie et al. [10] executed co-pyrolysis research on PET polyethylene terephthalate and rice straw through a fixed bed drop-type pyrolyzer. The research analyzed the product quantities from various pyrolysis temperature ranges from 450–600 °C using a steady 1:1 feedstock ratio. For determining the best temperature to produce maximum liquid fuel, the authors examined the products consisting of pyrolysis oil and biochar and non-condensable gases (NCG). A 550 °C temperature produced the highest oil yield during the process, yet elevated temperatures resulted in more gas formations along with vapor secondary cracking. The chemical evaluation of processed oil revealed the presence of aldehydes, ketones, acids, aromatics, and phenols functional groups. The research showed that co-pyrolysis allowed better energy recovery than single straw pyrolysis even though the oil's energy content was reduced by the present water. Integrated waste-to-energy approaches become more feasible through biomass-plastic waste co-processing because it enhances fuel properties in the final products.

Ahmad Zamri et al. [11] analyzed the applications and operational effectiveness of mesoporous catalysts during the biomass feedstock upgraded pyrolysis for biofuel manufacture. This review demonstrated how mesoporous catalysts can solve problems of raw bio-oil by addressing its composition-related issues including high oxygen content and acidity and poor stability. The study examined different types of mesoporous catalysts consisting of silica-based materials and metal-doped, waste-derived, and carbon-based forms which were evaluated based on their structural characteristics and catalytic activities. Studies focused on assessment of materials including SBA-15, MCM-41, KIT-6, hierarchical zeolites due to their ability to boost surface area characteristics and improve acidity and enhance diffusion pathways. This research examined both catalyst support materials and functionalization approaches that enhance product selectivity at the same time minimize coke production. Results from the review indicated that catalysts with mesoporous structures generate improved biofuel production rates through deoxygenation and hydrodeoxygenation procedures. Research going forward needs to address challenges that include unstable thermal characteristics and difficulties in controlling pore size and producing synreearch at reasonable costs.

Cao et al. [12] studied rice straw catalytic pyrolysis with metal salts together with basic oxides while also testing acidic metal oxides and zeolite catalysts to determine the impacts these catalysts had on yield output and bio-oil property transformation. The goal was to develop an efficient bio-oil production method by performing catalytic upgrading operations to improve its chemical features. Catalyst type testing used different reaction temperatures along with catalyst loadings to determine their effects on yield and resultant composition. Y-zeolite emerged as the best catalyst for bio-oil production in the conditions tested since it yielded 55.2 wt% at 450 °C while excelling at phenolic monomer production. Zeolite and MgO catalysts produced bio-oil with enhanced thermal properties shown by increased higher heating value (HHV) and carbon content values according to this research. Research findings demonstrated Y-zeolite stands as the best option for bio-oil production from rice straw because it optimizes yield alongside quality improvements through catalyst selection processes.

Alengebawy et al. [13] found in their work that sustainable rice straw recycling includes biogas production and creates valuable biofertilizers through their review. The research evaluated rice straw potential for global energy sustainability by examining its role in Chinese energy systems and worldwide energy systems. Rice straw presents three key challenges because of its low energy content and high ash concentration together with changing moisture levels. The study described how baling and ensiling and co-digesting rice straw with other materials improves its potential for anaerobic digestion. The creation of biogas from rice straw effectively lowers greenhouse gas emissions and serves as a climate change reduction method according to the authors. Integrating biogas production alongside digestate management together with life cycle assessment represents an essential framework for establishing a circular bioeconomy while ensuring sustainable rice straw use demands green technology combined with economic feasibility assessments.

Abd-Elgawad et al. [14] conducted a thorough research evaluation of rice straw's utilization potential for bioethanol fermentation through intensive examination of essential pretreatment processes. This research studied the complex structure of rice straw to show both obstacles and potential options for its use by analyzing its lignin and cellulose and hemicellulose and ash contents. Pretreatment of materials represents an essential step according to the authors because it helps break up lignin structures and makes cellulose crystals less densely packed. The research detailed physical, chemical, physicochemical, and biological pretreatment procedures from an efficiency and sugar recovery perspective. The review suggests that hydrolytic enzyme-based biological conversion represents the most sustainable method for the process. Optimizing all three stages from pretreatment to hydrolysis and fermentation will lead to economical and scalable bioethanol production methods. According to the authors rice straw has the potential to become a notable bioethanol provider in the transport sector due to advances in technology which include engineered enzymes and genetically modified yeast strains.

Gupte et al. [15] conducted a review on biotech approaches that transform different rice waste streams including straw, husk and broken and discolored rice for producing liquid and gaseous biofuels. This review tried to deliver an extensive analysis of microbial bioprocesses for bioethanol, biobutanol, biodiesel, biogas and biohydrogen production through microbial consortia or pure culture operations. The study examined physical and chemical and enzymatic and microbial pretreatments alongside thermochemical methods to improve fermentable carbon availability for biological processing unlike previous evaluations. The successful production of ethanol depends on effective pretreatment approaches and microbial fermentation techniques according to the authors. The study team established rice waste possesses unexploited capacity as a promising renewable energy source which helps combat climate change effects. Laboratory-scale success exists although the research organization acknowledges that both efficiency enhancement at commercial levels and development of biorefinery capabilities producing biofuels with high-value bioproducts are fundamental requirements for future development.

Domínguez-Escribà and Porcar [16] conducted research to evaluate rice straw's suitability for massive bioethanol manufacturing from renewable lignocellulosic sources plus they evaluated the solutions to present-day straw disposal problems. Greenhouse gas emissions consisting of CO_2 and methane occur at high levels through standard cultivation practices which use rice straw burning and soil incorporation. The authors described bioethanol production as a promising alternative to turn underutilized rice straw into a valuable renewable energy source. Research in pretreatment technology along with enzymatic development and fermentation processes holds promising results despite the initial technical obstacles faced when trying to transform lignocellulose into fermentable sugars. The research showed that innovative developments such as genetically modified organisms alongside enhanced cellulases and combined bioconversion methods would turn rice straw into an effective energy source for solving environmental pollutants and worldwide power needs.

Mohamed and Ashour [17] investigated bio-oil production from rice straw through a liquefaction process under modest thermochemical parameters. The researchers studied the effects of ethanol-to-solid ratio and retention time variations on bio-oil production under fixed pressure operation at 2.5 atm using 220 °C as the reaction temperature. The researchers separated the produced bio-oil into volatile components and two additional fractions of light and heavy type. Bio-oil production reached its peak under experimental conditions where ethanol solution amounted to 30 ml per gram solid and reaction duration reached 60 min. Instrumental analysis of the produced bio-oil through GC–MS and FT-IR showed that the main constituents were C_{16}–C_{19} chain carbon compounds and esters which indicate suitability for use as biofuel. Academic statistical software SAS together with MATLAB tools completed the optimization process by establishing 52.5 min and 35 ml/g as the most suitable conditions. Biofuels derived from rice straw through liquefaction under optimized conditions demonstrate excellent properties as usable biofuels according to the research findings.

Passoth and Sandgren [18] examined the biotechnological potential of straw feedstocks for biofuel and chemical production including cereal straw and rapeseed straw materials. The investigation evaluated the technical viability and sustained use potential of food crop byproducts known as straw for bioenergy systems while avoiding food crop farming conflicts and land utilization changes. The authors stressed that straw presents both advantages of abundant availability and net positive greenhouse gas emissions yet demands extensive heat and enzyme-based processing to unravel fermentable sugar molecules from its rigid lignocellulosic composition. The review analyzed current industrial practices of straw utilization in ethanol and biogas manufacturing while discussing research advancements for producing biobutanol and biodiesel and multiple valuable biochemical derivatives. The study team recognized pretreatment expenses as the biggest challenge that stands in the way of large-scale biofuel production. The economic viability of straw-based biofuel systems can be enhanced through innovations in microbial engineering together with process integration as well as by simultaneous production of valuable compounds. The authors emphasized complete utilization of straw elements including lignin because it supports sustainable diversified biorefinery systems.

2.3 *Summery*

In conclusion, the analysis presented in this chapter focuses on how scientists study the utility of pyrolysis to create biofuel from rice straw. The production of bio-oil from rice straw depends heavily on specific elements studied by investigators such as pretreatment approaches and heating rates as well as retention time along with particle size and pyrolysis temperature and catalyst selection and rice straw composition and type.

3 System Description and Methodology

3.1 Overview

A detailed investigation of rice straw pyrolysis experimentation for biofuel generation takes place in this chapter. The research investigates experimental setups and materials as well as methods and essential outcomes from chosen case studies. The research studies used for comparisons matched in terms of reaction parameters including temperature spans and catalyst applications together with reactor characteristics. This chapter offers a complete analysis of bio-oil yield and quality features during rice straw pyrolysis by studying experimental laboratory results.

3.2 Case Study One

Title: Production of Bio-oil from Rice Straw

Authors and year of publication: Mohamed, Olfat A.; Ashour, 2016.

3.2.1 Materials Used

The material under study in this work is rice straw, a lignocellulosic agricultural residue which is available in large quantities in rice producing regions. Initially, rice straw was collected, cleaned from debris, and dried to remove surface impurities and reduce its moisture content despite vague values in its moisture content. A uniform fine powder characterized by 0.2 to 1 mm particle sizes was produced after grinding, an amount highly useful and often used in liquefaction procedures. Ethanol was used in liquefaction process as both a solvent and hydrogen donor, a role which was essential in the liquefaction of biomass into smaller liquid hydrocarbons. There was experimentation of four ratios of ethanol to solids (10, 20, 30, and 40 mL/g) by weighing the feedstock to determine the effect of these ratios on bio-oil yield.

3.2.2 Methods and Experimental System Used

The safe processing, at increased temperatures and pressures, was achieved in these experiments using a lab-scale batch-type high-pressure liquefaction reactor. For every test, rice straw and ethanol were fed to the reactor in a known quantity. The experimental setup was leaked, and a reactor was heated to a well-controlled temperature of 220 °C with an internal pressure of 2.5 atm. Nitrogen injection was used to control oxidation by maintaining an inert reaction environment. Although the specific heating rate was not reported, it is quite commonplace for such a system to be heated at a rate of 10–20 °C/min. The time the samples spent in the reactor was fixed at 30, 60, or 90 min so the investigators could determine how time employed within the reactor influences product yields. After the reaction, the sample was cooled and then filtered to separate from its solid particles. The liquid products were divided into three separate fractions: bio-oil was defined as volatile and light or heavy based on its boiling point and molecular weight features. The yield of each fraction was determined by gravimetric analysis. The light and heavy oils were both subjected to GC–MS and FT-IR analysis, the former to identify their chemical constituents, while the latter provided fundamentals for characterization of the functional groups in the bio-oil. Further, the concept of response surface methodology was applied with the use of object software tools namely SAS 9.1.3 and MATLAB 7 to build statistical models. Using these tools, researchers were able to create 3D surface models of how ethanol ratio, reaction time, and product yield are interconnected. The optimization revealed the most efficient parameters of operation to improve the bio-oil production.

3.2.3 Results

From the study, it was found that there was a high impact of the ethanol to solid material ratio and retention time on the quantity and quality of bio-oil produced. The maximum combined yield of bio-oil fractions was obtained at an ethanol to solid ratio of 30 mL/g and retention time 60 min. Optimization of the process using a 30 mL/g ethanol proportion and a 60 min retention time generated light bio-oil yields of 0.2016 g/g, heavy bio-oil at 0.237 g/g, and volatile bio-oil at 0.0301 g/g, with a total bio-oil yield of approximately 0.47 g for every gram of the rice-straw (Fig. 2, 3, 4 and 5).

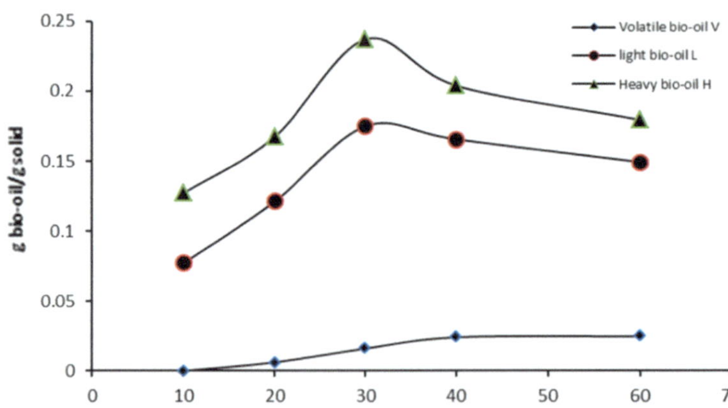

Fig. 2 Effect of ethanol to solid ratio in the amount of bio-oil produced at 15 min retention time [17]

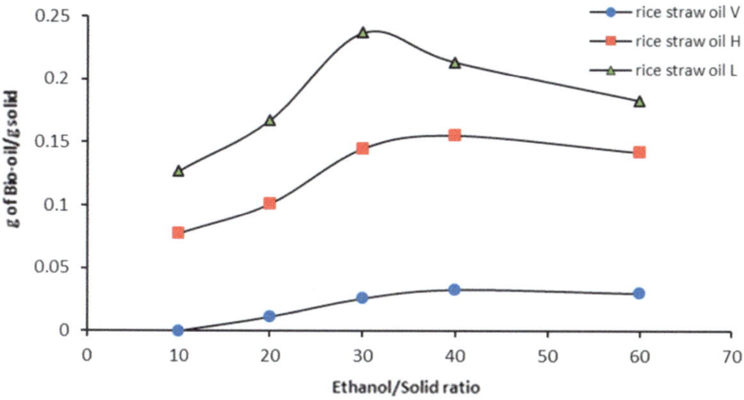

Fig. 3 Effect of ethanol to solid ratio in the amount of bio-oil produced at 30 min retention time [17]

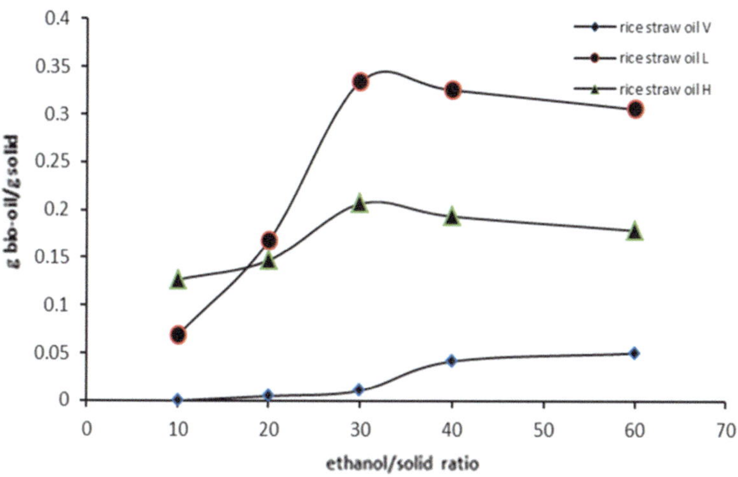

Fig. 4 Effect of ethanol to solid ratio in the amount of bio-oil produced at 60 min retention time [17]

Figures 6, 7 and 8 show how changing the amount of time the process goes on can affect the amount of bio-oil that is made when turning rice straw into bio-oil using 2.5 atm and 220 °C. The amount of volatile bio-oil peaks at around 60 min, after that longer holding times cause the amount to go down because the reactive pieces start to combine again and turn back into solids. The maximum amount of volatile oil comes out at 30 min, whereas the lowest amount is reached after 120 min. GC–MS results showed that both the light and the heavy bio-oils contained primarily esters, with most abundant type of esters being those with carbon chains length ranging from C_{16}

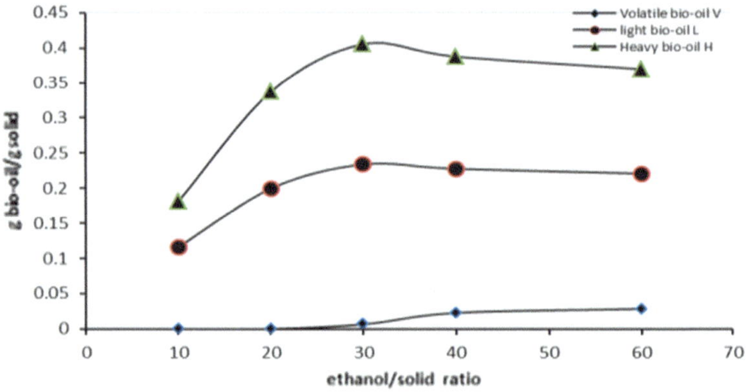

Fig. 5 Effect of ethanol to solid ratio in the amount of bio-oil produced at 90 min retention time [17]

to C_{19}. The existence of these esters like those found in the biodiesel indicates that the bio-oil is a good candidate for renewable fuel.

The data in Fig. 9 reveals that the main groups in rice straw bio-oil are ester and when it comes to benzene, the major carbon atoms are in C_{16} to C_{19}. As the volatile content is very small compared to the heavy molecules, it is also smaller. The bio-oil obtained from rice straw can be used as a biofuel. The generation of fatty acid ethyl esters proves to endorse the efficiency of liquefaction and esterification reactions in this process. The findings of these were confirmed by the FT-IR spectrum as Fig. 10 shows that exhibited strong absorption bands over the C=O (carbonyl),

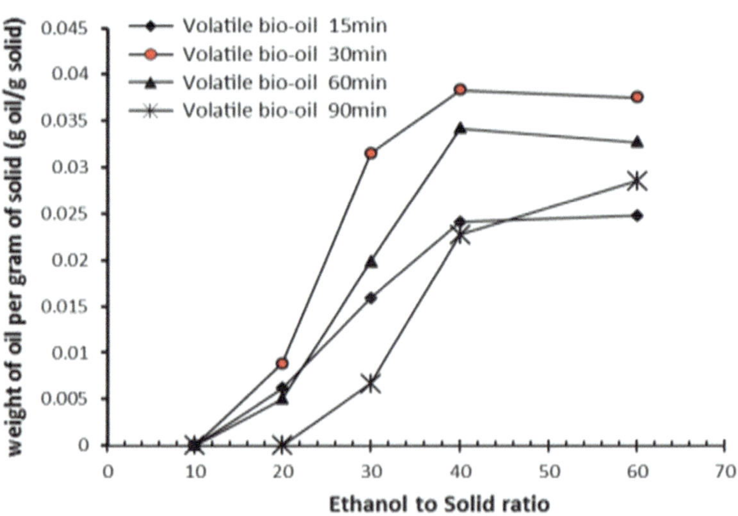

Fig. 6 Amount of volatile oil per gram of rice straw at different autoclaving time [17]

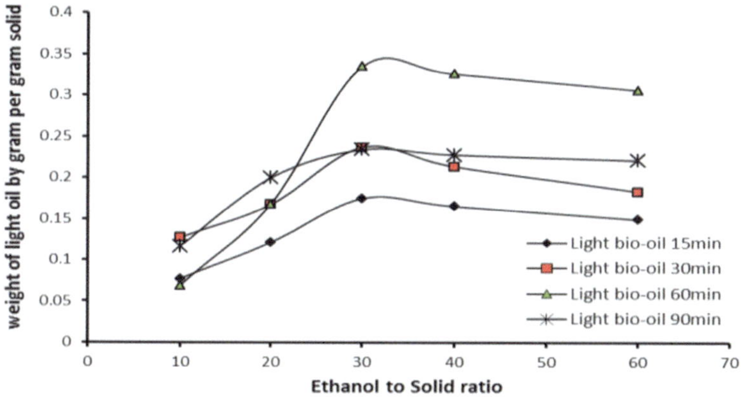

Fig. 7 Relation between ethanol solid ratio and amount of light oil produce per gram rice straw [17]

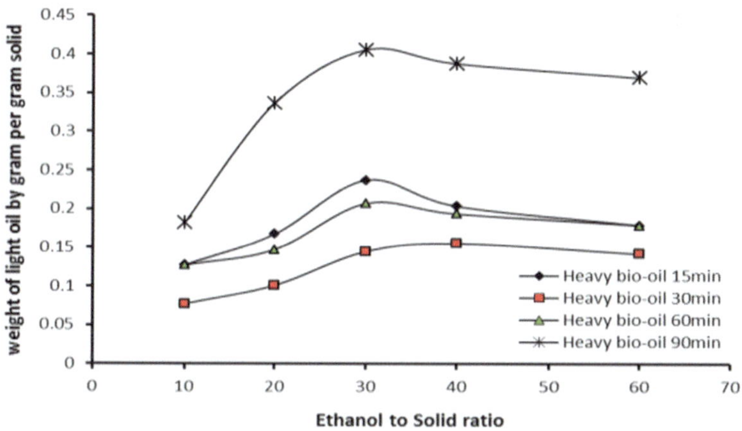

Fig. 8 Relation between ethanol solid ratio and amount of heavy oil produce per gram rice straw [17]

C–H (aliphatic hydrocarbons), and O–H bands representing the presence of alcohols and esters in the form of bio-oil.

Moreover, very precise forecasts for the response of yields were achieved by using statistical models using SAS and MATLAB. The models predicted optimal retention time as 52.5 min and an ethanol ratio of 35 mL/g, confirmed by experiment. Under these optimal conditions the model displayed the following yields: The yields achieved volatile bio-oil (0.1835 g/g), light bio-oil (0.7632 g/g), and heavy bio-oil (0.3853 g/g) provided a good fit with experimental data to attest reliability of model. Finally, our study was successful in demonstrating that rice straw can generate high-productivity energy-dense bio-oil via ethanol-based liquefaction under moderate operational conditions. Statistical and experimental optimization

RT	Area %	Compound Name	Chemical formula	Norm %
15.443	64.692	6-Octadecenoic acid, methyl ester	$C_{19}H_{36}O_2$	100
15.353	14.417	9,12-Octadecadienoic acid (Z,Z)-, methyl ester	$C_{19}H_{34}O_2$	22.28
14.123	10.548	Hexadecanoic acid, methyl ester	$C_{17}H_{34}O_2$	14.417
13.913	1.681	9-Hexadecenoic acid, methyl ester, (Z)-	$C_{17}H_{32}O_2$	2.6
15.548	0.972	Heptadecanoic acid, 16-methyl-, methyl ester	$C_{19}H_{38}O_2$	1.5
16.634	0.508	6,9-Octadecadienoic acid, methyl ester	$C_{19}H_{34}O_2$	0.79
16949	0.503	Z-8-Methyl-9-tetradecenoic acid	$C_{15}H_{28}O_2$	0.78
16.509	0.451	Oxiraneoctanoic acid, 3-octyl-, methyl ester, cis-	$C_{19}H_{36}O_3$	0.7
16.474	0.213	8,11-Octadecadienoic acid, methyl ester	$C_{19}H_{34}O_2$	0.33
4.264	0.151	2-Pentanol, 2,4-dimethyl	$C_7H_{16}O$	0.23
9.601	0.121	Benzene, (1-propyloctyl)-	$C_{17}H_{28}$	0.19
16.784	0.107	Methyl 18-methylnonadecanoate	$C_{20}H_{40}O_2$	0.17
9.231	0.094	Benzene, (1-butylheptyl)-	$C_{17}H_{28}$	0.14
11.817	0.085	Benzene, (1-pentylheptyl)-	$C_{18}H_{30}$	0.13
11.882	0.086	Benzene, (1-butyloctyl)-	$C_{18}H_{30}$	0.13
12.067	0.068	Benzene, (1-propylnonyl)-	$C_{18}H_{30}$	0.11
12.367	0.068	Benzene, (1-ethyldecyl)-	$C_{18}H_{30}$	0.10
12.882	0.07	Benzene, (1-methylundecyl	$C_{18}H_{30}$	0.11
13.142	0.07	Benzene, (1-pentyloctyl)-	$C_{19}H_{32}$	0.11
23.307	0.083	E-2-Methyl-3-tetradecen-1-ol acetate	$C_{17}H_{32}O_2$	0.13
25.127	0.067	9-Hexadecenoic acid, 9-octadecenyl ester, (Z,Z)-	$C_{34}H_{66}O_2$	0.1

Fig. 9 The main component of rice straw bio-oil from gas chromatography with Ms-detector [17]

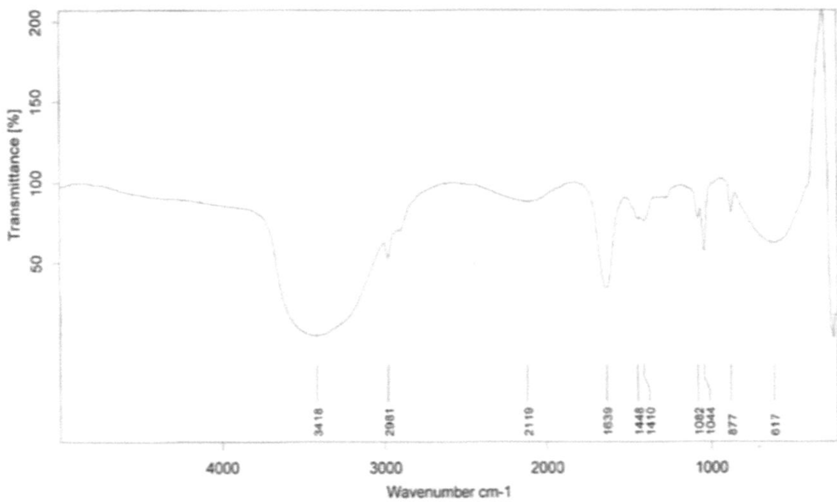

Fig. 10 FT-IR spectra for bio-oil produced from rice straw [17]

both demonstrated that solvent ratio and reaction time are significant variables in determining both yield and quality of bio-oil. The study emphasizes the appropriateness of rice straw as biofuel feedstock and promotes advancement in the development of sustainable energy technologies.

3.3 Case Study Two

Title: Catalytic Pyrolysis of Rice Straw: Screening of Various Metal Salts, Metal Basic Oxides, Acidic Metal Oxide, and Zeolite Catalyst on Product Yield and Characterization

Authors and year of publication: Cao, Zhenheng; Niu, Jiajia, [12].

3.3.1 Material Used

The feedstock for pyrolysis in this investigation was rice straw, which was obtained from Zhoukou, China. For the purpose of pyrolysis, the rice straw biomass was oven-dried at 105 °C for 24 h to remove moisture and obtain uniform moisture level before processing. Post-drying, the rice straw was ground and sieved to obtain particles of the size ranging from 0.3–0.45 mm reflecting optimum surface area for effective Pyrolysis. Various catalysts were used to determine their impact on the volume and composition of the rice straw-derived bio-oil produced by pyrolysis. The choice of the catalyst comprised $FeCl_3$, $CuCl_2$, $MgCl_2$, and $MnCl_2$ among the metal salts, MgO,

CaO, $MgCO_3$, and $CaCO_3$ as the basic metal oxides; ZnO, ZrO_2, CeO_2, and TiO_2 as the acidic metal oxides; along with zeolites ZSM-5, Y-zeolite, Mordenite, and SBA-15. All the catalysts used in this study were purchased from Sigma-Aldrich's China branch. The choice of such catalysts was targeted at understanding their efficiency in promoting bio-oil production, as well as changing the chemical nature of the bio-oil produced through pyrolysis of rice straw.

3.3.2 Methods and Experimental System Used

For pyrolysis experiments, a fixed bed-reactor 30 cm long and 4.0 cm (internal) in diameter was used. The reactor was operated at a heating rate of 10 °C/min, and it was maintained at four distinct settings of temperatures. Each experiment was carried out at four different temperatures (that is, 350, 400, 450, and 500 °C) for one hour. The experiment was carried out in an inert atmosphere produced by bubbling nitrogen gas through the reactor, with nitrogen used as the 40 ml/min carrier gas. The combined biomass was mixed with various catalysts levels: 5, 10, 15, and 20% (wt%). The bio-oil was condensed at 5 °C in a chilled condenser after pyrolysis, and non-condensable gases were also isolated in separate isolation. It also determined the mass of char residue separately so as to assess the yield of every product. Several approaches were used to learn about the composition and properties of the bio-oil under changing process conditions. The bio-oil was analyzed by GC–MS to quantify the presence of phenolic compounds, aromatic aldehydes, furan derivatives, and cyclic ketone. The FT-IR confirmed spectroscopic signatures of functional hydroxyl (–OH), ether (–O–), and carbonyl (C=O) groups in the bio-oil. We were able to elucidate details for the proton environments and the structure of the bio-oil using Proton Nuclear Magnetic Resonance (1 H-NMR) spectroscopy. Thermal stability and weight loss of the bio-oil were determined based on TGA, while elemental analysis (C, H, N, O) gave information on the bio-oil composition and higher heating value. Gel Permeation Chromatography (GPC) was used to determine the molecular weight distribution of the bio-oil, which is both reactive and stable.

3.3.3 Results

From the experiments, it was clear that the bio-oil yield was also very susceptible to changes in the pyrolysis temperature. As seen in Fig. 11, it was found that making the temperature 450 °C helped the process get the most amount of bio-oil, which ended up being 38.5 wt%. While lower such as 26.7 wt%, 33.2 wt%, and 32.5 wt% bio-oil yield was found when the temperatures used were 350, 400, and 500 °C respectively.

The use of metal salt catalysts accelerated the breakdown of biomass. As it observed in Fig. 12, $MgCl_2$ had the highest yield of bio-oil 48.4 wt%, followed by, $MnCl_2$ and $CuCl_2$ with yields of 45.3 wt% and 41.5 wt%. Simultaneously, $FeCl_3$ only generated 32.2 wt% bio-oil given its instability at such conditions.

Temperature, °C	Bio-oil, wt.%	Char residue, wt.%	Gas, wt.%	Conversion, wt.%
350	26.7	48.6	24.7	51.4
400	33.2	41.4	25.4	58.6
450	38.5	37.1	24.4	62.9
500	32.5	35.2	32.3	64.8

Fig. 11 Effect of different reaction temperature of rice straw on product yield [12]

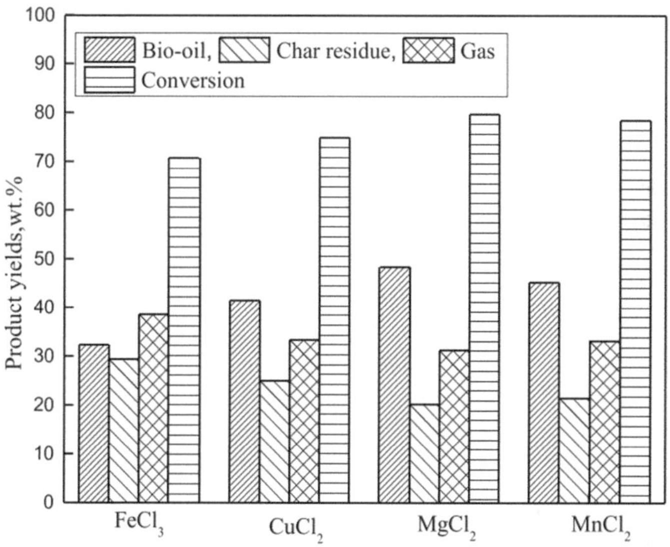

Fig. 12 Effect of metal salts catalyst on products yield [12]

The impact of acidic metal oxide catalyst on the products obtained during the pyrolysis of rice straw is presented in Fig. 13. MgO that is basic metal oxide catalyst gave 52.1 wt% bio-oil, while other basic oxides like CaO, $MgCO_3$, $CaCO_3$ generated much lower yields of bio-oil. The result of the acid metal oxide catalyst CeO_2 was 52.3 wt% bio-oil, indicating its capacity to increase yield of bio-oil.

The Y zeolite proved to be the best zeolite catalyst, yielding 55.2 wt% bio-oil followed by Mordenite (49.1 wt%), and ZSM-5 (47.4 wt%) as shown in Fig. 14. Bio-oil containing catalysts had increased levels of phenolic compounds (including alkyl phenols and guaiacol) compared to non-catalytic bio-oil that was rich in furan derivatives such as furfural and 2-furanmethanol. Higher heating values (HHV) of the bio-oil products obtained from catalytic pyrolysis were high. The highest HHV of 36.7 MJ/kg was recorded from the bio-oil produced using Y-zeolite as a catalyst, followed by those using MgO (36.2 MJ/kg) and CeO_2 (35.2 MJ/kg). At the same time, the non-catalytic bio-oil HHV value was only 30.7 MJ/kg. And further, the bio-oil

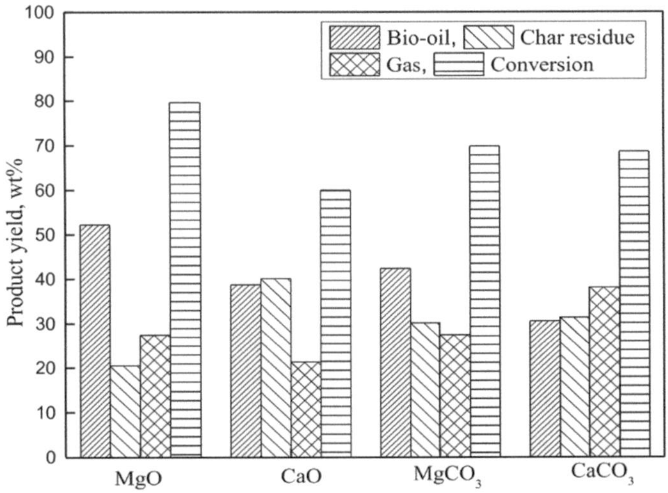

Fig. 13 Effect of metal oxide on products yield [12]

produced with the help of catalysts had a lower molecular weight than that obtained without catalysts implying the catalysts facilitated the breakdown of biomass to smaller, more volatile species. Y-zeolite-transformed bio-oil was the least heavy molten (690 g/mol) and therefore attracted attention as refined and quite reactive product. These results show how catalysts greatly increase the yield, chemical profile, and energy content of the bio-oil, with Y-zeolite proving superior for pyrolysis of rice straw.

3.4 Case Study Three

Title: Pyrolysis Pretreatment and Low-Temperature Liquefaction: Developing Rice Straw Biofuels

Authors and year of publication: Zhou, Yuqian; Tan, Zhongchao, [7].

3.4.1 Material Used

The focus for this research was on rice straw as the primary raw material for biofuel production. Rice straw was collected from an experimental field at Huazhong Agricultural University in Wuhan, Hubei, then naturally dried, ground before being sieved through a 60-mesh sieve and dried for 24 h at 105 °C. For pyrolysis temperatures of 300, 400, and 600 °C, different types of biochar have been produced. Which were labeled RSC3, RSC4, and RSC6. The elemental composition and higher heating

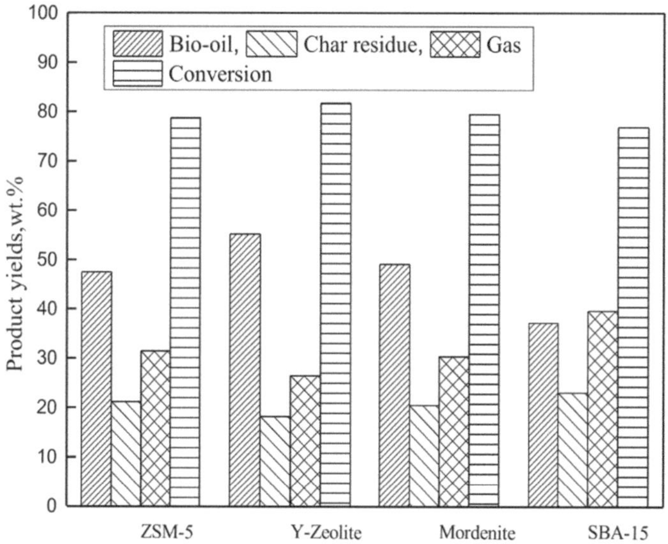

Fig. 14 Effect of zeolites based catalyst on products yield [12]

value (HHV) of this biochar was determined. Tetralin (THN) was chosen as the hydrogen-donor solvent for liquefaction due to its high hydrogen supply and excellent dispersion qualities that are essential to the transformation of solid fuels into liquid biofuels. Furthermore, use of acetone was necessary to promote recovery of solvents in the liquefaction.

3.4.2 Methods and Experimental System Used

Pyrolysis pretreatment of rice straw was carried out in a unique pyrolysis furnace. The pyrolysis of rice straw was done with the use of nitrogen (N_2) as a carrier gas at a flow rate of 20 ml/min, under three different temperature conditions: The pyrolysis of rice straw was performed at 300, 400, and 600 °C with residence time at 20 min each. Elemental analysis, determination of higher heating value (HHV) and g thermogravimetric characterization were conducted on the biochar samples obtained—RSC3, RSC4, and RSC6. After pyrolysis, the biochar was liquefied using tetralin (THN) as the hydrogen donor solvent. Liquefaction with biochar was performed in a HPR at temperatures between 25 and 280 °C and a hydrogen pressure of 5 MPa. Liquefaction was performed at 5°C/min heating rate and 15 min residence time. RSC4 was chosen as the primary precursor for synthesizing rice straw-biochar-liquid fuel (RS-C-LF) due to the unique properties of the RSC4 Biochar (higher HHV and lower oxygen levels than the alternative biochar obtained through a pyrolysis process at 400 °C). Biofuels were first fractionated using filtration, and further purification by means of

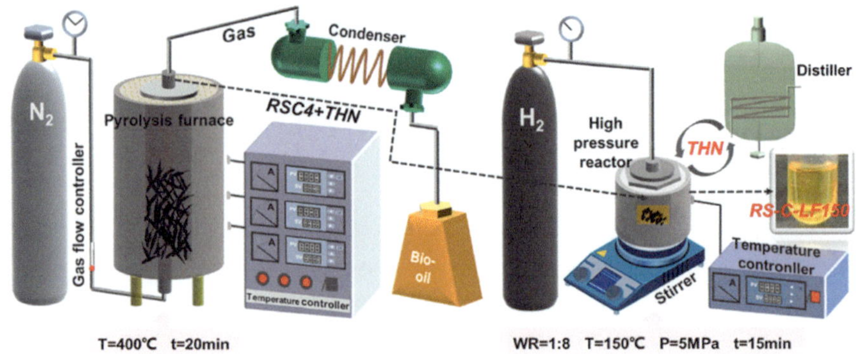

Fig. 15 The preparation process of rice straw-biochar-liquid fuel at 150 °C [7]

acetone in a rotary evaporator gave solvents back and reduced waste. The elemental properties and heating capacities of the end products were determined (Fig. 15).

3.4.3 Results

The results emphasized the effectiveness of pyrolytic pretreatment for optimization of characteristics of rice straw biochar and optimization of liquefication process. As the pyrolysis temperature increased carbon content and HHV of biochar increased and its oxygen content therefore reduced. Among the biochar, RSC4 produced at 400 °C resulted in best outcomes in terms of carbon (54.92 wt%), higher heating value (19.51 MJ/kg), and lower oxygen content than raw rice straw, which is suitable for the following liquefaction as shown in Fig. 16.

The liquefaction process was characterized by significant changes in yield in terms of temperature increases as it observed in the chart (Fig. 17). As the liquefaction temperature increased, the yield of RS-C-LF enhanced, and the maximum yield (67.41%) was at 150 °C. The yield of RS-C-LF decreased when temperature rose to 200 and 280 °C indicating the breakdown of unstable functional groups including carboxyl groups into gaseous compounds.

A shift in composition of the biofuels was observed upon functional group analysis, in that at higher liquefaction temperatures, more aromatic compounds were

Liquefied raw materials	Element analysis (wt.%, daf[a])					Atomic ratio		HHV (MJ/kg)	Reference
	C	H	O	N	S	H/C	O/C		
RS	39.34	5.84	41.47	0.75	0.44	1.80	0.79	15.22	This work
RSC3	44.17	4.87	33.96	0.78	0.48	1.32	0.58	18.40	This work
RSC4	54.92	3.60	17.19	1.00	0.47	0.72	0.24	19.51	This work
RSC6	56.83	1.55	8.00	0.88	0.45	0.36	0.11	20.24	This work
Tunçbilek Lignite	67.13	4.68	21.93	2.52	3.64	0.84	0.29	21.80	[3]
Inner Mongolia Lignite	66.15	4.30	25.88	1.21	2.46	0.78	0.29	19.18	[30]
Huolinhe Lignite	67.67	4.62	25.44	1.05	1.21	0.81	0.28	24.95	[31]

[a] Dry and ash-free basis.

Fig. 16 Element analysis and HHV of several types of biomass, biochar, and lignite [7]

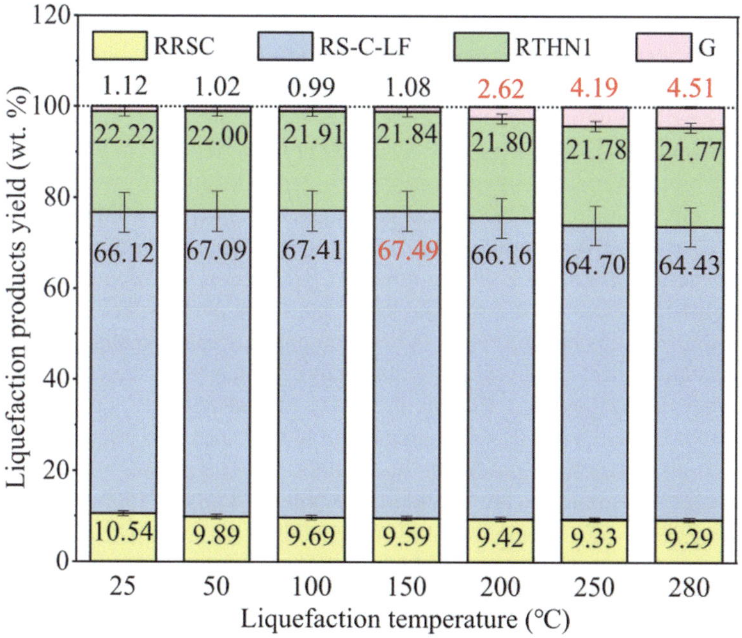

Fig. 17 Effect of liquefaction temperature on the yield of liquefaction products [7]

present whereas the number of carboxyl groups were reduced. Figure 18 illustrated the HHV for the biofuels produced was at about 42 MJ/kg and still within the European biodiesel standard (EN 14214, and RS-C-LF150 recorded the highest HHV of 42.16 MJ/kg). This verification demonstrated that the resulting biofuel was of a superior quality and similar to commonly used biofuels.

Results of energy balance during liquefaction showed excellent efficiency as the recovery rates (η) were greater than 95% at all the temperatures and the highest in 97.95% at 150 °C. The material-energy ratio (ε) was high, and this implies that the input materials were well used in the manufacture of the biofuel products. RS-C-LF yield outweighed 65% at any temperature tested, which indicated that the liquefaction process was efficient. Economic review to produce RS-C-LF150 revealed a total

Liquid Fuels	Element analysis (wt.%, daf[a])					H/C	HHV (MJ/kg)
	C	H	O	N	S		
RS-C-LF150	86.97	8.68	0.23	0.78	0	0.10	42.03
CLF10	82.68	11.28	0.82	0.89	0	0.13	42.16

Fig. 18 Element analysis and HHV of RS-C-LF150 and CLF10 [7]

production expenditure (TPE) of $264.89/bbl., which is competitive with the conventional biodiesel and diesel market prices. The wasteful costs that resulted from tetralin and solvent recovery presented critical reasons for the increased TPE, showing the pressing need for optimization to increase the economic efficiency. Using zero-waste as well as self-circulating preparation systems, TPE was reduced to $171.29/bbl. for CLF10, a biofuel produced by means of an ecologically optimized process. The RS-C-LF system is shown in Fig. 19 and is designed for preparing lithosphere fluids using environmentally friendly methods such as distillation and concentration.

The benefits for RS-C-LF and its means of production were found to be on par with worldwide trends toward renewable energy, while barriers including technical barriers that need to be overcome and outcompeting rivals remain. The possibilities of technological progress and wider market penetration may make or break this biofuel, in the long run. Finally, the study demonstrated adequately that rice straw biochar, after Pyrolysis pretreatment and low-temperature liquefaction, could be used as a feedstock for producing liquid biofuels. The process yielded biofuels of better standards in line with the industry standard; RS-C-LF150 was most superior in biofuels quality, production, and energy efficiency. Economic analysis of the biofuel production process was also done, and it was predicted that implementation of optimization strategies would lead to a cost reduction. By this work, we learn

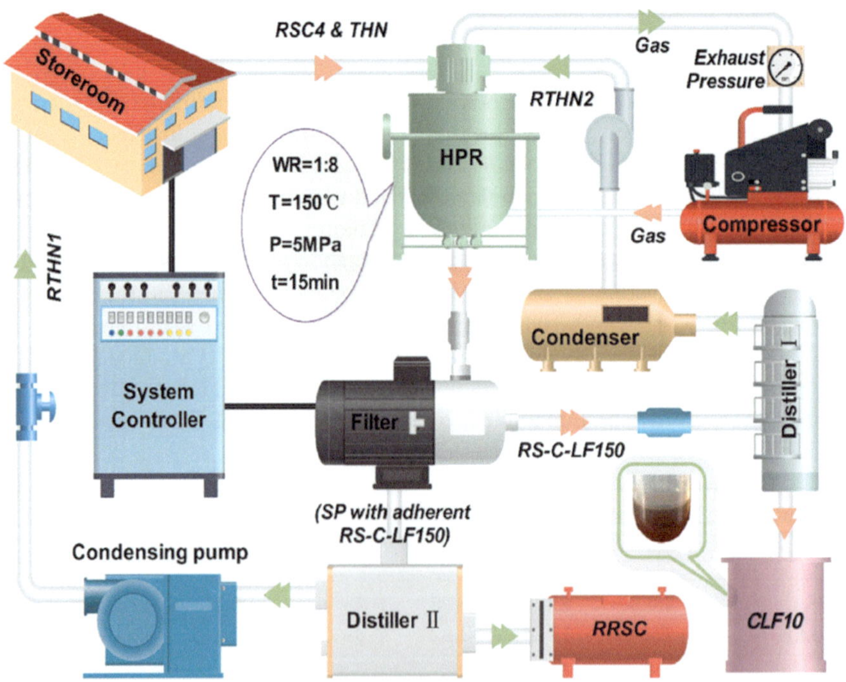

Fig. 19 Optimized zero-waste and self-circulating system for CLF10 [7]

very important things about the sustainable use of rice straw for biofuel production, advances the cause of renewable energy technologies.

3.5 Case Studies Summary

This section gives a summary of the critical findings and experimental values obtained on the basis of the three selected case studies in relation to development of bio-oil from rice straw. The three case studies, through experimentation, reviewed several thermochemical processes seeking to formulate bio-oil from rice straw. Using ethanol-based liquefaction in Case Study One [17] achieves a maximum bio-oil yield (0.47 g/g) at a temperature of 220 °C after a reaction duration of 60 min. In Case Study Two, Cao and Niu [12] applied catalytic pyrolysis with various catalysts, being Y-zeolite the most effective, that produced 55.2 wt% bio-oil which contained a high energy content. Case Study Three by Zhou and Tan [7] showed the greatest biofuel yield (67.41%) and energy efficiency (at 150 °C) as a result of combining pyrolysis and low-temperature liquefaction. Together, this research supports the notion that rice straw is a viable feedstock for production of efficient and sustainable bio-oil (Table 1).

4 Results and Discussion

4.1 Overview

This chapter encapsulates selected research of the experimental results and comparative analyses of rice straw conversion to bio-oil. Different types of thermochemical conversion methods were used as a part of each experiment: ethanol-assisted liquefaction, catalytic assisted pyrolysis, and pyrolysis followed by liquefaction pretreatment. The study investigates how key elements of the processes such as temperature, establishment of reactor, types of the catalyst, heating rate, and pressure influenced the yield, nature of the bio-oil, and the energy content. The comparative analysis reported in this chapter is a goal to provide useful practical information on how different approaches to rice straw bio-oil production compare in terms of effectiveness and sustainability, and thereby guiding biofuel development field.

4.2 Yield Performance Across Techniques

Clear evidence has emerged in the data that achieving high yield of bio-oils is predicated on adoption of the right thermo-chemical approach as well as modification of the operating conditions. In ethanol liquefaction at 220 °C and 2.5 atm, Mohamed

Table 1 A comparison table for all 3 case studies

Aspect	Case study one [17]	Case study two [12]	Case study three [7]
Title	Production of bio-oil from rice straw	Catalytic pyrolysis of rice straw	Pyrolysis pretreatment and low-temperature liquefaction: developing rice straw biofuels
Material used	Rice straw, dried and ground into a fine powder (0.2–1 mm)	Rice straw, oven-dried and ground to 0.3–0.45 mm particles	Rice straw, dried, ground, and sieved through a 60-mesh sieve
Methods and experimental system	Ethanol-based liquefaction in a high-pressure batch-type reactor	Catalytic pyrolysis using a fixed-bed reactor and various catalysts	Pyrolysis pretreatment followed by liquefaction with tetralin (THN) in a high-pressure reactor
Experimental conditions	220 °C, 2.5 atm, varying ethanol:solid ratio (10–40 mL/g) and retention time (30–90 min)	Pyrolysis at 350–500 °C, nitrogen atmosphere, different catalyst concentrations (5–20%)	Pyrolysis at 300, 400 and 600 °C is performed before a liquefaction process from 25 to 280 °C
Catalysts used	None	Metal salts ($FeCl_3$, $CuCl_2$, $MgCl_2$), metal oxides (MgO, CaO), zeolites (Y-zeolite, ZSM-5)	None (focus on pyrolysis and liquefaction)
Bio-oil yield	Maximum total yield of 0.47 g/g at 30 mL/g ethanol ratio and 60 min retention time	Y-zeolite catalyst produced 55.2 wt% bio-oil	Maximum yield of 67.41% biofuel at 150 °C liquefaction temperature
Energy content (HHV)	Not explicitly stated	Y-zeolite: 36.7 MJ/kg, MgO: 36.2 MJ/kg, CeO_2: 35.2 MJ/kg, non-catalytic: 30.7 MJ/kg	RS-C-LF150 produced biofuel with 42.16 MJ/kg HHV
Key findings	High ethanol ratio and retention time positively impacted yield and quality	Catalysts (especially Y-zeolite) enhanced yield, chemical profile, and energy content of bio-oil	Pyrolysis temperature increases carbon content and HHV, liquefaction optimized at 150 °C

on the other hand, Cao and Niu [12] used catalytic pyrolysis at 450 °C and they were able to increase the bio-oil produce up to 55.2 wt% the use of the Y-zeolite catalyst. Zhou and Tan [7] with a precursor utilizing pyrolysis and then low-temperature liquefaction had the best yield of 67.41% exemplifying the advantages of a combined process strategy.

4.3 Effect of Key Parameters

All processes were highly sensitive to temperature for highest yield. Biomass degradation was significantly promoted even in pyrolysis (450–600 °C), but volatiles were lost; liquefaction was carried out at 150 °C in turn protected biomass structure and tailored yield. The effect of catalyst use was notable, particularly in Case Study Two where Y-zeolite not only increased yield but also improved the higher heating value and chemical quality of the oil. Reactor design varied: Batch processing in high-pressure reactors regulated and prolonged reaction times in liquefaction whereas fixed-bed reactors enhanced catalyst contact in pyrolysis. Case Study Three used a hybrid reactor that involved combination of pressure and temperature control to provide both high yield and a minimal rate of energy recovery process.

4.4 Practical Implications and Scalability

The three approaches reveal that rice straw is a highly promising resource for sustainable feedstocks. Catalytic pyrolysis yields rich outputs with high yields as opposed to liquefaction that maximizes consistent product output at mild temperatures. Case Study Three emphasizes the unique benefit of integrating pyrolysis pretreatment with low-temperature liquefaction, with its impressive yield and quality, and high economic value for scalability of the process. This helps to greatly enhance the applicability of solvent recycling and energy recovery systems.

4.5 Conclusion and Economic Outlook

Analysis of the case studies shows that rice straw is an ideal renewable feedstock for making bio-oil. Ethanol-based liquefaction allows clean processing with relatively light conditions, catalytic pyrolysis boosts both product reach and energy content and pyrolysis-treated liquefaction has the advantage of being the most productive and energy efficient. Looking at the economics of treatment, Case Study Three provided the most efficient results. After process improvement and recycling solvents, the TPE for each barrel was reduced from $264.89 to $171.29. This means that bio-oil from rice straw would be as competitive as current fossil fuel prices if production was scaled with suitable technology. At the same time, high energy recovery and efficiency mean the process is both efficient and good for the economy and the environment. It is particularly significant to me as a process engineering student that rice straw which is normally burned or thrown away can now supply energy. Thus, it suggests that using waste-to-energy strategies could be very helpful in places where agricultural residues are common and rarely put to good use. Emphasizing the integration of solvent recovery, more efficient catalyst reuse and cutting environmental emissions should be made in future studies. Running simulation experiments on an industrial scale and in Aspen Plus or SuperPro Designer might give a clearer picture of feasibility. Using renewable energy not only cuts use of fossil fuels but also helps Oman remain sustainable for the future.

5 Conclusion

It examined if bio-oil could be made from rice straw through thermochemical conversion processes. Because it's easy to find and has a lot of lignocellulose, rice straw makes a suitable raw material for eco-friendly energy. The research team studied three different cases to look at various approaches: making ethanol, using catalytic pyrolysis and including a pyrolysis preliminary stage in low-temperature liquefaction. From these studies, we found that the temperature, type of catalyst used, heating speed, time retained in the reactor and the design of the reactor all change the amount produced, its makeup and the energy it contains. When compared to other techniques, the pyrolysis-pretreated liquefaction process resulted in greater bio-oil output (67.41%) and its heating value (42.16 MJ/kg) and recovery rate (above 95%) were superior. Catalytic pyrolysis performed well and Y-zeolite was found to be the best catalyst, leading to high yields and better chemistry in the bio-oil. Simpler ethanol-based liquefaction showed that the best ester-rich bio-oil was obtained when perfecting the ratio of solvent to solid and the period of reaction time. Economic benefits came from the third case study, with solvent recovery and system tuning lowering costs from $264.89 to $171.29 per barrel. They show that by scaling up and combining different systems, rice straw bio-oil might provide a suitable alternative for traditional fossil fuels. The results of this study are that rice straw can both be a waste product that is not used enough and a great resource for making biofuels. The lessons we gain from these case studies make it clear that efficient conversion parameters and effective economic recovery systems play a big role. Proposing new experiments at pilot plants, examining the environmental impact, and finding better ways to manage catalysts and cut waste will help make bio-oil production from rice straw viable for industry. The process used here is helpful to the global environment effort and holds promise for sustainable growth in regions such as Oman that produce a lot of rice farming biomass but mostly leave it behind or burn it.

References

1. Das PP, Sontakke AD, Purkait MK (2023) Rice straw for biofuel production. In: Green approach to alternative fuel for a sustainable future, Elsevier, pp 153–166. https://doi.org/10.1016/B978-0-12-824318-3.00034-5
2. Vamvuka D (2011) Bio-oil, solid and gaseous biofuels from biomass pyrolysis processes—an overview. https://doi.org/10.1002/er.1804

3. Sharma A, Singh G, Arya SK (2020) Biofuel from rice straw. Elsevier Ltd. https://doi.org/10.1016/j.jclepro.2020.124101
4. Rodionova MV et al (2017) Biofuel production: challenges and opportunities. Elsevier Ltd. https://doi.org/10.1016/j.ijhydene.2016.11.125
5. Bhaskar T, Balagurumurthy B, Singh R, Poddar K, Thermochemical route for biohydrogen production
6. Uddin I, Sohail M, Hussain MI, Alhokbany N, Amaro-Gahete J, Estévez R (2023) Probing the pyrolysis process of rice straw over a 'dual-catalyst bed' for the production of fuel gases and value-added chemicals. Sustainability (Switzerland) 15(14). https://doi.org/10.3390/su151411057
7. Zhou Y, Tan Z, Zhou T (2024) Pyrolysis pretreatment and low-temperature liquefaction: developing rice straw biochar-based clean liquid biofuel. J Energy Inst 117. https://doi.org/10.1016/j.joei.2024.101710
8. Seo YD, Seo TC, Oh SY (2022) Co-pyrolysis of rice straw with industrial wastes: waste disposal and environmental remediation. Waste Manage Res 40(3):339–348. https://doi.org/10.1177/0734242X211027066
9. Moholkar VS, Singh S (2014) A Corrigendum to feasibility of rice straw as alternative substrate for biobutanol production by Ranjan et al. [Appl Energy 103:32–38 (2013)]. Elsevier Ltd. https://doi.org/10.1016/j.apenergy.2013.10.022
10. Izzatie NI, Basha MH, Uemura Y, Hashim MSM, Amin NAM, Hamid MF (2017) Co-pyrolysis of rice straw and Polyethylene Terephthalate (PET) using a fixed bed drop type pyrolyzer. J Phys: Conf Ser. https://doi.org/10.1088/1742-6596/908/1/012073
11. Ahmad Zamri MFM et al (2024) Progress and challenges of mesoporous catalysts in upgraded pyrolysis of biomass for biofuel production. Elsevier B.V. https://doi.org/10.1016/j.jaap.2024.106651
12. Cao Z, Niu J, Gu Y, Zhang R, Liu Y, Luo L (2020) Catalytic pyrolysis of rice straw: screening of various metal salts, metal basic oxide, acidic metal oxide and zeolite catalyst on products yield and characterization. J Clean Prod 269. https://doi.org/10.1016/j.jclepro.2020.122079
13. Alengebawy A, Ran Y, Ghimire N, Ahmed, Osman I, Ai P (2023) Rice straw for energy and value-added products in China: a review. Environ Chem Lett 21:2729–2760. https://doi.org/10.1007/s10311-023-01612-3
14. Abd-Elgawad AMM (2023) Biofuel extraction methods from rice straw: review. J Int Soc Sci Eng 5(4):90–97 [Online]. Available: www.jisse.journals.ekb.egwww.isse.org.eg
15. Gupte AP, Basaglia M, Casella S, Favaro L (2022) Rice waste streams as a promising source of biofuels: feedstocks, biotechnologies and future perspectives. Elsevier Ltd. https://doi.org/10.1016/j.rser.2022.112673.
16. Domínguez-Escribà L, Porcar M (2010) Rice straw management: the big waste. Biofuels, Bioprod Biorefin 4(2):154–159. https://doi.org/10.1002/bbb.196
17. Mohamed OA, Ashour I (2016) Production of bio-oil from rice straw. [Online]. Available: https://www.researchgate.net/publication/326584369
18. Passoth V, Sandgren M (2019) Biofuel production from straw hydrolysates: current achievements and perspectives. Springer. https://doi.org/10.1007/s00253-019-09863-3

Production of Biofuel from Co-pyrolysis of Algae and Plastic Wastes

Muna Abdullah Musabah Al Risi and Seyed Mojtaba Sadrameli

Abstract Research of sustainability biofuel production methods has accelerated due to the world's transition to renewable energy. This study explores the possibility of co- pyrolysis plastic waste and algae, two abundant yet environmentally problematic materials to produce biofuel. The combination of plastic trash as an energy-rich co-feedstock and algae's high lipid and protein content makes them ideal for thermochemical conversion. In this study, several experimental case studies involving different algae species including Nannochloropsis sp, Ulva intestinalis, Ulva lactuca, and Chlorella vulgaris co-processed with synthetic polymers including LDPE, PET, and HDPE are analyzed and contrasted. The impact of operating parameters, including reactor type, temperature, catalyst use, and the particles size, the resultant bio-oil in the yield and energy content is systematically examined in this study. According to the results, the fixed-bed reactor in Case Study 1 produced the most bio-oil (approximately 82 wt%) at 600 °C and 30 min for reaction, although microwave-assisted systems provided benefits in terms of energy content and bio-oil quality. Catalysts like silicon carbide and activated carbon improved conversion and increased thermal efficiency. Additionally, greater heating values of the generated oils varied among the investigations, Case Study 4 achieved 35.23 MJ/kg, whereas Case Study 2 (microwave vacuum reactor) had the highest value (36.6 MJ/kg). The technologies used for condensation, which included solvent wash and ice-water, also contributed to the recovery and purity of the final product. The study evaluates yield and energy output while also emphasizing the environmental significance of turning waste into energy and proving that the co-pyrolysis method is a significant step in decreasing waste accumulation and lowering reliance on fossil fuels. The potential of algae–plastic co-pyrolysis in creating a sustainable, circular bioeconomy is further supported by this study.

Keywords Co-pyrolysis · Algae · Plastic waste · Sustainable fuel production · Reactor type · Thermochemical conversion · Catalysts · Biofuel production

---

M. A. M. Al Risi · S. M. Sadrameli (✉)
Department of Engineering, German University of Technology in Oman, Muscat, Oman
e-mail: seyed.sadrameli@gutech.edu.om

A. Azizi (ed.), *Mechatronic Innovations in Sustainable Energy*, Emerging Trends in Mechatronics, https://doi.org/10.1007/978-981-95-5863-6_4

1 Introduction

1.1 Overview

Searching for sustainable and renewable energy alternatives has grown more urgent due to growing global worries about climate change, environmental deterioration, and the limited supply of fossil fuels. Making biofuels by thermochemically converting waste-derived feedstocks is one promising tactic. In this context, the co-thermal decomposition of algae and trash which is made of polymer (plastic) has emerged as an approach which is worthy of note. Through the simultaneous thermal breakdown of both materials in an oxygen-limited environment, energy-rich products including biochar, syngas, and bio-oil are produced. Co-thermal decomposition is a potentially useful strategy for integrating disposal of trash (plastic waste), and energy recovery since it can enhance the standard and quantity of bio-oil in comparison to the pyrolysis of algae alone. It also makes it possible to value plastic waste.

The following chapter presents a summary of recent experimental studies on the co-pyrolysis of algae and the trash of the polymer (plastic). It studies how important factors influence fuel properties and product distribution, including feedstock composition, blending ratio, reactor design, pyrolysis of heating value, boiling rate, catalyst use, and the time spent on the experiment. In this chapter, the reviewed studies are arranged in chronological order, starting with the most current to earlier works throughout the years.

1.2 Review of Recent Research

In 2024, Farobie et al. [1] co-pyrolyzed Ulva lactuca macroalgae with polyethylene terephthalate (PET), a form of plastic waste, to examine the generation of biochar and bio-oil at different blend ratios using a pyrolysis reactor. The experiment was conducted at temperatures as high as five hundred degrees Celsius, and forty percent Ulva lactuca and sixty percent PET were shown to be the ideal conditions. The higher value of bio-oil production up to 37.91% was attained to be lower ideal circumstances, with the hydrocarbon content reaching 57.16%, indicating better fuel quality, according to their analysis of the impacts of PET on product yield. The bio-oil's unwanted components were likewise inhibited by the addition of PET. When compared to raw biomass, a bio charcoal formed by a seventy percent of U. lactuca, and thirty percent of PET blend demonstrated a noticeably higher region size of the outer layer (20.18 m2/g). With the value of activating power round to Forty-three point three five kilojoules per mole, kinetic analysis revealed the initial-order structure for the material reaction replica. The study demonstrated how plastic trash, and marine biomass can be combined to provide sustainable energy.

The study was led by Tan et al. [2] in the year 2024, used HZSM-5 support on limestone (Zeolite Catalysts/LS) as catalytic substance, researchers investigated the catalytic co-thermal decomposition of microalgae with plastic waste (polyethylene component which is distinguished with high-density (HDPE) and polyethylene component which is distinguished with low-density (LDPE)). The ideal thermal decomposition temperature was found to be 873 K after the studies were conducted at different heating ratios. PDSE, or Progression Deep Swarm-Evolution, a neural-evolutionary optimization strategy, was used to identify the ideal feedstock mix and operational parameters. They discovered that the presence of plastic trash, particularly LDPE, greatly accelerated the thermal decomposition of microalgae and produced biofuel through co-pyrolysis. When make a comparison for the thermal decomposition of pure microalgae, the power activated which used in the experiment is decreased to around 20 kilo joule per mole, indicating a beneficial synergistic impact. According to kinetic modeling, the combination of microalgae and LDPE used 23% less energy activation than microalgae by themselves. Under catalytic circumstances, the ideal blend, which included 0.13% LDPE, 0.77% HDPE, and 0.11% microalgae, produced a 99.6% degradation rate. The co-pyrolysis process was endothermic and non-spontaneous, according to thermodynamic study, but it produced a higher-quality bio-oil product and increased fuel conversion efficiency.

This paper was directed by Fan et al. [3] in the year 2024. A microwave-assisted reactor operating at about 500 °C was used to study the co- thermal decomposition of a type of microalgae called Chlorella vulgaris with a plastic waste that's polyethylene which is distinguished with higher value density (HDPE) waste at a one-to-one mixing ratio. Bimetallic catalysts supported by activated carbon (AC) cerium–iron on activated carbon, cerium–nickel on activated carbon, and cerium–copper on activated carbon were also used. To assess the impact of different catalyst loadings (30–50%) on reaction time, product yield, and bio-oil quality, tests were conducted. When they examined the biofuel made by co-pyrolysis, they discovered that adding a catalyst greatly improved the pyrolysis process's efficiency by shortening reaction times and encouraging larger yields. 30% of Cerium–iron on activated carbon produced the best bio-fuel production (25.8%), whereas Forty percent cerium–copper on activated carbon produced higher value of hydrocarbons components reach to (67.43%) and aromatic fraction (50.75%). Additionally, the catalysts helped the bio-oil undergo efficient deoxygenation and denitrification, up to 44.38% and 58.63%, respectively. This enhanced fuel quality by lowering undesirable molecules that included nitrogen and oxygen and boosting hydrocarbon selectivity.

This paper was led by Hou et al. [4] 2024 witnessed research on the co- thermal decomposition of plastic trash, specifically high-density polyethylene (HD) and protein-rich microalgae, utilizing microwave-assisted thermal decomposition in a semi-batch reactor. The four hundred degrees Celsius operating temperature was used for the experiment. The ratio of microalgae to polyethylene was kept constant at 1:1. The findings demonstrated that the co-pyrolysis strategy limited gas which is generated from the process while increasing the production of biofuel. When we do a comparison to solitary microalgae thermal decomposition, the bio-fuel generated from this co-pyrolysis showed a notable decrease in the value of the N_2 which is produced around zero point eight five to one point six eight weight percent and a rise in the value of the O_2 which is produced around thirty-three point seven five to forty-seven point four zero weight percent because Ester compounds and alcoholic beverages with long chains were formed. This experiment provided important insights into the possibility of improving the biofuel qualities from microalgae and polyethylene co-pyrolysis by demonstrating the efficiency of the PE component in lowering nitrogen concentration while raising oxygen content in bio-oil.

This experiment was published by Ormanci et al. [5] in the year 2023. In a semi-batch reactor, the co-thermal decomposition of the plastic waste which is called polystyrene (PS) with a type of microalgae (Spirulina sp. (SP)) was investigated at heating value between 350 and 450 °C. The experiment sought to determine how co-pyrolysis affected the quality and dispersion of the final product. To examine their influence on the productivity and makeup of the biofuel product, the feedstock mixture (PS and SP microalgae) was exposed to varying temperatures. The quantity produced and make-up of the liquid, gas, and solid which is produced from the experiment considerably changed by the method, according to their analysis of the biofuel generated by co-pyrolysis. A variety of aromatic substances like C_8H_8 and C_7H_8, ethyl benzene, and alpha-C_9H_{10}, were added to the liquid bio-oil that was generated. These compounds were mostly created by the synergistic interaction between PS and SP at higher temperatures. Particularly, when doing a comparison for the pyrolysis for each feedstock alone, co-thermal decomposition produced a noteworthy rise in the yield of toluene, styrene, and ethyl benzene at 400 °C. Furthermore, oxygenated molecules like phenol were found in the liquid product, suggesting that the co-pyrolysis had altered the chemical makeup. The study emphasized how the PS and SP feedstocks worked in concert to raise the hydrocarbon which is the main component of the biofuel and alter the distribution of chemical components in the liquid bio-oil, increasing its fuel value.

In 2023, Chen et al. [6] examined the using of graphite/alumina (GP/AlO_3) compound additions in a microwave thermogravimetric reactor to co-pyrolyze Chlorella vulgaris (CV) microalgae and polypropylene (PP) plastic waste. With total additive loading levels of 10, 20, 30, and 40%, the experiments were carried out at different biomass-to-plastic ratios (CV: PP=10:0 to 0:10) and additive compositions ($GP:Al_2O_3$=10:0 to 0:10). After characterizing the biofuel which is a liquid product from the co- thermal decomposition process, they discovered that adding 40% GP/Al_2O_3, particularly at a GP: Al_2O_3 ratio of 7:3 (G7A3), greatly rise the value of production and biofuel qualities. With alkyl-cyclohexane making up 65.95% of the total hydrocarbons, the bio-oil's alcohol and hydrocarbon concentrations in this configuration were 29.56% and 34.60%, respectively, underscoring its potential for chemical uses in organic synresearch and antistatic coatings. Additionally, the additive mixture reduced the production of unwanted chemicals including acids, amines, and nitriles while promoting efficient denitrification and deoxygenation. When employed alone, alumina had no effect, while graphite showed considerable catalytic stimulation. But when used together, they showed a synergistic impact that improved product quality and pyrolysis efficiency without using more energy.

This article was carried out by Jiang et al. [7] in the year 2022. Thermal decomposition gas (Py-GC/MS) and temperature-gravimetric analysis (TGA) which are used to study the co-thermal decomposition of the microalgae which is called Chlorella vulgaris (C.V.) with the polymer (plastic) waste, namely polypropylene (PP), with tests conducted at a final pyrolysis temperature value of 500 °C. The composition of the pyrolysis products and their kinetic behavior were evaluated by varying the blending ratios of PP to C.V. Comparing the co-pyrolysis reaction to individual feedstock pyrolysis, the former produced biofuel with much improved fuel characteristics, including gaseous hydrocarbons and liquid bio-oil. By lowering energy activation and raising hydrocarbon yields, the procedure demonstrated the greatest synergistic effect when PP was added at a 0.75 blending ratio. The bio-oil's peak hydrocarbon content was 39.06%. Furthermore, PP facilitated the synresearch of aliphatic hydrocarbons while preventing the synresearch of aromatics, nitrogen compounds, and oxygenated species. In addition to improving deoxygenation and denitrification, the synergistic interaction shifted the peak pyrolysis temperature of C.V. downward, improving the overall quality and energy content of the fuel that was produced.

The study was carried out by Chen et al. [8] in the year 2022. High-density polyethylene (HDPE) waste and Chlorella vulgaris (microalgae) co-pyrolysis were investigated at different mixing ratios utilizing a semi-batch microwave reactor with activated carbon (AC) acting as a microwave absorbent. The operating circumstances included different AC loadings (10–50%) and microalgae-to-HDPE mass ratios (1:0, 4:1, 2:1, 1:1, 1:2, and 1:4). The ideal setting, according to their analysis of the biofuel generated by co-pyrolysis, was at a 1:1 ratio (C1HP1) with 40% AC. The hydrocarbon content and bio-oil production rose under this configuration, and the reaction rate and conversion were greatly enhanced. When we made a comparison to the thermal decomposition of pure microalgae, the co-thermal decomposition procedure also caused a noteworthy decrease in N_2-containing chemicals (by 75.75%) and oxygen-containing compounds (by 42.46%). While AC further encouraged deoxygenation and the creation of aromatic hydrocarbons, HDPE's involvement improved the development of long-chain Alcoholic substances and hydrocarbons, thus raising the bio-fuel's overall quality.

In 2022, Chen et al. [9] used the carbon atom which is used to activate reaction speed (AC)-supported the single-metal catalysts like cerium on activated carbon, iron on activated carbon, and nickel on activated carbon to co-pyrolyze the microalgae which is name is chlorella vulgaris and a polyethylene component which is distinguished higher value of density (HDPE) in a microwave reactor at an operating temperature of 550 °C. Co-pyrolysis was performed with different catalyst loadings (40–50%) to examine the impact of catalysis on the production of biofuel. The ratio of chlorella vulgaris microalgae to the plastic waste which is a polyethylene with high density was one to one (co-pyrolysis process (C1HP1)) for biomass to plastic. The generated biofuel was examined using GC-MS and chemical composition analysis, among other techniques. The findings demonstrated that by increasing bio-oil yields and changing the bio-oil's chemical makeup, the application of catalysts greatly enhanced the co-pyrolysis process. Fe/AC had the best bio-oil production (25.6%) with a 40% loading and the quickest reaction time (2901 s) among the catalysts evaluated. The production of useful bio-oil molecules, especially hydrocarbons, was encouraged by Ce/AC and Fe/AC. Aromatic hydrocarbons have the highest rate around thirty-six-point sixty-four percent was produced by Fe/AC, but the maximum hydrocarbons (66.68%) were produced by Ce/AC. The catalysts significantly reduced the quantity of O_2 and N_2 in the biofuels, increasing its deoxygenation and denitrification efficiency, in addition to increasing bio-oil yields. Ce/AC exhibited a deoxygenation efficiency of 47.33% and a denitrification efficiency of 42.28%. The production of higher-quality biofuels with fewer unwanted substances, such as oxygenates and nitrogenates, depends on these advancements.

The paper was done by Chen et al. [10] in the year 2021. Dunaliella salina microalgae were co-pyrolyzed with several plastic wastes (polypropylene PP, polystyrene (PS), polyethylene terephthalate (PET), and polyvinyl chloride (PVC)) using an FTIR spectrometer and thermogravimetric analyzer. For PVC, the kinetic analysis was especially concentrated between 410 and 535 °C, although the temperature range extended up to about 600 °C. Through a coating effect, the molten plastics facilitated the microalgae's pyrolysis, according to their analysis of the interaction between the two during thermal degradation. The hydrogenation reactions between unsaturated pyrolysis products and bio charcoal caused a decrease in the quantity of the remains solid precise by one point fifty-five weight percent, 1.39 weight percent, and 1.69 weight percent for the combination of DS-Polypropylene, DS-Polystyrene, and DS-Polyethylene Terephthalate, respectively. In contrast, DS-PVC displayed an increase in residue because of interactions with HCl. The DS-PET blend improved the quality of the bio-crude by releasing more CO_2, according to FTIR measurements. Kinetic research showed that co-pyrolyzing plastics with D. salina greatly decreased their activation energies, suggesting energy savings and synergistic effects during conversion.

This article was directed by Abomohra et al. [11] in the year 2021 who investigated the co-thermal decomposition by the microwave vacuum of the polyethylene which is distinguished from a lower density (LDPE) and Ulva intestinalis, a macroalgae species, using a microwave vacuum reactor set to 500 °C. Ulva intestinalis was the best option for co-pyrolysis out of all the macroalgae that were examined since it produced the most biomass and bio-oil. Using a blend ratio of 25:75 (Ulva intestinalis: LDPE), the procedure produced 68.9 weight percent bio-oil. The macroalgae and LDPE had a synergistic effect during the co-thermal decomposition technique, which improved the bio-fuel product and how much it has a good quality. By supplying hydrogen radicals, LDPE increased the hydrogen-to-carbon ratio and decreased molecules that included nitrogen and oxygen. A considerable rise in hydrocarbon content (up to 51.2%) and a decrease in acids, phenols, and other undesirable substances were revealed by GC-MS analysis. According to the findings, this co-pyrolysis method has the potential to generate cleaner, more energy-dense bio-oil, which could cause production of scalable, environmentally friendly fuel.

This paper was published by Xu et al. [12] in the year 2020. To improve the generation of biofuel, the type of marine macroalgae which is called Enteromorpha prolifera and type of plastic waste which is a polyethylene which distinguished with a higher value of density were catalytically co-pyrolyzed at 550 °C in a type of reactor which calls it fixed bed and thermogravimetric analyzer (TGA). Zeolite catalysts were utilized to identify the effect of the non-isothermal procedure on reaction performance and product quality. Analysis of the biofuel generated by the co-pyrolysis revealed that the mutually beneficial relationship between E. Prolifera and HDPE considerably decreased the bio-oil's concentration of unwanted substances such acids, oxygenates, and nitrogen-containing species. At the same time, the hydrogen-rich HDPE feedstock caused a rise in the amounts of aromatical content and light hydroponic content. By reducing the activated power of the reactant feedstocks with no alert in the reaction process, HZSM-5 further improved the synresearch of aliphatic hydrocarbons over aromatics, expanded product yields, and enhanced catalytic activity. A significant drop in energy activation during catalytic co-pyrolysis was confirmed by kinetic analyses based on TGA data using five distinct techniques. These results were corroborated by (Fourier Transform Infrared Spectroscopy and Gas Chromatography-Mass Spectrometry) studies of the thermal decomposition oils, which revealed increased hydrocarbon content and general biofuel quality. Additionally, the co-pyrolysis decreased the production of solid residues and hindered catalytic coking, suggesting enhanced process effectiveness and cleaner product profiles.

This paper was authored by Tang et al. [13] in the year 2019 who investigated the co-thermal decomposition of plastic waste which is distinguished for a lower value of density (LDPE), and a type of microalgae which is called Nannochloropsis sp. (NS), in a fixed bed reactor set at 550 °C. Different LDPE and NS mixing ratios were examined to see how they affected fuel quality and product dispersion. Gaseous hydrocarbons and liquid biofuel were the byproducts of the co- thermal decomposition process. Heating value which is at low level (LHV) of the gas which are produced was considerably raised and gas yields were enhanced using LDPE. With a maximal C_2+ production of eighty-four point eighty-six milliliters per gram at seventy-five percent LDPE, co-pyrolysis notably aided in the creation of light hydrocarbons like CH_4 and C_2+ compounds, particularly ethylene. Aliphatic hydrocarbons rose starting from twenty-two-point sixty-three percent in pure Nannochloropsis sp thermal decomposition to seventy-seven-point four percent with twenty-five percent LDPE, improving the liquid bio-oil's composition. In the meantime, there was a significant decrease in the amounts of nitrogenous and foreign chemicals. Nitrogen was mostly liberated into the gas phase, while oxygen in Nannochloropsis sp leaned to change into water and carbon monoxide (rather than of CO_2), enhancing the overall quality and energy content of the biofuel that was created.

In 2019. Cao et al. [14] evaluated the synergistic effects on product distribution by co-pyrolyzing the marine macroalgae Enteromorpha clathrate (EN) and the plastic waste, polyvinyl chloride (PVC), in a reactor which is fixed bed running at five hundred fifty degrees Celsius. Special attention was paid to the characteristics of biofuel and biochar in addition to the dynamics of Cl convert. Different PVC fractions of the mass (5–25 weight percent) were present in the feedstock mixes. In their analysis of the biofuel produced by the co-pyrolysis, they discovered that while the overall yield of the bio-oil dropped little (by 0.43–2.57 weight percent) from theoretical values, its composition greatly improved. According to GC-MS data, co-pyrolysis increased the production of energy-rich phenolic, aromatic, and aliphatic hydrocarbons while decreasing the amounts of unwanted oxygenated and nitrogenated chemicals, improving the bio-oil's calorific value and chemical stability. Synergistic processes between the macroalgae and PVC, including deoxygenation and denitrogenating, which were aided by HCl liberated during PVC breakdown, were credited with these changes. Overall, the study showed that co-pyrolysis of PVC and Enteromorpha clathrate can stabilize toxic components like chlorine, enhance the properties of the biofuel including the quality of it, and provide a practical way to turn marine biomass and plastic trash that contains chlorine into cleaner and more valuable biofuels.

The paper was published by Dai et al. [15] in the year 2018. The Dai examined the co- thermal decomposition of type of microalgae with a type of plastic (polyvinyl chloride) a waste product of plastic, in a microwave reactor set at 605 °C. To assess the impact of power input and feedstock ratio on product yields and biofuel composition, co-pyrolysis was conducted using different MA to PVC proportion (five to five and three to seven) and the microwave machine energy levels of eight hundred watts and 1000 W. Higher microwave power was shown to boost bio-oil yields in the biofuel production process. The highest temperature value and heating ratio were significantly higher at 1000 W, resulting in more effective pyrolysis. At 1000 W, the maximum value of bio-oil output was recorded at a 5:5 MA to PVC ratio. At 800 W, a notable synergistic impact was detected at a ratio of 5:5 and 3:7, indicating the potential of PVC as a substance to catalytic the reaction to increase bio-oil yield. In bio-oil, the nitrogenous and exogenous chemicals that usually degrade the quality of biofuel were greatly diminished by the rise in PVC concentration. According to GC-MS analysis, the proportion of aromatic hydrocarbons increased significantly as the PVC concentration increased, getting to ninety-two-point seven percent at eight hundred watts and ninety-one point nine percent at one thousand watts, whereas the content of aliphatic hydrocarbons declined. The number of organic molecules that contained nitrogen and oxygen decreased significantly; at 800 W, the oxygen content reduced from 29.65 to 0.38%. This decrease in unwanted substances emphasizes how much better the bio-oil made in these circumstances is. By encouraging the production of valuable hydrocarbons, PVC also served as a hydrogen donor, improving the bio-oil's quality even further.

This experiment was led by Qi et al. [16] in the year 2018. Using HZSM-5 zeolite catalysts, the study aims to investigate the catalytic co-thermal decomposition of microalgae (Nannochloropsis sp (NS)) and polypropylene within a bench-scaled reactor with bed that is fixed. NS to PP mass ratios ranged from 1:0 to 0:1, and the operational the value of the temperature is ranged between 500–900 °C. The effect of varying the raw material-to-the mass catalyst rate from one per zero to one per ten on the synresearch of aromatic hydrocarbons was also examined. Comparing the catalyst co-thermal decomposition of NS and polypropylene to their solo pyrolysis, the study of the biofuel produced illustrates a notable combined impact on the generation of permutational hydrocarbons. At 800 C, the mass ratio of biomass to polypropylene, which is one to one, produces the maximum monoaromatic compounds. The methods by which the main aromatic hydrocarbons are formed during the co-pyrolysis process were also covered in the study. According to the findings, catalytic co-pyrolysis of

Nannochloropsis sp and polypropylene are viable methods for producing aromatic hydrocarbons efficiently. It may be optimized in subsequent research by designing a catalyst that lowers pyrolysis temperatures and increases energy efficiency.

2 Summary

In conclusion, a comprehensive revision of the existing literature on the co-pyrolysis of algae and plastic waste has demonstrated how important a variety of pyrolysis parameters are to the creation of biofuels. Numerous important variables have been carefully examined, including pyrolysis temperature, reactor type, blend ratios, catalyst use, residency timing, and heated rates. The results show that these elements have considerably impact on the products' yield, composition, and dispersion. Maximizing biofuel yields requires optimizing pyrolysis settings, with temperature ranges being found to be crucial for improving product results. The findings also highlight the intricate thermochemical relationships between algae and plastic trash, specifically regarding the contents of biofuel, synergistic effects, and the processes of deoxygenation and denitrification. In sum, the study has provided insightful information about the potential of co-thermal decomposition of algae and plastic waste as an effective and sustainable process for producing biofuel, highlighting the necessity of meticulous parameter optimization to attain desired fuel qualities.

3 Methodology

3.1 Introduction

This chapter presents a thorough summary of the experiment's setups, supplies, and strategies used in four chosen investigations that investigate the co-thermal decomposition of the waste of plastic materials and algae to produce biofuel. To assess their potential for synergy in enhancing the producing and the properties of biofuel, experimental studies concentrate on the thermochemical transformation of macroalgal and microalgal biomass when combined with synthetic polymeric substances, specifically polyethylene (PE) and polyethylene terephthalate (PET). The algae species employed in these investigations include a wide range of macroalgae (Ulva lactuca, Ulva intestinalis, Sargassum polycystum, and Hypnea valentiae) and microalgae (Chlorella vulgaris, Nannochloropsis), all of which were selected for their biochemical makeup, accessibility, and significance in bioenergy research. The goal of co-pyrolyzing these algae feedstocks with PE and PET is to investigate the interplay between chemicals obtained from biomass and hydrocarbons derived from plastic during thermal breakdown. The chapter is organized to provide a detailed explanation of each study's experimental design, including how raw materials are prepared and characterized, how mix ratios are created, and how pyrolysis equipment and operating parameters are described. It also describes the analytical techniques used to describe the final products, with an emphasis on the compositional analysis of bio-oil. Furthermore, a few findings from each study are combined and examined to shed light on how feedstock composition and process variables affect fuel quality and product distribution. By consolidating the methods and important experimental results presented in this chapter provide a crucial basis for assessing the feasibility of co-pyrolysis as a sustainable technique of producing biofuel by valuing plastic waste and algae biomass.

3.2 *Case Study I: Co-pyrolysis of Microalgae and Plastic*

Characteristics and Interaction Effects [13]

Authors: Ziyue Tang, Wei Chen, Yangquan Chen, Haiping Yang, Hanping Chen

Year: 2019

3.2.1 Materials Used in the Experiment

NS (Nannochloropsis sp algae) was obtained from the company of biology technology co, Ltd. and polyethylene which distinguished of low level of density (LDPE) was made from wrapping of plastic which we can buy it from the shops. Polyethylene with a low level of density and nannochloropsis sp algae were crushed and sieved to result in pieces smaller than three hundred eighty micrometers and one hundred twenty micrometers, respectively. After that, they dry it under fifty-five degrees Celsius for twenty-four hours. Thermal decomposition approximation of the plastic trash that's from the polyethylene which featured a low value of density conducted with a type of algae which call it nannochloropsis sp by used an analyzer (thermogravimetric analysis -two thousand). Materials are carried out in the final assessments by using a rudimentary analyzer. Using the calorimeter to evaluate the minimum heating value for each testing sample. Checking the quantity of protein in the algae which is (Nannochloropsis sp) by utilizing a smart technique. Determining the content of the lipid which is in the algae by extractable technique. Moreover, calculating the nannochloropsis sp algae carbohydrate amounts by identifying the differential value which can be calculated by subtracting the weight quantity of the moisture, ash, lipids and proteins from 100%. Because NS is mostly made up of thirty percent and proteins contents are forty-point eight percent, with a little quantity of carbohydrates (19.20 wt%), it possesses high amounts of oxygen atoms and nitrogen. The NS's oxygen and nitrogen concentrations (twenty-four point seventy-five and six points sixty-eight weight percent, respectively) and hydrogen percentage (7.31%) are expected to result in pyrolysis oil having considerable oxygen and nitrogen contents.

Conversely, oxygen and nitrogen were very low in LDPE, which is primarily made up of carbon to be eighty-four-point thirty two percent and hydrogen being fifteen point forty seven. Because of this, low density polyethylene is one of the best additives for co-pyrolysis with biomass. It was decided to co-pyrolyze LDPE and NS using a fixed bed approach. The unit of condensation value, value of temperature control point, feeding unit, flowing of gas in meters, using a reactor which is fixed bed type which distinguished a property including height of sex hundred-millimeter, diameter with forty-five millimeter in the inert side of reactor, and a unit for cleaning and drying jobs (filled with allochronic silicagel) comprised the system [13].

3.2.2 Experimental Set-Up and Methodology

Nannochloropsis sp algae and polyethylene plastic which has low density were ground in the appropriate rate to produce a uniform mixture in an agate mortar before each trial. Nannochloropsis sp algae/low density polyethylene combinations were created with various weight percentages of low density polyethylene (zero, twenty five, thirty three, fifty, sixty seven, seventy five, and one hundred). After that, a bunch of quartz filled with around two gram particles mixtures and Argon ninety-nine point ten percent functioning as a container for gases with a 150 mL/min rate of discharge was placed on top of the reactor. After reaching the desired value of temperature around sex hundred degree Celsius, the sample combination was immediately positioned within the reactor's steady temperature portion, where it remained for thirty minutes. When the sample materials start to heat the volatility of it increases a few times. Following the trapping of the condensation of volatility in a condenser unit submerged in a solution of icy water, the gas which can't condense was collected and purified by utilizing a gas tank. After heating for thirty minutes, the pyrolysis experiment was taken off, and the char was continuously cooled by Ar purging. As demonstrated in the co-pyrolysis system in Fig. 1, each attempt is repeatedly by utilizing liquid content of N_2 rather than mixture of icy water to be the oil under condensation system, with the production of gas being vented immediately to the air, to precisely measure the components of the oil. This was done because the oil might not be completely captured by the ice-water mixture. The charcoal product was calculated using the differential of weight between the prior to and following the combustion experiment, a quartz container. In the same way, the biofuel (liquid) production was calculated using the weight differential value of N_2 liquid phase through the condensation process. The gas product (wt%) was identified the value of it by doing calculation process by utilizing the sum of gas volumes and the using volume fraction for every component. Reactor time and carrier gas flow were used to compute the cumulative volume, and micro-GC was used to measure the volume of all the gases that's produced from the experiment. For comparison, polyethylene which distinguished from low level of density and Nannochloropsis sp algae were also thermal decomposition independently. The mass balance was precise, the experiments repeated a minimum of 3 times, and the average value of all the results was calculated. The standard deviation was less than 5% [13].

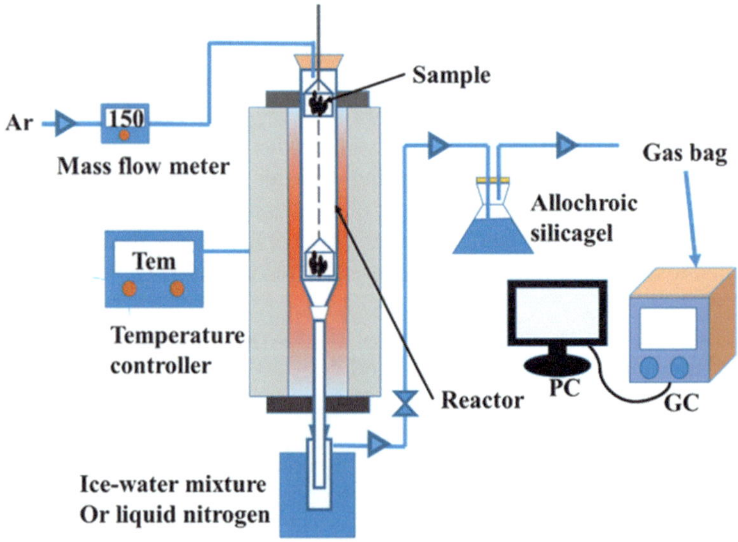

Fig. 1 The co-pyrolysis experimental system [13]

Gas productions were examined by utilizing a two-channel micro gas chromatograph apparatus with thermal conductivity detectors from Agilent Technologies, USA. Two columns were utilized to evaluate carbon dioxide, ethane, ethylene, and acetylene at sixty degrees Celsius. The minimum heating value for the gases which produced from the experiment were calculated by the following formula:

$$\text{LHV}\left(\text{MJ/Nm}^3\right) = \sum_{i=1}^{n} (v_i \times \text{LHV}_i)$$

V_i and LHV_i are the volume fraction, the LHV of gas species *i*, respectively.

The potential of hydrogen (PH) of thermal decomposition fuel was calculated by utilizing the measured using the potential of hydrogen instrument beginning two thousand one hundred over three c pro. Identifying the waiting content in the liquid production by utilizing a karl fischer moisture analyzer. Then, liquid product analysis by using was finally analyzed using a basic analyzer. Using a type of detector (Gas Chromatography-Mass Spectrometry) to identify the essential component of the fuel and a column of capillaries with a temperature ranging between zero degree Celsius to three hundred fifty degrees Celsius. The heat of the injection is three-hundred-degree Celsius, and the oven was kept up to three hundred degrees Celsius for 10 min. The weight spectrometry was operational in the range between 30 to 500 micros. The atoms sensitivities quotient was used for determining each element's contents based on its associated peaked region, with carbon utilizing as reference. By using the scanning electron microscopy of the field -emitting to examine the char's shape [13].

3.2.3 Results and Discussion

Figure 2 displays the dispersion of products that were produced by co-pyrolyzing LDPE and NS at various ratios. NS exhibited similar pyrolysis characteristics to other biomass samples, resulting in a significant output of liquid oil (65.17 weight percent), the creation of some gases (about 15 weight percent), and a solid residue (19.53 weight percent).LDPE, on the other hand, demonstrated pyrolysis characteristics that were significantly different, with the same gas output around sixteen point twelve weight percent and a larger liquid yield (81.7 wt%), but no solid residue [13].

The gas output rose as LDPE was added to co-pyrolysis, peaking at 20.10% at 25% LDPE. This amount (about 5 weight percent) exceeds the products for the independent thermal decomposition of low-density polyethylene and Nannochloropsis sp algae as well as the projected value determined by the rate of mixing. However, when the low density of polyethylene rose, the rate of gas products decreased. Additionally, the char output gradually decreased while the product of liquid oil rose correspondingly with the low-density polyethylene concentration because LDPE contains over 80% volatile and breaks down to the liquid materials and gases contents through the thermal decomposition process. In contrast to the liquid yield, which is smaller and does not differ significantly for char, the value of the gas product which produced from the experiment is continuously larger the calculated value. Because co-pyrolysis promotes the subsequent breakdown of the volatiles which is heavy into particles of the gases which are lighting, more gas species might have emerged [13].

The LHV and gas compositions of the gas products generated by co-pyrolyzing Nannochloropsis sp algae and polyethylene plastic which has low density mixes are displayed in Fig. 3a and b. The principle of gas products of pyrolysis of NS alone are carbon dioxide (thirty-one-point eighty-six percent) and carbon monoxide (twenty-point ninety-seven percent), which may be produced by breaking down the carbon = oxygen and carboxylate compounds because of the higher level of proteins and carbohydrates in Nannochloropsis sp algae. Furthermore, some methane (seventeen-point sixty-nine percent) and C_2+ (seventeen-point eighty percent) are produced via

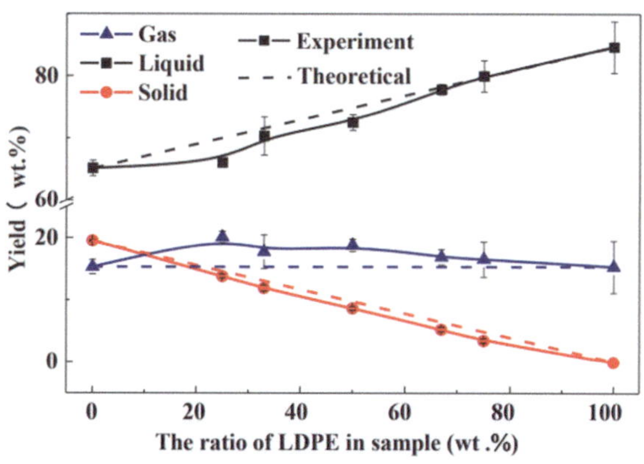

Fig. 2 Production values from LDPE and NS co-pyrolysis [13]

nannochloropsis sp algae pyrolysis. Algae's Dem ethoxylation and end-chain fission processes mostly yield CH_4, whereas long-chain fatty acid scission and cyclization events produce C_2+. Thus, at 23.56 MJ/Nm^3, the minimum heating value of the nannochloropsis sp algae gas production was greater than that of the lignocellulosic biomass. Therefore, it should enhanced before it can be utilized as a natural gas particles. Methane (twenty-one-point twenty-two percent) and C_2+, especially ethylene (thirty-two-point sixty-four percent), were the main constituents of the pyrolysis gas that was recovered from LDPE. These components could have been created by alkylation procedures, LDPE depolymerization, or the atomic structure of hydrocarbon chains with a length at the final stage. The cyclization and aromatization of the alkenes also resulted in the production of some H_2 (3.19%). Consequently, the lower heating value of the plastic trash which is polyethylene which distinguished lower level of density thermal decomposition gas was sixty points ninety-one megajoules per normal cubic meter. Co-pyrolyzing of Nannochloropsis sp algae and polyethylene plastic which has low density resulted in a much lower carbon dioxide content (18.98%) than pyrolyzing NS alone. The trends for the CO and H_2 concentrations were similar, with the CO_2 content decreasing in direct rate to rise in the level of polyethylene low density. On the other hand, the amount of C_2^+ rose significantly while the CH_4 concentration increased little as the LDPE level rose. When NS and LDPE are co-pyrolyzed, the predicted percentage of C_2^+ which includes ethylene, propylene, and ethane is above thirty percentage, as opposed to twenty-five percentage when LDPE is pyrolyzed alone. This might make it possible to use the product to make simple synthesized substances that are now acquired through the petrochemical sector. The generated pyrolysis gas may be employed as a natural gas substitute in urban and industrial settings because its LHV (thirty-six point twenty-five Megajoules per normal cubic meter) was comparable to that of the gas which is natural [13].

The functional groups that contain O or N may be the origin of this phenomenon, as they promote the removal polymers of the polyethylene which distinguish a low density to generate light vapor of the molecules. As the mixture's NS quantity decreased, so would the quantity of nitrogen and oxygen which have a group that functions interact with lower density of polyethylene, resulting in less contact with C_2+. Additionally, when interact a quantity of twenty five percent of the plastic which is the lower density polyethylene that's doing a large impact on the generation amount of methane and hydrogen gas that's around twenty mile per gram methane and ten mile per gram hydrogen [13].

GC/MS analysis was used to identify the constituents of the pyrolytic oil (bio-oil) that are comparatively bright. Moreover, some oxygen or nitrogen contains a faction group that are invisible to GC/MS may be present in the heavier components. Therefore, the various elements of the fuel were further identified using proton nuclear magnetic resonance spectroscopy (Table 1). Incomplete pyrolysis of proteins and lipids resulted in long-chain esters and peptides, which caused the large proportion of protons in the zero to one point five part per million range in the NS oil. The hydrocarbons which are fat in the thermal decomposition oil increased with the addition of LDPE, as demonstrated by the GC/MS results, and this ratio increased rapidly as

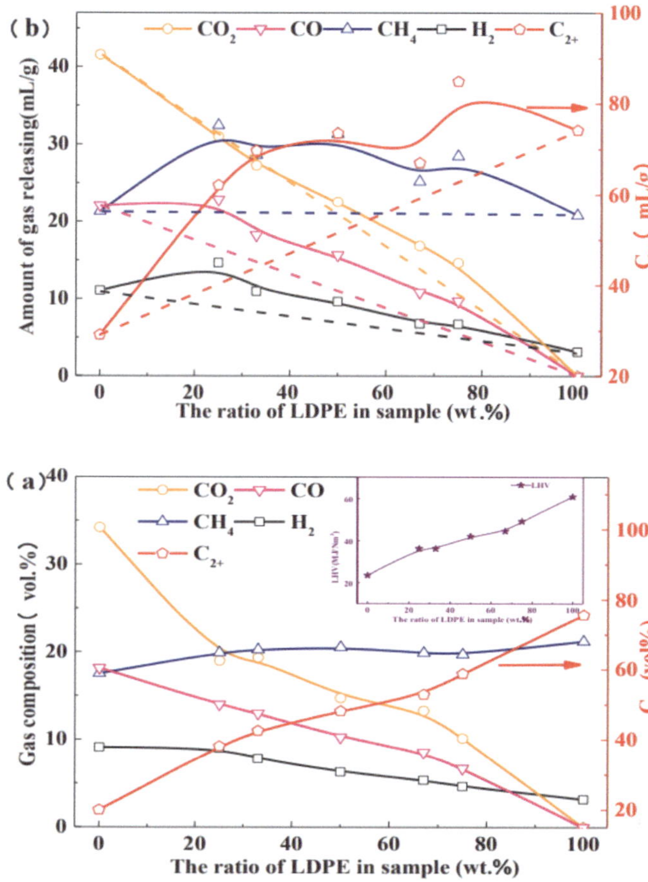

Fig. 3 The gas produced through the co-thermal decomposition of LDPE and NS is defined by **a** the yield of gases and **b** their composition. Theoretical outcomes are shown by dashed lines, whereas experimental outcomes are represented by solid lines [13]

well. The four point sex to five-point sex part per million regions, which comprised twenty-one-point seventy eight percent from the protein contents in the fuel which is from the thermal decomposition of nannochloropsis sp, was also considerably raised to a minimum value of 8.82% by the addition of 25% LDPE. The Hydroxyl groups in phenolic substances produced during the pyrolysis of carbohydrates being hydrogenated and dehydrated may account for this action [13].

Figure 4a and c displays the primary constituents of liquid fuel obtained from Nannochloropsis sp and low-density polyethylene thermal decomposition. In addition to several organic chemicals that contain oxygen (such as carboxylic acids (18.66%), alcohol (16.21%), and other oxycompounds (7.69%), the primary constituents of Nannochloropsis sp. oil consist of hydrocarbons, including $C_{19}H_{40}$ (twenty-two point sixty-three percent), $C_{66}H_6$ (fourteen point eighty-four percent),

Table 1 H-NMR integrates for the bio-oils produced at 600 °C in particular chemical-based range [13]

Chemical shift region (ppm)	Type of protons	Hydrogen content (% of all hydrogen)					
		NS	25% LDPE	33% LDPE	50% LDPE	66% LDPE	75% LDPE
0.5–1.5	$-CH_3-$, $-CH_2-$	43.56	62.75	69.70	67.33	64.36	69.00
1.5–2.1	RC = C–CH_2–; ROH	6.93	7.84	6.06	5.94	4.95	6.00
2.1–3.0	RCOCH, CH–Ar, NC–CH, ROH	12.87	7.84	5.05	3.96	1.98	1.00
3.0–4.6	RO–CH; CH–OH; RCO_2–CH; PhO–CH	3.96	3.92	2.02	0.99	0.99	0.00
4.6–5.6	Ph–OH; HC = C	21.78	8.82	12.12	15.84	23.76	20.00
5.6–6.5	RCO–NH; Ph-OH	2.97	1.96	2.02	2.97	1.98	2.00
6.5–8.0	RCO–NH, ArH	6.93	5.88	3.03	2.97	1.98	2.00
8.0–10.0	–CHO; –COOH	0.99	2.98	0.00	0.00	0.00	0.00

and $C_{19}H_{30}$ (six point ten percent). Notably, a lot of N-containing substances (17.2%) are also present, such as nitriles, nitrogen heterocycles, and amides/amines. Hexadecanamide and tetradecanamide are examples of long-chain amides that may arise because of interactions between long-chain aliphatic acids and amino compounds that are produced by protein pyrolysis. On the other hand, the LDPE oil had no molecules that contained O or N and was primarily composed of high value of hydrocarbons which is aliphatic around ninety eight point eighteen percent. As the LDPE concentration in the mixture increased for co-pyrolysis process, a decline in Organic substance that contains oxygen was noted (Fig. 4c) [13].

An overview of the characteristics of the Charcoal created via co-thermal decomposition is given in Table 2. The amounts of carbon, nitrogen, oxygen and hydrogen gradually decreased as the amount of LDPE in the co-pyrolysis chars rose. This pattern indicated that the char and LDPE might have interacted to promote the decomposition of nannochloropsis sp algae the substance tar and boost the producing of the volatility. Through co-thermal decomposition, nannochloropsis sp algae particles that has a melting low-density polyethylene around it. Could speed up the secondary processes by returning the volatility from the nannochloropsis sp algae (NS) pyrolysis under high value of temperatures for a long duration [13].

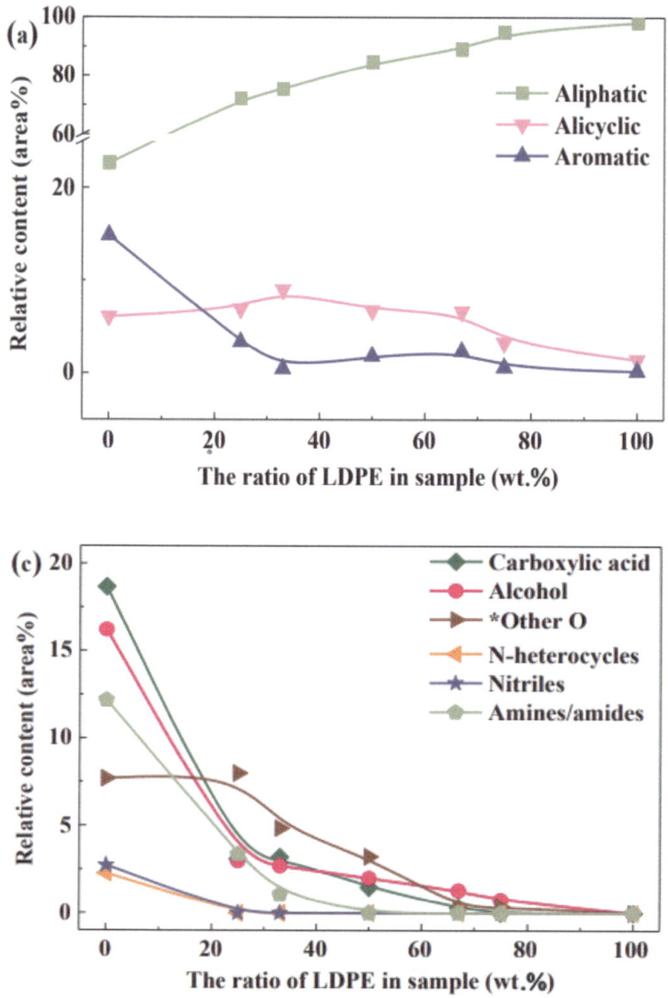

Fig. 4 Liquid product properties derived from Co-pyrolysis of LDPE and NS: **a** components that contain hydrocarbons, and **c** components that contain nitrogen and oxygen [13]

Figure 5 illustrates the distributed percentage of oxygen and nitrogen in the thermal decomposition process. Following NS pyrolysis, the oil's O production was a very high 29.46 weight percent. Moreover, gas and water contained the majority of microalgae O. The gas' calorific value decreased because most of the O was present as CO_2. The oil's distribution quickly dropped to 4.41 weight percent when LDPE was added during co-pyrolysis, greatly enhancing the oil's calorific value and quality. Despite no appreciable change of the oxygen component in gas contents, the quantity of the oxygen raised noteworthy in the water, recommending that oxygen which is from the microalgae evacuated as H_2O rather than CO_2. The more CO_2 content in the

Table 2 Characterization of char properties (wt%) [13]

Sample	C	H	N	S	O	Volatile	Fixed carbon	Ash, d	atomO/C	atomH/C	LHV (MJ/kg)
NS	54.41	1.73	6.53	0.10	5.77	13.44	55.10	31.47	0.0796	0.3810	19.51
25%	52.88	1.49	6.37	0.13	5.80	12.61	54.06	33.33	0.0823	0.3370	18.69
33%	52.92	1.47	6.50	0.16	5.09	14.17	51.97	33.86	0.0721	0.3325	18.82
50%	51.95	1.44	6.29	0.17	4.26	12.89	51.21	35.90	0.0615	0.3316	18.60
67%	50.55	1.39	5.98	0.23	3.91	13.46	48.60	37.94	0.0580	0.3294	18.14
75%	50.45	1.32	5.91	0.20	3.38	19.43	41.83	38.75	0.0503	0.3148	18.11

gas has caused a better gas and a good quantity of calorific material. Nannochloropsis sp thermal decomposition resulted in the majority of N (56.43 weight percent) being released into the oil, which caused inclined in the characterization of the bio-oil which a product from the thermal decomposition process (pyrolysis) and make it unfit for the environment. Among addition, there was some N scattered among the charcoal and gas contents. For co-pyrolysis, the production of nitrogen in the oil contents steadily dropped to thirty sex point forty nine weight percent at seventy five percent of low density polyethylene as the amount of LDPE increased [13].

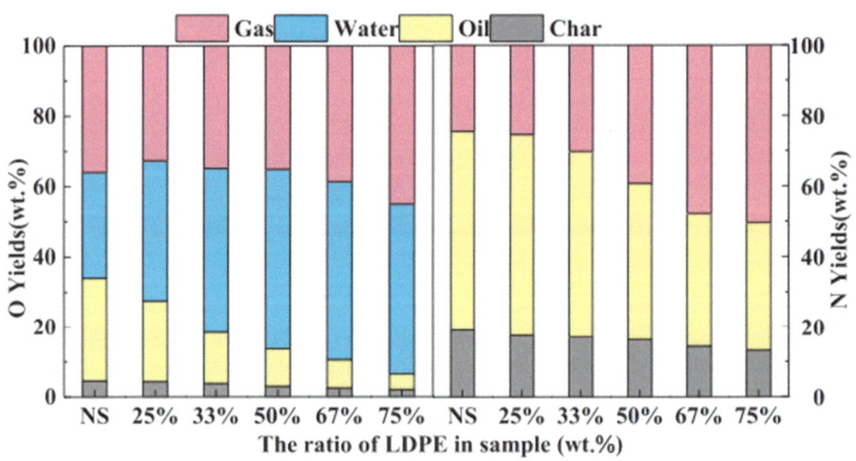

Fig. 5 The co-pyrolysis products' O and N distributions are as follows: Ooil = 100 weight percent − Ochar − Ogas − Owater; Ngas = 100 weight percent − Noil − Nchar [13]

3.3 Case Study II: Microwave Vacuum Co-pyrolysis of Waste Plastic and Seaweeds (Macroalgae's) for Enhanced Crude Bio-oil Recovery: Experimental and Feasibility Study Toward Industrialization [17]

Authors: Abd El-Fatah Abomohra, Huda M.A. Sheikh, Amal H. El-Naggar, Qingyuan Wang

Year: 2021

3.3.1 Materials Used in the Experiment

In April 2018, seaweeds belonging to the Hypnea valentiae macroalgae, Ulva intestinalis, and Sargassum polycystum were collected along the Jeddah, Saudi Arabia's red sea coastline. The collection area was set up using a steel quadrate (1 × 1 m) to measure the growth of the vegetation areas. There were five quadrates gathered at each of the five collecting sites. Sand and contaminants were eliminated using seawater after the seaweed was collected from each quadrate. Each species was transported to the lab and stored in sterile containers packed with ice after being thoroughly separated. To remove any leftover salt, we should wash all the test materials with fresh water and put the air under drying process on absorbent paper at room temperature. After determining the dry weight (dw), the product which is a biomass material was measured as drier wt/m^2. The overall common particles of enormous size of algae, which stand for rhodophytes, phaeophytes, and chlorophytes, were selected for this investigation. The components which are biological chemical analyze it in a proximately, and carried out the end data analysis were on the 3 chosen particles. Grain size 36 sic nanoparticles were used as a microwave absorbent. Bags which are made from plastic material which we found on the local markets were used to make low-density polyethylene, or LDPE. After the plastic was sliced into tiny pieces, LDPE and seaweed were ground up in a lab mill. After that, they were dried at 60 °C until their weight was constant after passing through a sieve with pores smaller than 3 mm [17].

3.3.2 Experimental Set-Up and Methodology

To find the appropriate seaweeds (algae) with a high value of biofuel product, the three seaweeds were microwave vacuum pyrolyzed at various temperature ranging between those values (four hundred, four hundred fifty, five hundred, five hundred fifty, and sex hundred degree Celsius. When the reactor unit achieved the desired value of heating, then the chemical reaction will continue for thirty minutes until the microwave unit shut down. After that, the cooling system is in working in the next step by condensing the tube units and utilizing a flask for the collecting process being cleaned by dichloromethane material to extract the biofuel. Then, the solvent liquid being under evaporating process for around forty-five-degree Celsius vacuum oven, the biofuel product was measured by gravimetric analysis. In addition, the growth of the weight values of the reactors materials was used to calculate the weight of the biochar. Eqs were used to calculate the production values yields and quality of the conversational materials as a proportion of the drier raw materials that is inserted

$$Y_o(\%) = \left({}^{M_o}/_{M}\right) \times 100 \tag{1}$$

$$Y_c(\%) = \left({}^{M_c}/_{M}\right) \times 100 \tag{2}$$

$$Y_G(\%) = \left(M - \frac{M_o + M_c}{M}\right) \times 100 \tag{3}$$

$$\text{Conversion efficiency}(\%) = 100 - Y_c \tag{4}$$

In this case, Y_O, Y_C, and Y_G stand for syngas, biochar, and bio-oil yields. The crude bio-oil, biochar, and feedstock have corresponding weights of M, M_O, and M_C. According to the biomass which produced and the biofuel product, the seaweed (algae) for further study was selected as having the best bio-oil areal productivity. By mixing various quantities of low-density polyethylene that's (twenty five percent, fifty percent, and seventy five percent) mixed with the algae feedstock for co-pyrolysis, which produced a final feedstock weight of 40 g beside the independent thermal decomposition of algae (seaweed) with around zero percent and lower density polyethylene which is hundred percent. The screening tests determined the optimal temperature for co-pyrolysis, and various pyrolysis yields were measured. The impact of the synergistic of lower density polyethylene at various rates blend with algae through co-thermal decomposition was assessed utilizing Eqs and the calculated value of biofuel production (Y_{th}).

$$Y_{\text{th}} = (Y_s \times W_s) + (Y_L \times W_L) \tag{5}$$

$$\text{Synergestic efffect}(\%) = \left[\frac{Y_{\text{ex}} - Y_{\text{th}}}{Y_{\text{th}}}\right] \times 100 \tag{6}$$

The resulting products from the experiment of biofuel from the independent pyrolysis of algae (seaweed) and lower density polyethylene are denoted by Y_S and Y_L, whereas W_S and W_L indicate the weight ratio of algae (seaweed) and the polyethylene plastic, which is lower density, respectively. Y_{ex} is the symbol for the biofuel which is produced from the experiment. Additional research was done on some of the properties and chemical contents of the biofuel (bio-oil), as detailed in the next section [17] (Fig. 6).

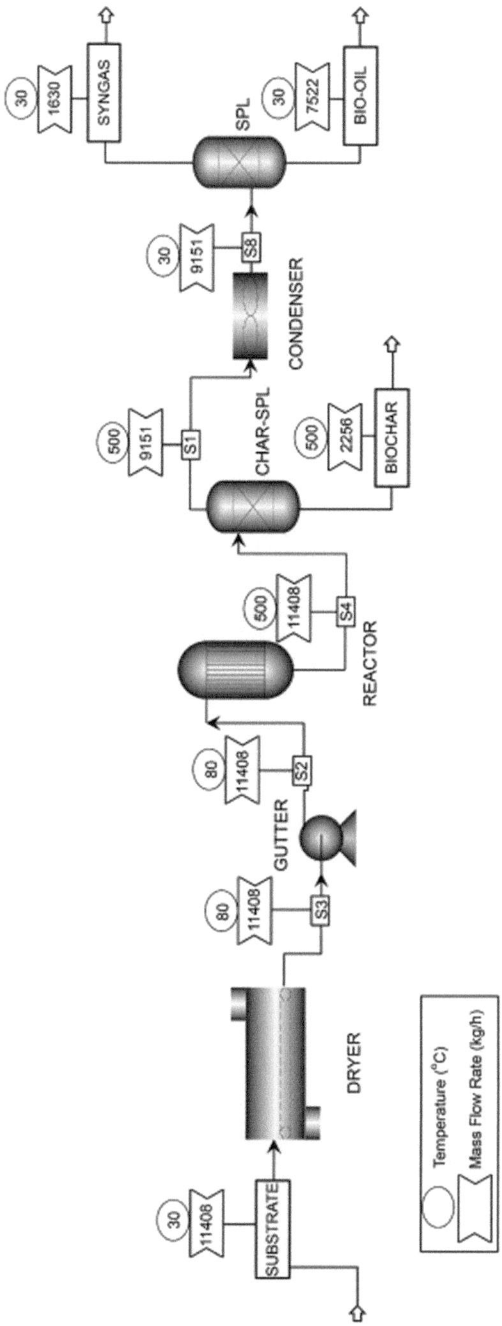

Fig. 6 Aspen Plus flowsheet uses 75% waste plastic in the vacuum of microwave co-pyrolysis of Ulva intestinalis [17]

3.3.3 Results and Discussion

Table 3 lists the salient properties of the 3 macroalgae and the polyethylene plastic which has low density that are being examined. Of the seaweed collected, Ulva intestinalis, Sargassum polycystic, and Hypnea valentiae illustrated the biggest growth. At 25.17 dw%, H. valentiae possessed the higher value of protein which contains compared to other algae type, surpassing that of S. polycystum by 1.4 times and U. intestinalis by 57.4%. H contains a lot of protein. Of the species studied, Valentiae has the highest nitrogen concentration (7.95%). On the other hand, U was used to identify the highest levels of fat and carbohydrates. Intestinalis makes up 45.7 weight percent of the whole. Consequently, of all the species studied, it contained the highest quantities of carbon and hydrogen (forty sex point twenty seven percent and seven-point thirty nine percent, respectively). Sargassum polycystum had the largest quantity of ash reaching around twenty-nine-point eighty five percent as resulted to its less volatility, however U. intestinalis had the less percent of ash quantity around (13.66%) and, thus, the highest reported volatile (71.83%). On the other hand, because LDPE is mostly composed of hydrogen and carbon, its volatile component was 98.84% [17].

Figure 7 illustrates the (Thermogravimetric Analysis/Derivative Thermogravimetric) curves for LDPE and the 3 type of algae. The breakdown of algal biomass

Table 3 The biochemical makeup of the three algae (seaweeds) under study (Ulva intestinalis, Sargassum polycystum, and Hypnea valentiae) and waste plastic (low density polyethylene) as well as their proximate and final analyses [17]

Parameters	Ulva	Sargassum	Hypnea	LDPE
Biochemical composition				
Biomass (g m^{-2})	59.85 ± 1.60^{a}	32.65 ± 1.31^{b}	41.96 ± 1.84^{c}	na
Carbohydrates (dw %)	36.47 ± 1.33^{a}	33.81 ± 1.70^{b}	29.14 ± 1.25^{c}	na
Lipids (dw%)	9.18 ± 1.07^{a}	5.24 ± 0.20^{b}	2.91 ± 0.24^{c}	na
Proteins (dw%)	10.55 ± 0.71^{a}	15.99 ± 1.87^{b}	25.17 ± 1.78^{c}	na
Proximate analysis (%)				
Moisture	6.73 ± 0.13^{a}	11.15 ± 0.13^{b}	9.80 ± 0.34^{c}	1.14 ± 0.41^{d}
Volatiles	71.83 ± 1.71^{a}	53.42 ± 2.44^{b}	59.75 ± 2.10^{c}	98.84 ± 0.39^{d}
Fixed carbon	7.78 ± 1.55^{a}	5.58 ± 1.28^{b}	8.85 ± 1.93^{a}	0.02 ± 0.02^{c}
Ash	13.66 ± 0.71^{a}	29.85 ± 2.71^{b}	21.60 ± 1.49^{c}	nd^{d}
Ultimate analysis (%)				
C	46.27 ± 1.42^{a}	37.63 ± 3.07^{b}	32.55 ± 2.59^{c}	85.91 ± 2.46^{d}
H	7.39 ± 0.40^{a}	3.94 ± 0.80^{b}	4.44 ± 0.33^{b}	14.09 ± 2.45^{c}
N	2.14 ± 0.14^{a}	3.58 ± 0.47^{b}	7.95 ± 0.96^{c}	nd^{d}
S	3.41 ± 0.29^{a}	2.11 ± 0.13^{b}	2.14 ± 0.15^{b}	nd^{c}
O	27.13 ± 2.96^{a}	22.89 ± 2.18^{b}	31.32 ± 0.34^{c}	nd^{d}

was commonly observed to occur in three main phases. A weight loss of about 8% and evaporate the lighter volatility components and the wet of it (up to ≈187 °C) marked the start of the first stage. However, most of the biomass degradation occurred in the next stage, with U exhibiting the highest value which is lost around eighty-seven percentage. Most organic macromolecules break down in the intestinalis around about 670 °C. As fragile connections in the polymer's feedstock components are heated broken down, stronger and stronger chemical connections are created during this step. Lastly, the very modest weight loss seen in the final stage can be attributed to the comparatively sluggish rate of decomposition of carbonaceous wastes. Moreover the 3 seaweeds (algae) illustrate the same value of weight, which is losing profile, the charcoal remains changed when S was present. The proximate analysis showed that Polycystum has the highest value. The LDPE thermal breakdown started at a temperature higher than seaweed about four hundred forty-five degrees Celsius, and the 98% of the weight loss was achieved in an individual step between 445 and 675 ∘C. While the DTG curve revealed a single peak for LDPE, different seaweeds showed many peaks due to the breakdown temperatures of various macromolecules, including proteins, carbohydrates, and lipids. Additionally, LDPE broke down thermally without producing any residues, which is in line with proximate analysis. It's crucial to keep in mind that the amounts of various additions, such as plasticizers, stabilizing elements, and the fullering, may have an impact on how various waste plastics break down [17].

The algae (seaweeds) was selected according to the production area of the biofuel, which is defined as high quantity of biomass output or a high value of biofuel product, which cause a high value of biofuel which recovered per unit area. Figure 8 illustrates the conversion efficiency and yield distribution of algae (seaweed) thermal decomposition at various values of temperatures. At 50.22%, the United States produced the crudest biofuel. At five hundred degrees Celsius °C the intestinalis contained the least quantity of biochar. Thus, increasing the thermal decomposition temperature to five hundred degrees Celsius further raised the converted quality, and it thereafter illustrates a constant at higher temperatures. Sargassum polycystum algae as well as hypnea valentiae algae. The pyrolysis process of Valentiae was similar, with the highest noticeably biofuel generation taking place at five hundred degrees Celsius. However, the biofuel produced in the U.S. Intestinalis was 34.4 and 53.8% greater than H. valentiae and S. polycystum, respectively. Protein and carbohydrate breakdown may occur before algal lipids decompose, which could account for the increased crude bio-oil output at higher temperatures. S. polycystum had the highest biochar content across all temperatures examined, which was in line with the findings of proximate thermal decomposition and TGA. The durability maybe a little decline in the biofuel product at higher temperatures of five hundred fifty and sex hundred degree Celsius could be due to the second cracks of the condensation volatility, which was found to enhance gas generation [17].

Selecting the appropriate algae (seaweed) for biofuel product is contingent upon the specific conditions. Because U. intestinalis generated the highest value of biomass and biofuel product in this investigation, also it is recording the thirty gram per meter square is the greatest predicted bio-fuel areal production, which is seventy

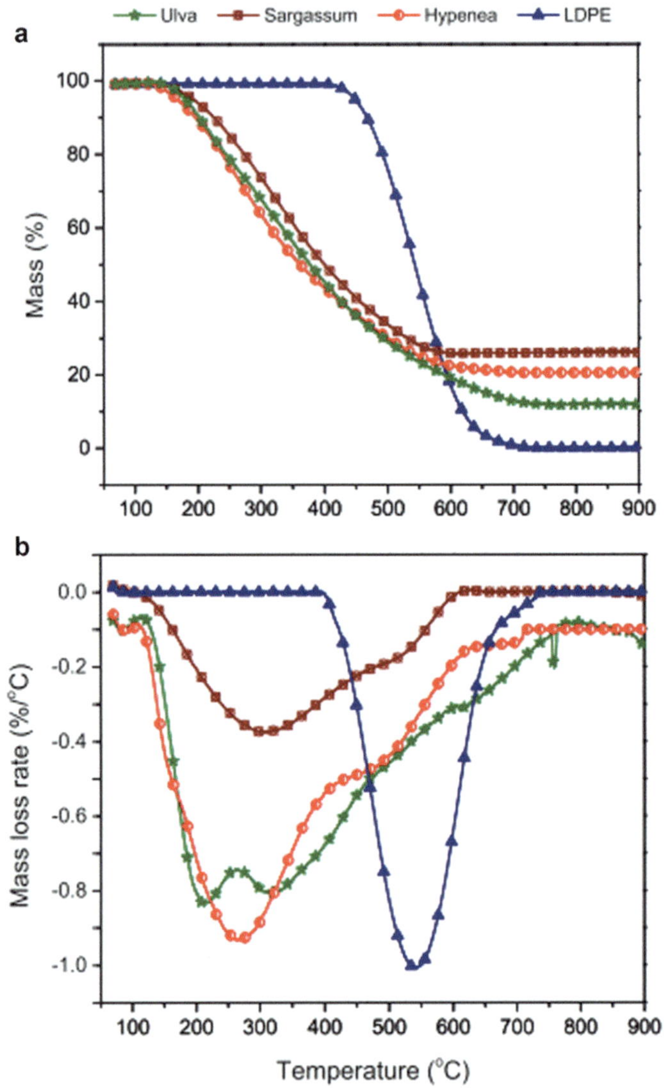

Fig. 7 Thermogravimetric analysis at temperatures up to 900 °C displaying thermogravimetric analysis (**a**) and derivative thermogravimetric (**b**) curves of the 3 seaweeds under study (Ulva intestinalis, Sargassum polycystum, and Hypnea valentiae) and waste plastic (Low density polyethylene) [17]

one percent and eighty-point three percent largest than that of H. valentiae and S. polycystum, respectively (Fig. 9). As a result, U. intestinalis was chosen for co-thermal decomposition with lower density polyethylene in this study based on its potential as algae (seaweed) for producing crude biofuel [17].

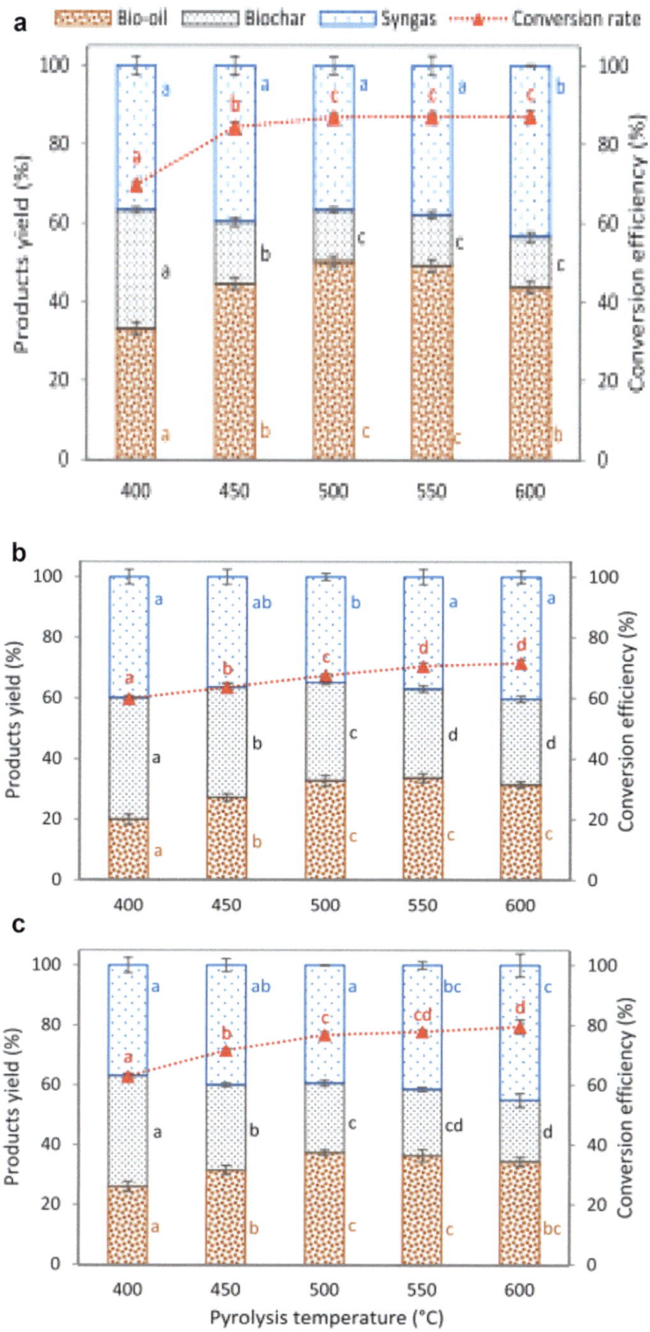

Fig. 8 Evaluation of the three seaweeds' capacity to produce bio-oil and their conversion efficiency at varying pyrolysis temperatures: Ulva intestinalis (**a**), Sargassum polycystum (**b**), and Hypnea valentiae (**c**). The same-letter values in the same series did not differ significantly (at $P < 0.05$) [17]

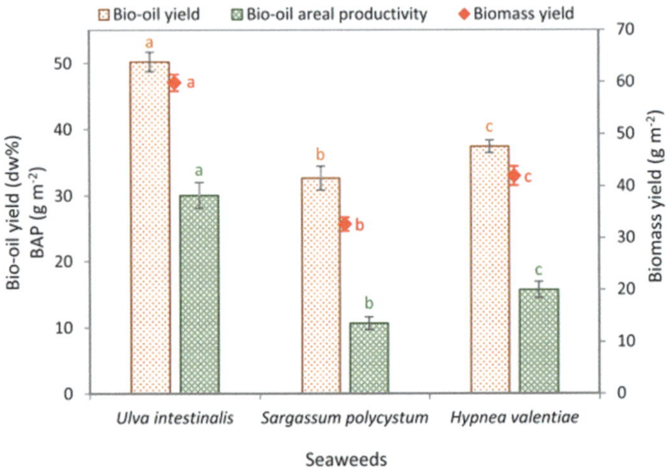

Fig. 9 Biomass production, bio-oil production, and bio-oil areal productivity (BAP) from the three seaweeds under study that microwave vacuum pyrolyzed at 500 °C. Significant variations were seen between seaweeds and different letters for the same measured parameter ($P < 0.05$) [17]

The impact of various LDPE ratios on the allocation of the production was examined at 500 °C after co-pyrolysis with U. intestinalis (Fig. 10a). The result's values illustrate that, with a small quantity of biochar (2.1%), the pyrolysis of LDPE mostly produced gas products (26.3%) and bio-oil (71.6%). The production of biochar may be the result of following volatile reactions since LDPE pyrolysis yields carbonaceous compounds with a rather extended pyrolysis duration. compared with U's individual pyrolysis. Intestinalis, raising the LDPE ratio in feedstock increased the product amount of crude biofuel but significantly reduced the product of bio charcoal (Fig. 10a). The cause might be the loss of ash in low density polyethylene and the impact of ash in quantity of biomass, which has been demonstrated to raise the thermal decomposition volatility generation at high value of heating (about eight hundred degree Celsius) while decreasing the ratio of converted at low temperatures. It's interesting to note that the bio-oil output of LDPE individual pyrolysis and that of a 75% LDPE mix ratio did not differ much. Furthermore, the synergistic impact of different ratios of LDPE mix and algal biomass was investigated (Fig. 10b). All biofuels produced from the experiment were largest than the values calculated at various blend percentage, suggesting a cooperative effect caused by the interacting the biomass lower density polyethylene was through co-thermal decomposition. In this situation, Low density polyethylene DPE was considered as a contributing source of the hydrogen, raising the hydrogen to carbon ratio of biofuel and promoting the production of hydrocarbon contents. This notion is supported by the outcomes of the biofuel(bio-oil) qualities. Compared to individual algal pyrolysis, LDPE pyrolysis produces less hydrogen radicals, which aid in the breaking down the ponds of the polymers group in the material which is biomass and reduce the quantity of biochar produced. The maximum synergy impact of ten percent was seen at a 25% lower density polyethylene mix rate (Fig. 10b). The synergistic effect dropped to 4.1% at a 75% LDPE mix ratio as the LDPE ratio rose. As a result, there were very little differences in the conversion efficiency and bio-oil output at 75 and 100% LDPE (Fig. 10a) [17].

By further examining the bio-oil's key attributes, its quality was evaluated (Table 4). Following LDPE individual pyrolysis, the oxygen content dropped from 18.6 to 9.0%, while the C and H mass fractions markedly increased relative to U. intestinalis. Deoxygenation increases the amount of bio-oil produced, increases its hydrophilic properties, and facilitates its isolation from volatile substances. The hydrophobicity of biocrude is demonstrated by the confined water's tiny mass portion in the current study, which declined from four-point one percent in the biofuel algae to zero-point three percent in the low-density polyethylene bio-oil (Table 4). The HHV of low-density polyethylene bio-oil was fifty-seven-points two percent larger than the biofuel algae due to the concurrent decrease of O and rise of H and C. The properties of the biofuel are impacted by the percentage of LDPE mixing. When comparing 75% LDPE blend ratio to 100% LDPE, the slight differences in oxygen, sulfur, hydrogen to carbon rate, oxygen to carbon ratio, and water that binds may be due to highlights the mutually beneficial relationship between biomass and plastic waste which is lower density polyethylene. Additionally, at a 75% LDPE mix ratio, the carbon, hydrogen, and sulfur proportion of the mass of the biofuel are comparable to those of petroleum oil and diesel (Table 4). One of the initial variables impacting the characteristic of bio-oil is widely thought to be HHV, which is affected by the combination of elements. The findings show that the HHV of bio-oil which as outcome from a seventy five percent polyethylene which distinguished with a lower density mix rate is two-point three percent and ten percent larger than the highest amount of diesel and petroleum oil, respectively (Table 5). Additionally, it exceeded the maximum HHV that could be attained by co-pyrolyzing HDPE with seaweed by 2.1%. Additionally, raising the LDPE blend ratio enhanced hydrocarbon production [17].

In Fig. 11, the Van Krevelen plot shows the hydrogen to carbon and oxygen to carbon ratios of the bio-oil made by empty space co-thermal decomposition using microwaves of LDPE and U. intestinalis. It expedited the hydrogenation processes as a hydrogen donor, increasing the hydrogen to carbon ratio and decreasing the oxygen to carbon ratio at the same time. These finding expectations that low density polyethylene improves hydrogenation processes rather than employing direct deoxygenation to eliminate oxygen. Furthermore, by reducing molecules that include nitrogen and carboxylic acids, it is possible that LDPE triggered the deamination and decarboxylation reactions. This could account for the bio-oil blends' increased concentration when compared to seaweed. This activity was heightened by the singular lower density polyethylene thermal decomposition, which had the higher recording of hydrogen to carbon ratio of any treatment. The present findings further expect that polyethylene which distinguished with a lower density may generate the hydrogen radical when biomass is co-pyrolyzed, increasing the hydrogen to carbon ratio and decreasing the oxygen to carbon ratio of the biofuel constituents. This explanation is

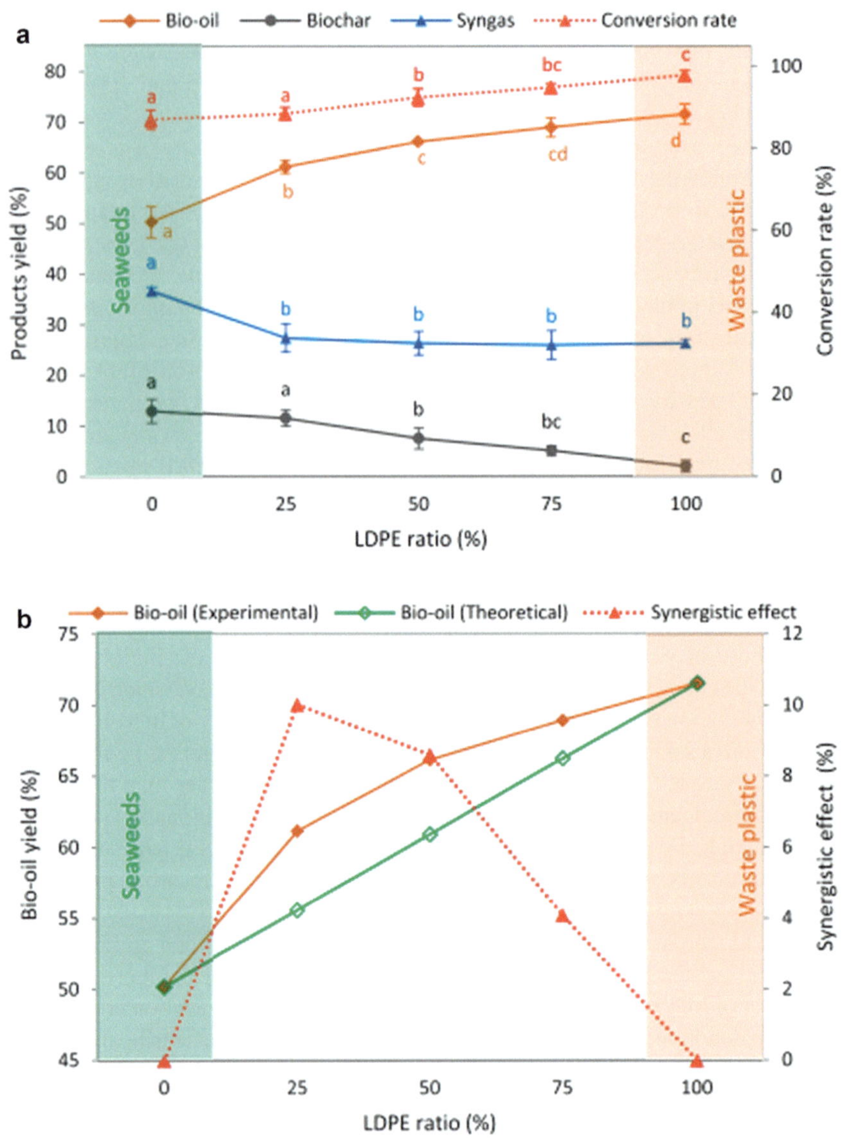

Fig. 10 LDPE/seaweeds' synergistic effects on the production of bio-oil during co-pyrolysis (**b**) and the impact of varying waste plastic (LDPE) mix ratios on Ulva intestinalis conversion efficiency and microwave vacuum co-pyrolysis products at 500 °C (**a**). The same-letter values in the same series did not differ significantly (at $P < 0.05$) [17]

Table 4 Characteristics of bio-oil made by microwave vacuum co-pyrolyzing waste plastic (LDPE) and Ulva intestinalis in contrast to traditional fuel [17]

Parameters	LDPE ratio (%)					Diesel	Petroleum oil
	0	25	50	75	100		
C	62.56 ± 1.45^{a}	71.45 ± 1.69^{b}	78.84 ± 1.73^{c}	85.15 ± 1.60^{d}	89.44 ± 1.45^{e}	86.2–86.3	85.00
H	8.32 ± 0.45^{a}	9.64 ± 0.19^{b}	11.24 ± 0.41^{c}	12.95 ± 0.50^{d}	14.22 ± 0.73^{e}	13.1–13.3	11.00
N	3.25 ± 0.20^{a}	2.51 ± 0.06^{b}	1.94 ± 0.50^{c}	0.54 ± 0.23^{d}	0.02 ± 0.02^{e}	na	0.30
O	18.55 ± 1.57^{a}	16.21 ± 0.63^{b}	12.14 ± 2.00^{b}	11.15 ± 3.16^{b}	8.95 ± 6.81^{b}	na	1.00
S	0.16 ± 0.05^{a}	0.05 ± 0.03^{b}	nd^{c}	nd^{c}	nd^{c}	0.034–0.039	na
H/C ratio	0.13 ± 0.01^{a}	0.13 ± 0.01^{a}	0.14 ± 0.01^{ab}	0.15 ± 0.02^{b}	0.16 ± 0.01^{c}	na	na
O/C ratio	0.30 ± 0.02^{a}	0.23 ± 0.01^{a}	0.15 ± 0.02^{b}	0.13 ± 0.03^{b}	0.10 ± 0.07^{b}	na	na
HHV (MJ kg^{-1})	30.03 ± 0.90^{a}	34.85 ± 0.27^{b}	39.71 ± 0.09^{b}	43.99 ± 0.87^{c}	47.21 ± 1.08^{c}	42.7–43.0	40
Bound water (%)	4.14 ± 0.20^{a}	3.26 ± 0.32^{b}	1.82 ± 0.27^{c}	0.45 ± 0.06^{c}	0.31 ± 0.03^{d}	na	na

na—Not applied, nd—Not detectable (it was considered 0 in statistical analysis)
Values in the same raw with the same letter showed insignificant difference (at $P \leq 0.05$)

Table 5 The raw material's elemental and chemical analysis [1]

No.	Feedstock	Chemical analysis (wt%)				Elemental analysis (wt%)					HHV (MJ/kg)
		MC	VM	FC	AC	C	H	N	S	O	
1	*U. lactuca*	7.2	49.7	1.0	42.1	39.1	6.2	4.5	7.3	42.9	12.04
2	PET	0.4	82.1	16.7	0.8	63.4	4.3	0.1	0.0	32.2	45.81

supported by the results. Who verified that when LDPE and lignin were co-pyrolyzed, hydrogen radicals were produced [17].

Figure 12 lists the initial groups of biofuels produced by co-pyrolyzing U. intestinalis at various ratios of lower density polyethylene mix. The findings verified that whereas hydrocarbons were relatively low (16.7%), molecules with N_2 and carboxylic acids dominated the individual pyrolysis of algal biomass (38.7% and 23.7%, respectively). Nitrogen and oxygen make up most of the chemicals in the biofuel made by algae thermal decomposition, with a lower percentage of hydrocarbons. The high quantity of C_6H_6O in the bio-oil made by biomass thermal decomposition is caused by. The heat transformation of aryl-ringed amino acids. The hydrocarbon content rose to 51.2% at a 75% blend ratio when the LDPE blend ratio was raised. At the same time, the levels of carboxylic acids and substances that include N_2 were dramatically reduced to 30.5% and 6.8%, respectively. There was also a notable drop in furans, the form of C_6H_6O carbohydrates, and aldehydes to ketones ratios. Consequently, the reaction between lower density polyethylene and biomass (algae) through co-pyrolysis prevented the synresearch of unwanted components that is as experiment production through biomass singular thermal combustion. Interestingly,

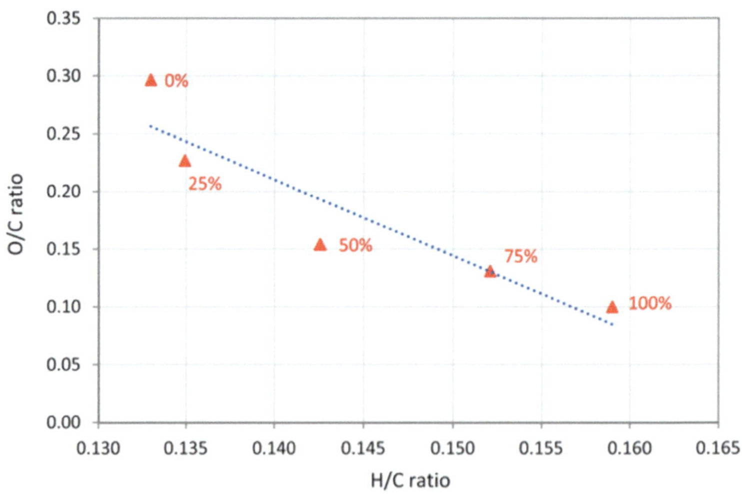

Fig. 11 A Van Krevelen plot of the bio-oil made from Ulva intestinalis microwave vacuum co-pyrolysis at 500 °C with varying proportions of waste plastic (LDPE) [17]

the proportion of carboxylic acids revealed very little change at seventy five percent and one hundred percent of the polyethylene which distinguished a lower density pyrolysis. The secondary reactions brought on by the addition of LDPE to the reactor and the volatile substances of the process of forming oligomers during the phase of condensing may account for these insignificant differences. This outcome shows that biomass and LDPE working together in a synergistic manner through a co-thermal decomposition in the vacuum of microwave that's results not producing only bio-oil but also instead of that is improved the quality of it [17].

In addition to the range of chemical components, LDPE and seaweed biomass were combined to increase the bio-oil's carbon chain length range. While solo pyrolysis of U. intestinalis produced bio-oil notable in thermal decomposition or co-thermal decomposition with low density polyethylene expanded the carbon chain length to twenty-seven carbon atoms from two carbon atom to ten carbon atom compounds. (Fig. 13a). The number of carbons of the constituents can be utilized to discriminate between the crude bio-oil's viability as fuel and the size of its organic molecules. Depending on the length of the carbon chain, the substances which are organic can be separated into four types of fuels: (light fuels < C5), diesel which contain carbon atom ranging between thirteen to twenty-four carbon atoms, gasoline which is contain carbon atom ranging between five to twelve carbon atoms, and (heavy oil contains > C24). With asphaltic bitumen's low economic worth coming in last, followed by tar and paraffin wax, the latter has a stable structure. According to Fig. 13b, most of the chemicals from U. intestinalis pyrolysis were light fuels around fifty sex point

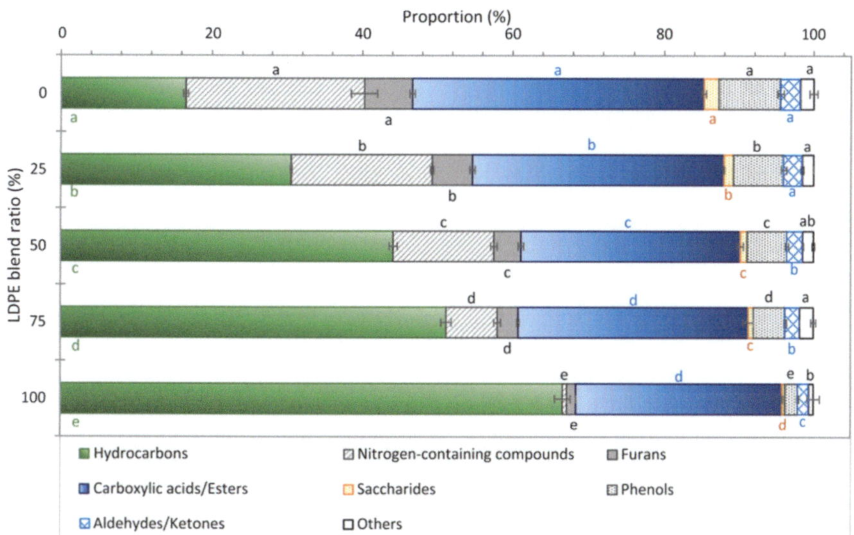

Fig. 12 The primary ingredients of the bio-oil made by microwave vacuum co-pyrolysis of Ulva intestinalis with various proportions of LDPE waste at 500 °C. There were negligible variations (at $P < 0.05$) between values in the same series that had the same letter [17]

one percent and gasoline around thirty-one-point five percent, with a little amount of diesel (10.6%). LDPE and U are combined. The intestinalis biomass improved the chemicals in the diesel and heavy fuel categories while reducing the use of light fuel and gasoline. This study shows that lower density polyethylene increased the components with long chains of carbon atoms in the biofuel, Whereas the biomass's exceptionally large ash content facilitated the breakdown of low-density polyethylene into substances with smaller carbon chains. Interestingly, there were comparable amounts of diesel and gasoline ranges for 75% LDPE (Fig. 13b). Accordingly, additional research on the catalyst co-thermal decomposition of low-density polyethylene with seaweed and the improvement of conditions for thermal decomposition would raise the quality of the bio-oil which is produced [17].

3.4 *Case Study III: Co-pyrolysis of Plastic Waste and Macroalgae Ulva Lactuca, Sustainable Valorization Approach Toward the Production of Bio-oil and Biochar [1]*

Authors: Obie Farobie, Apip Amrullah, Widya Fatriasari, Asep Bayu Dani Nandiyanto, Lusi Ernawati, Surachai Karnjanakom, Seng Hua Lee, Rangabhashiyam Selvasembian, Nur Izyan Wan Azelee, Muhammad Aziz

Year: 2024

3.4.1 Materials Used in the Experiment

Lombok Island in Indonesia is the source of the Ulva lactuca (ULV) macroalgae employed in this investigation. The algae were thoroughly cleaned to remove any impurities, then left to air dry before being roasted at 110 °C for 12 h. In Pradha Karya, Perkasa, Indonesia, we bought polyethylene terephthalate (PET) flakes from PT. That had been ground and filtered until the pieces were 0.5 mm in size. Five various mixes of U were made with the following weight proportions. 70% of the Ulva Lactuca (algae): 30% of polyethylene terephthalate, 60% of the Ulva Lactuca (algae): 40% of polyethylene terephthalate, 50% of the Ulva Lactuca (algae): 50% of polyethylene terephthalate, 40% of the Ulva Lactuca: 60% of polyethylene terephthalate, and 30% of the Ulva Lactuca: 70% of polyethylene terephthalate. While blended samples were created by combining the proper proportions of PET and algae, the 100ULV sample was manufactured using 50 g of pure U. Lactuca. A maximum thermal decomposition of the samples was carried out with a thermogravimetric analytical device. To test the amount of wetness, the samples were heating to one hundred ten degrees Celsius in the atmosphere nitrogen; to identify the subject of volatility, they were heated to 900 °C; and to assess the ash content, they were kept at 900 °C in air for 45 min. The

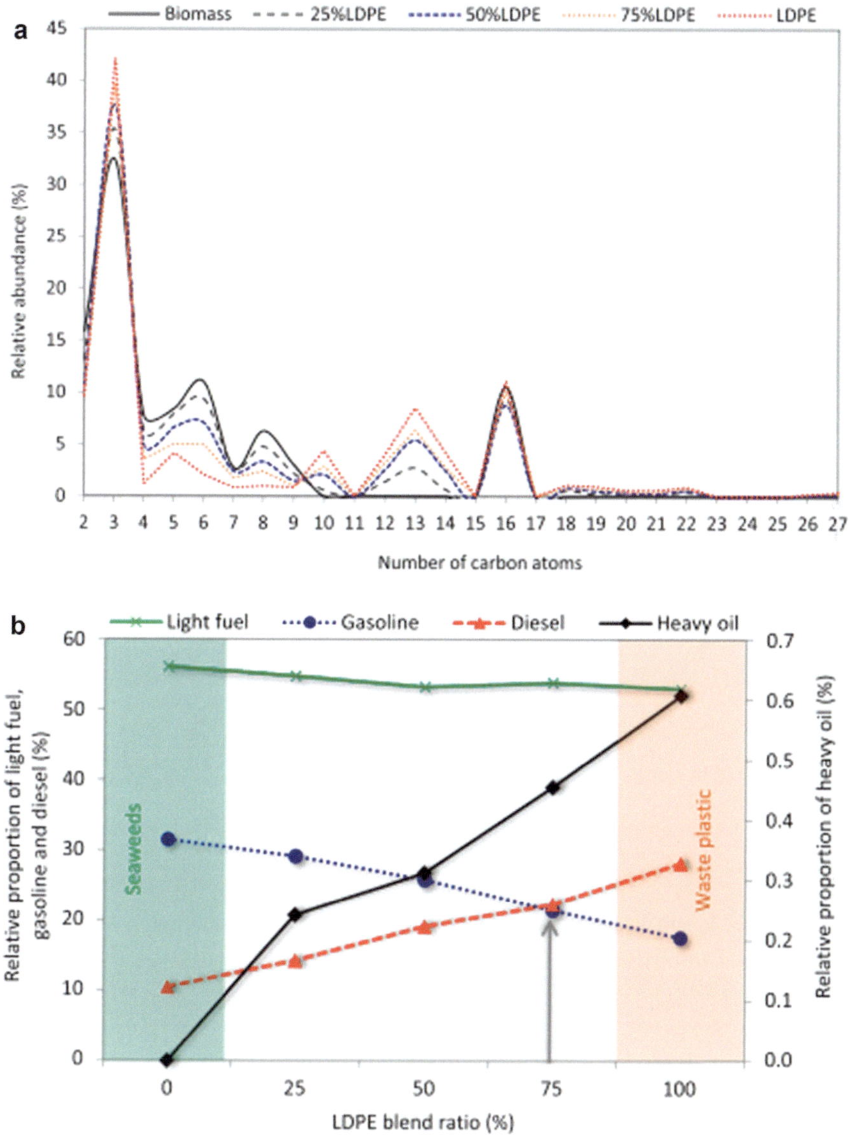

Fig. 13 Comparative amounts of light fuel, gasoline, diesel, carbon chains (**a**), and heavy oil (**b**) in the crude bio-oil made from Ulva intestinalis and various ratios of waste plastic (LDPE) at 500 °C using microwave vacuum co-pyrolysis [17]

carbon which is fixing carbon content was then measured by utilizing the difference. All analyzing measurements were repeated three times to ensure correctness [1].

3.4.2 Experimental Set-Up and Methodology

Experiment was conducted in a group jar with a thermoelectric and condensing fitted (Fig. 14). A cylindrical cylinder that is 0.50 m tall serves as the experiment's reactor. The reactors inside diameter (ID) are zero point zero three five meters, while it's the diameter which is on outside the cylindrical (OD) is zero point zero five meters. It is built of stainless steel. Because of the location of the reactor inside a furnace that runs on electricity, its surface will experience constant circulation of heat through the experiment. The tank was filled with fifty grams of dry feedstocks, nitrogen-purged and steamed from ambient range of temperature that's between four hundred to sex hundred degree Celsius at a ratio of twenty degree Celsius per minute. The vapor was collected using a liquid sample port after these temperatures were reached. The mixture's water content was extracted utilizing Na_2SO_4. Filters by vacuuming were utilized to extract the liquid. Using gravity methods, the products of biofuel and bio charcoal were measured; three experiments were carried out to guarantee accuracy. To calculate the product yield, the following formulas were applied.

$$Y_{\text{bio-oil}} = \frac{m_{\text{bio-oil}}}{m_{\text{feedstock}}} \times 100\% \tag{7}$$

$$Y_{\text{biochar}} = \frac{m_{\text{biochar}}}{m_{\text{feedstock}}} \times 100\% \tag{8}$$

$$Y_{\text{gas}} = 100\% - (Y_{\text{bio-oil}} + Y_{\text{biochar}}) \tag{9}$$

$Y_{\text{bio-oil}}$, Y_{biochar}, and Y_{gas} are the product value of bio-oil, biochar, and gases, respectively; $m_{\text{bio-oil}}$ and m_{biochar} are their corresponding masses; and $m_{\text{feedstock}}$ is the initial mass of the feedstock [1].

3.4.3 Results and Discussion

Significant differences were found between the preliminary and final assessments of the raw materials properties of PET and U. lactuca. Important details on the behavior and potential of these materials during pyrolysis are revealed by these variances. The elemental and proximate raw materials components utilized in this investigation are provided in Table 5. The composition analysis shows that U. Lactuca has a significantly higher ash content (forty-two-point one percent) than polyethylene terephthalate which is zero-point eight percent. The high content of ash of macroalgae from the oceans, which is composed of minerals and trace elements that result in increased residue after pyrolysis, is a common characteristic of these algae. U. Lactuca has a

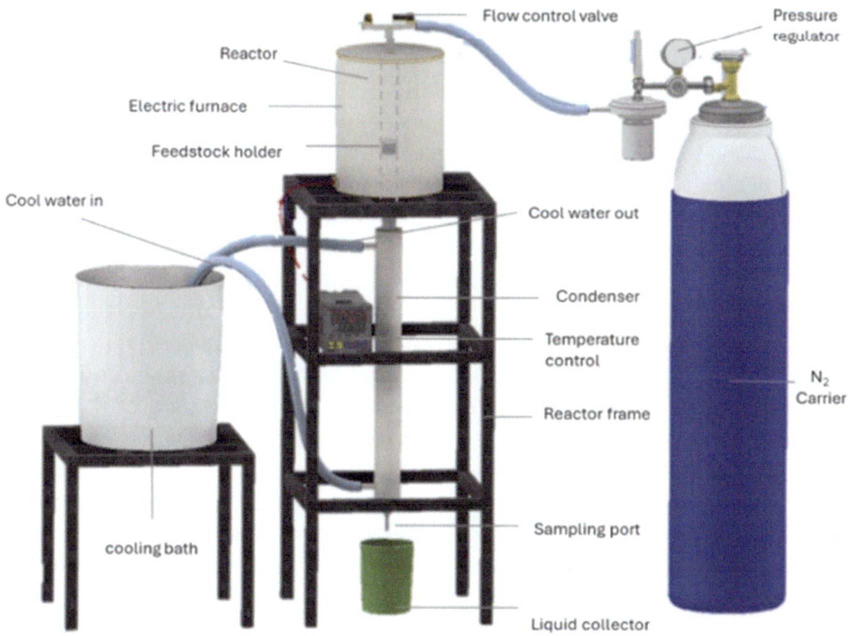

Fig. 14 Schematic of the pyrolysis reactor [1]

lower volatile matter concentration (49.7%) than polyethylene terephthalate (82.1%). This implies that a high percentage of U. Lactuca are not volatile, which means it is bonded as residual carbon or solid ash. The final analysis revealed significant differences in the elemental compositions of U. Lactuca and polyethylene terephthalate. The U. Lactuca has 42.9% oxygen, thirty-nine-point one percent of carbon, sex point two percent of hydrogen, four point five of nitrogen, and seven point three of sulfur. A lower value of C which contains, and a higher value of O_2 concentration combined with a high oxygen-to-carbon proportion. A high oxygen-to-carbon proportion is detrimental to fuel uses due to the significant amount of oxygen lowering the heating value. Sulfate polysaccharides, which are abundant in marine algae, supply a significant amount of sulfur. In contrast, polyethylene terephthalate has a high value of carbon which is sixty-three-point four percent, a low of oxygen level (32.2%), no sulfur, and very little nitrogen (0.1%). This composition of elements is linked to a large heating value that as result of high concentration of carbon suggests that additional energy can be produced through burning. PET's high energy potential has been confirmed by other tests that have revealed similar compositions [1].

Crucial details regarding the producing of gas, which is produced, biochar, and bio-oil were revealed by examining the manner of PET and U. Lactuca through thermal decomposition at various degrees and raw materials percentage. Figure 15 illustrates the production value of co-pyrolysis production at 400, 500, and 600 °C and various U. Lactuca to PET ratios. Pure U. Lactuca produced a notably greater amount of biochar at all temperatures examined. At a temperature of four hundred degree Celsius, the product was eighty two point thirty one percent. But when the temperature rose, the product yield declined (it was seventy-point twenty nine percent at five hundred degrees Celsius and sixty sex point forty nine percent at sex hundred degree Celsius). Nonetheless, the product was still noticeably larger than combinations that included polyethylene terephthalate. These increased solid remains are consistent with research findings showing that the ash content is increased by the natural minerals and trace elements found in macroalgae, increasing the yields of biochar. The addition of polyethylene terephthalate to the mix of raw materials consistently declined the formation of bio charcoal whereas concurrently boosting bio-oil manufacturing across all the ranging of values temperatures studied. For example, when 70% PET was added to the feedstock, the amount of biochar produced at 400 °C declined from eighty-two-point thirty-one percent (pure U. Lactuca) to thirty-two-point seventy eight percent. In comparison, within the same range of raw material composition, the output of biofuel rose from twelve-point forty-nine percent to twenty-seven-point eleven percent. This negative association is most likely caused by no contents of ash in polyethylene terephthalate and the catalyst action of U. Lactuca volatility of minerals production, particularly at high values of heating like sex hundred degree Celsius. Additionally, hydrogen radical generated throughout the polyethylene terephthalate thermal decomposition process assists in the breaking down of groups of polymers in biomass, improving the synresearch of biofuel. At five hundred degrees Celsius, the highest biofuel output (37.91%) was obtained with a raw materials percentage of forty percent U. 60% Lactuca PET. The combination of this ideal component and heating temperature imply that the largest yields of liquid products are produced at moderate temperatures, which encourage the components' synergistic effects. Moreover, as the temperature raised from five hundred to sex hundred degree Celsius, the producing quantity of biofuel decreased slightly, causing in the product of thirty-five-point sixty four percent for the same raw materials proportion. These declines are caused by the fact that more non-condensable gases are created because of more cracking processes occurring at higher temperatures. The increasing gas emission at sex hundred degree Celsius confirms the improved of the cracked down activity. Additionally, research demonstrating that thermal decomposition charcoal product dramatically dropped from fifty-four-point eighty-three weight percent to twenty six point sixty-seven weight percent when temperatures raised from three hundred to nine hundred degrees Celsius was in line with the loss in biochar production with rising temperature and PET concentration [1].

The analysis of the bio-oil content generated by co-pyrolyzing PET and U. lactuca is significantly influenced by temperature and feedstock ratio. As illustrated in Fig. 16, the GC-MS analysis reveals notable variation in the structure of chemicals of bio-oils generated from the thermal decomposition of U. Lactuca alone as opposed to a co-thermal decomposition linked of U. lactuca and PET. The primary components of the bio-oil are generated by U's autonomous pyrolysis. N-aromatic chemicals, carboxylic acids, amines, and phenolics were all found in lactuca. These compounds are common when macroalgae undergo pyrolysis. But adding PET to

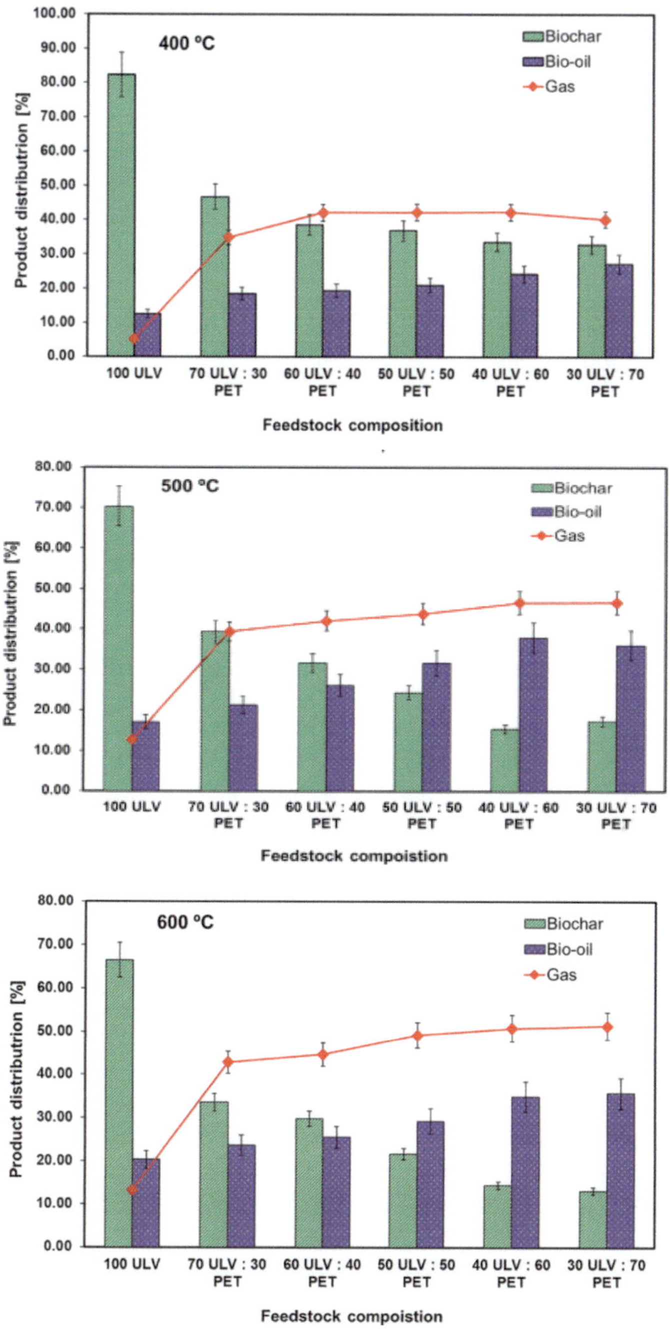

Fig. 15 Percentage of U. lactuca and PET products that co-pyrolyze at different temperatures [1]

the thermal decomposition process significantly changed the makeup of chemical structure of produced bio-oil. This transformation in chemical structure expects that co-thermal decomposition with polyethylene terephthalate increases the bio-oil's chemical variety and contributes significant aromatic hydrocarbons [1].

The distribution of bio-oil structure obtained from the co-thermal decomposition of Ulva Lactuca: polyethylene terephthalate at various values of temperatures is displayed in Fig. 17. Pure U. Lactuca pyrolysis produced bio-oils that were mostly high value of amines/amides (15.33%), phenolics (31.89%), and carboxylic acids (22.63%). The heat degradation of marine biomass is commonly associated with these factors. Similar patterns are seen in other marine biomasses, and the carboxylic acids collected from the breakdown of lipids are consistent with earlier observations. Phenolics are anticipated to occur as nitrogen-containing compounds, including $C_4H_{11}N$, $C_4H_{10}N_2$, and $C_6H_4NH_2$ decompose and aromatic molecules steaming reform, indicating that proteins are broken down during pyrolysis [1].

Figure 18 depicts a potential set of the pathways of the reactions for the co-thermal decomposition of polyethylene terephthalate and U. Lactuca. The hydrocarbon content rose significantly at 400 °C, reaching as high as 53.32%, when the blend ratio was raised from thirty percent U. Lactuca to seventy percent polyethylene terephthalate. At the same time, the amounts of $C_4H_{11}N/C_4H_{10}N_2$, $C_7H_6O_2$, and $C_4H_4O_2$, phenolics, N-aromatics, and ketones significantly decreased. Co-thermal decomposition with polyethylene terephthalate significantly alters the distributed of the products by reducing the formation of organic substances, potentially enhancing the properties of the biofuel. The suppression of unwanted components is assigned of this. The thermal decomposition temperature was raised to five hundred degrees Celsius, which notably changes the characteristic contents of the bio-oil. Pure U. lactuca nevertheless contained notable quantity of phenolics (20.16%), amines/amides (23.31%), and carboxylic acids (28.02%). But when PET was added, the hydrocarbon content rose dramatically, peaking at 57.16% for the highest PET mix (30% U. Lactuca and 70% PET). Because PET oligomerization influences secondary reactions that impact the entire chemical pathway during condensation, higher PET concentrations may result in consistent amounts of carboxylic acids. The significant drops in phenolics, N-aromatics, and amines/amides that coincided with the enrichment of hydrocarbons at 400 °C mirrored similar trends. At 600 °C, the bio-oil's composition keeps shifting in favor of hydrocarbons, which now account for most of its content. The highest amount of hydrocarbon (63.62%) was produced by a mix ratio of 30% U. Lactuca and 70% PET. The synresearch of volatile hydrocarbon fractions is enhanced by this temperature because it allows for higher thermal and secondary cracking, which further reduces the amount of phenolics, carboxylic acids, amines/amides, N-aromatics, and minor components like ketones and furan derivatives. PET and U. lactuca work in concert during co-pyrolysis, as evidenced by the steady increase of hydrocarbons with increasing temperature and PET %. The amount of bio-oil is increased, its chemical composition is improved to boost energy-rich hydrocarbons, and the amount of detrimental nitrogenous and oxygenated components decreased [1].

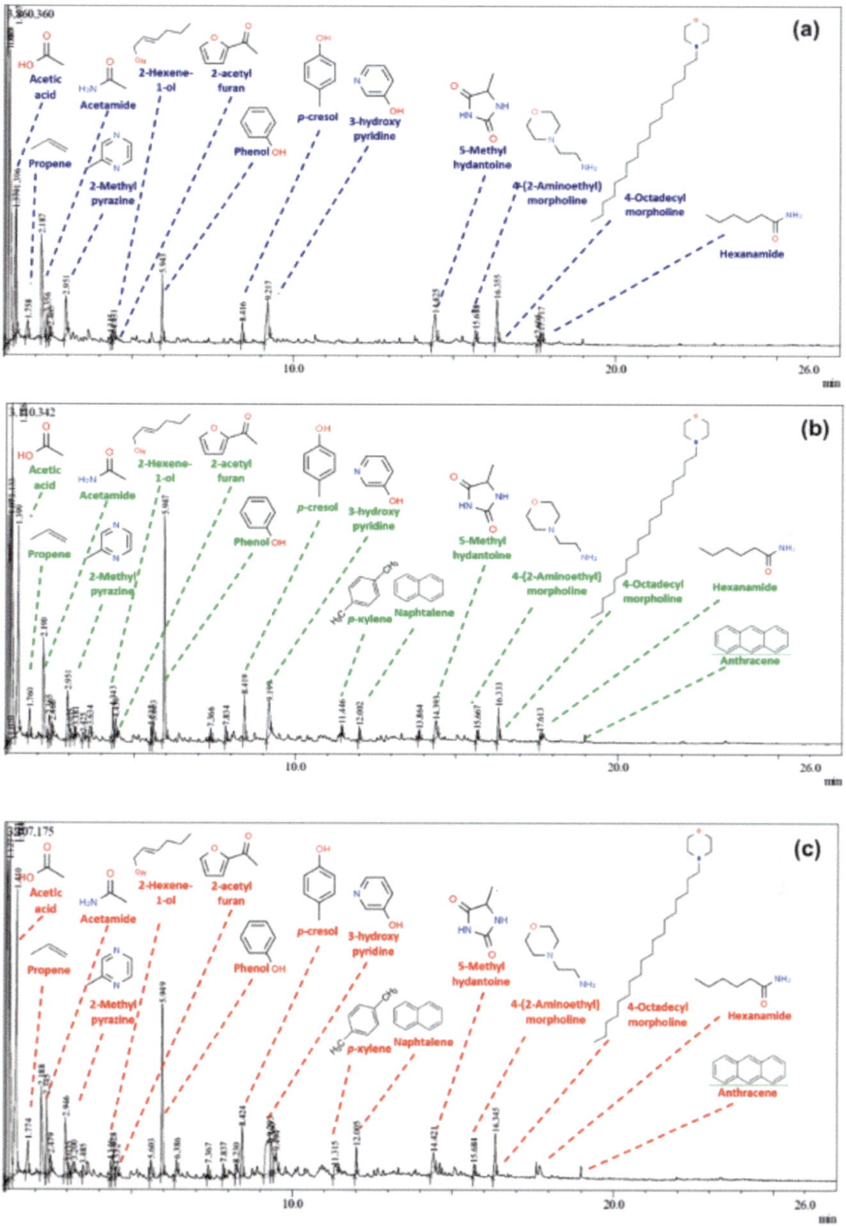

Fig. 16 Bio-oils from the pyrolysis of **a** individual U. lactuca, **b** 70 ULV: 30 PET, and **c** 30 ULV: 70 PET at 500 °C were analyzed using GC–MS [1]

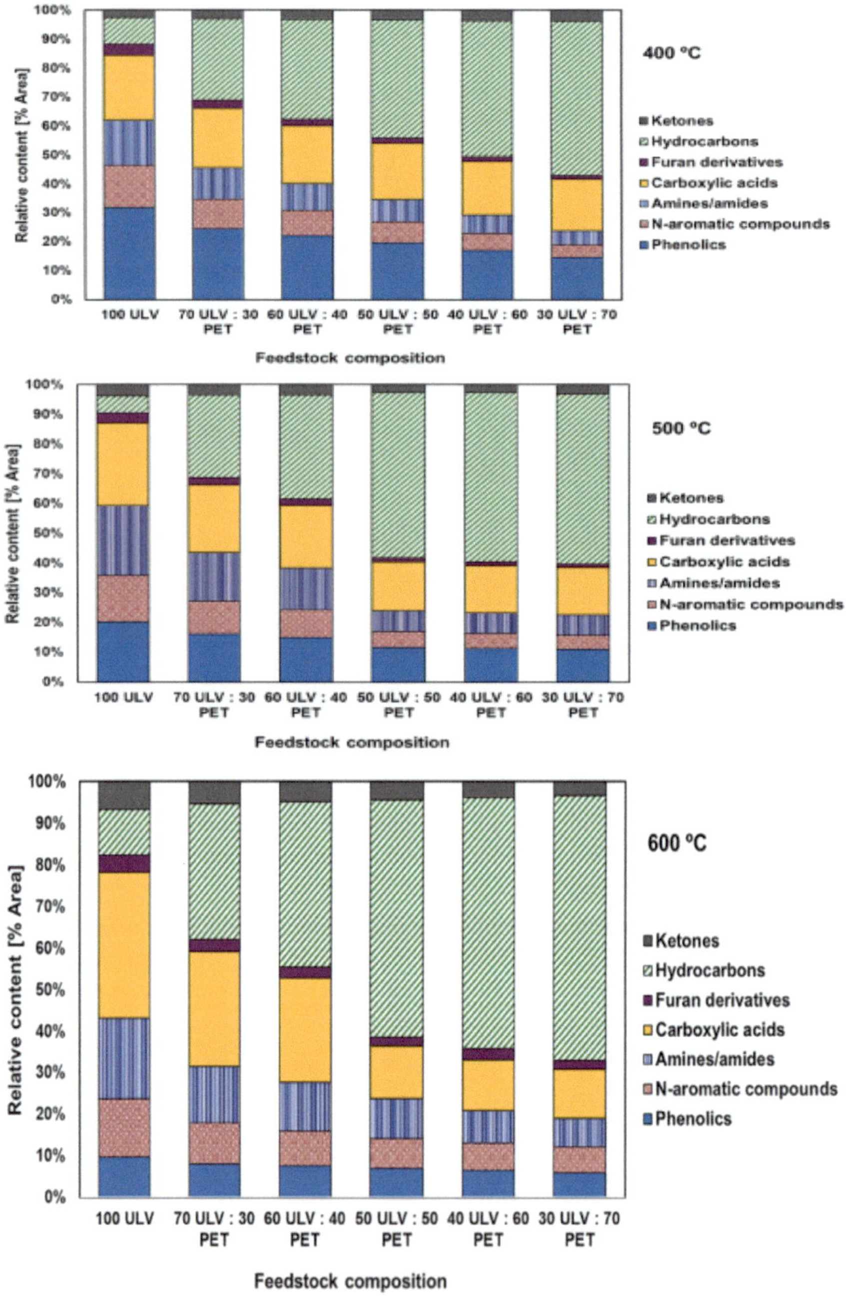

Fig. 17 Distribution of the bio-oil compositions produced at various temperatures by co-pyrolyzing ULV and PET [1]

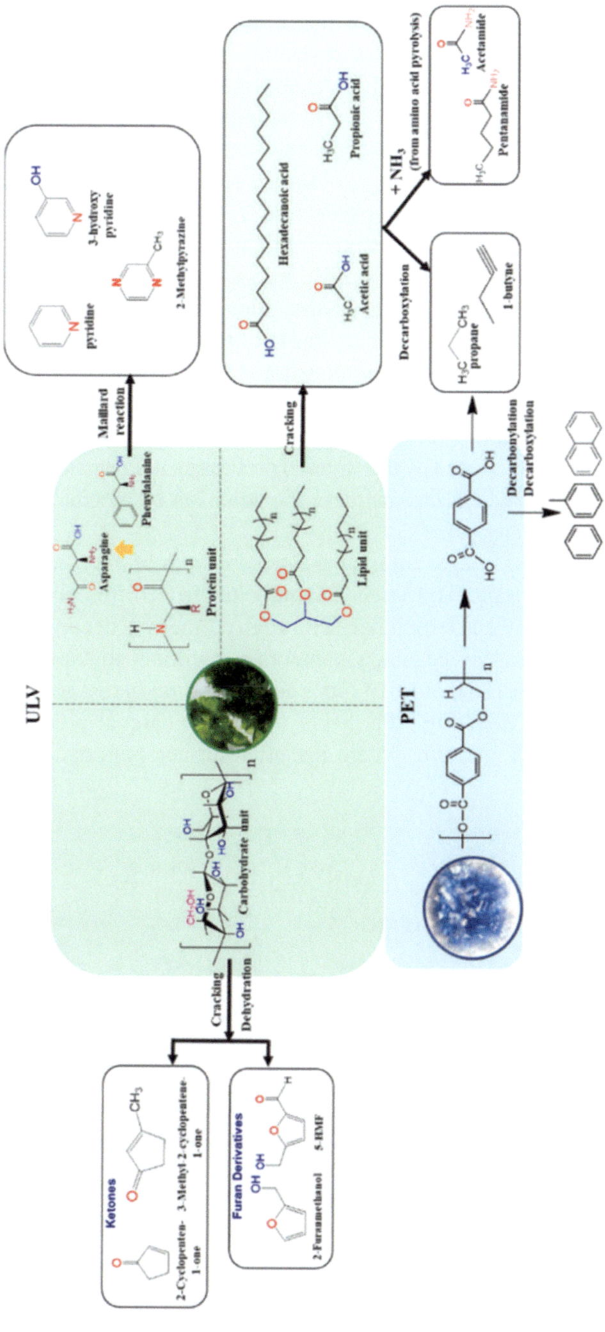

Fig. 18 A believable reaction mechanism for PET and U. lactuca co-pyrolysis [1]

SEM was used to analyze the biochar created by co-pyrolyzing PET and U. Lactuca to investigate any potential transformations in morphology that may have taken place following thermal decomposition. The pictures from scanned electron microscopic in Fig. 19 are essential for revealing both the structure and surface alterations of U. lactuca and its related bio charcoal at different raw materials rate, particularly at a pyrolysis temperature of five hundred degree Celsius. The images from scanned electron microscopic showed a substantial difference according to the surface characteristics, differences can be observed between the original U. Lactuca raw material and the resulting bio charcoal derived from various biomass proportion. The initial U. Lactuca biomass had a somewhat flat surface and was the largest in terms of size. However, the biochar's surface warped during pyrolysis, displaying numerous cracks and holes. The rapid volatilization of organic molecules, which produces a porous structure, is the primary cause of this deformation. Biochar with increasing PET levels (60 and 70%) demonstrated significant agglomeration and compaction, as well as increasingly larger physical sizes, when compared to bio charcoal with a small quantity of polyethylene terephthalate concentration. This combination is most likely caused by the sintered process characteristics of the plastic composed of polymeric, which can increase the adhesion of biochar particles and reduce their overall porosity [1].

In Fig. 20, the HRTEM images offer an analysis in depth of the characteristics of the structure of U. Lactuca and bio charcoal manufacture from different combinations of U. Lactuca and polyethylene terephthalate through co-thermal decomposition. The U. Lactuca (Fig. 20a) has a surface which is smooth and without any obvious porosity or graphitic formations, it is a feature of unprocessed natural material before undergoing heat therapy. However, Fig. 20b displays the bio charcoal produced from a 70% ULV and 30% PET mixture, where irregular surface patterns and graphitic

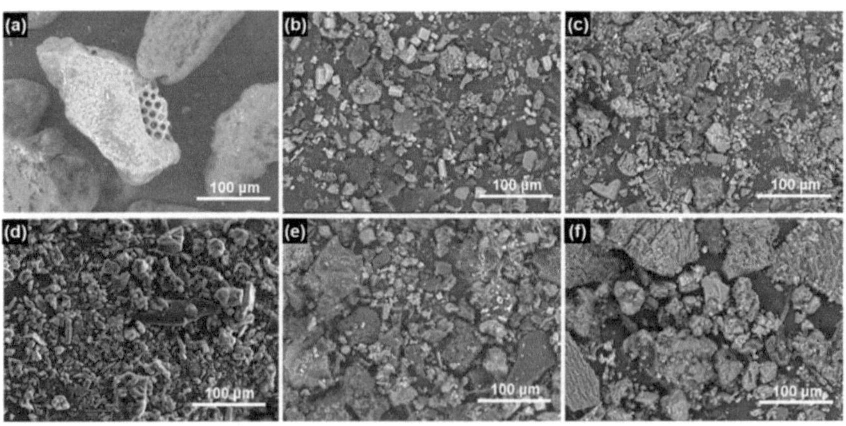

Fig. 19 **b** Biochar from 70 ULV: 30 PET, **c** biochar from 60 ULV: 40 PET, **d** biochar from 50 ULV: 50 PET, **e** biochar from 40 ULV: 60 PET, **f** biochar from 30 ULV: 70 PET, and **a** U. Lactuca all are shown in SEM pictures [1]

Fig. 20 **a** U. Lactuca, **b** biochar from 70 ULV: 30 PET, **c** biochar from 60 ULV: 40 PET, **d** biochar from 50 ULV: 50 PET, **e** biochar from 40 ULV: 60 PET, and **f** biochar from 30 ULV: 70 PET are included in the HRTEM photos [1]

carbon layers are clearly visible. The layers of graphitic graphite became more obvious, and a more uneven layer arrangement emerges when 60% ULV and 40% PET are combined, as seen in Fig. 20c. Higher graphitization and carbon restructuring are shown by the presence of microscopic holes and better lamellar structures in the biochar containing fifty percent of polyethylene terephthalate (Fig. 20d) and sixty percent of polyethylene terephthalate (Fig. 20e).The bio charcoal well-formed graphitic strata at 70% PET show a notable improvement in the carbon structure (Fig. 20f) [1].

To provide light on how varying ratios of Ulva Lactuca and PET blend impact the final biochar's morphological characteristics, Table 6 compares polyethylene terephthalate area of the exterior measurements. The U. Lactuca raw material had a relatively less polyethylene terephthalate surface area of one point thirty-eight-meter square per gram, having a one point fifty-four nanometer indicate pore diameter. However, the thermal decomposition and co-thermal decomposition processes caused a significant increase around the surface and pore diameter of the bio charcoal, particularly when PET was added. The 70% U.S. produced biochar. For instance, the Lactuca and thirty percent polyethylene terephthalate linked demonstrated a much larger mean pore width of 12.41 nm and a significantly higher polyethylene terephthalate of the surface of area of 20.18 m^2 g^{-1}. This significant increase is caused by the organic materials' vaporization components at a higher level of temperature, which facilitates the structure of more permeable and complex surface structures. However, the amount of PET in the feedstock increased. A mean pore width of 11.47 nm and a reduced the size of polyethylene terephthalate area of 18.56 m^2 g^{-1} were acquired from a combination of sixty percent U. Lactuca and forty percent of polyethylene terephthalate. With a mean pore width of 6.48 nm and the size of the area surface of polyethylene terephthalate of 10.64 m^2 g^{-1}, a significant drop was seen when the PET concentration was raised to 70%. The reason for this decreasing trend is most likely the pyrolysis-induced sintering and agglomeration of plastic. The partial melting and flow of PET may lead to surface structures clumping together and blocking pore channels, resulting in a thicker, less porous biochar matrix [1].

Table 7 highlights the evaluation of elements of the bio charcoal made by co-pyrolyzing PET and U. Lactuca. One significant discovery is the quantity of the C component in the bio charcoal raised notably and exceeds that of the pure Ulva Lactuca raw material. However, when the rate of the polyethylene terephthalate within the blend increases, the amount of carbon, hydrogen, nitrogen, and sulfur in the bio charcoal usually declines [1].

To confirm the precision of these examples, the yields of the experiment were compared to the expected values, as illustrated in Fig. 21. The products are correctly predicted by the models, as evidenced by their high values of correlation of zero point nine zero nine for gas, zero point nine zero sex for bio charcoal, and zero point nine two five for bio-oil. A large degree of agreement compared to the experiment results and expected data is indicated by plotting the points close to each other following the line that is diagonal. This demonstrates that the RSM accurately forecasts outcomes that fall within the experimental range [1].

The relationship between theoretical predictions and experimental data is shown in Fig. 22, emphasizing how temperature affects the product of biofuel. The experimental findings and the theoretical model showed a high degree of agreement [1].

The corresponding parity diagrams provided more evidence for this, which displayed a substantial factor of association of zero point nine nine nine five. Arrhenius plots illustrate the percentage of breakdown of the co-thermal decomposition of polyethylene terephthalate and Ulva Lactuca, is shown in Figs. 23 and 24. This graph was utilized to determine the repeated factors and the activated power. It is significant to highlight that the power which is activated was measured under initial conditions, for example under five hundred degrees Celsius of temperature using a raw materials mixture of forty percent U. Lactuca and sixty percent polyethylene

Table 6 Polyethylene terephthalate surface area of biochar and Ulva Lactuca at different percentages of feedstock [1]

Feedstock	BET surface area [m^2 g^{-1}]	Mean pore diameter [nm]
U. lactuca	1.38	1.54
Biochar 70 ULV: 30 PET	20.18	12.41
Biochar 60 ULV: 40 PET	18.56	11.47
Biochar 50 ULV: 50 PET	15.36	9.48
Biochar 40 ULV: 60 PET	12.81	8.39
Biochar 30 ULV: 70 PET	10.64	6.48

Table 7 Analysis of the elements in the biochar produced by co-pyrolyzing PET and U. Lactuca [1]

No	Sample	C	H	N	S	O[a]	Atom O/C	Atom H/C	HHV (MJ/kg)	LHV (MJ/kg)
1	*U. lactuca*	39.1	6.2	4.5	7.3	42.9	1.0972	0.1586	16.61	15.26
2	Biochar 70 ULV: 30 PET	47.3	3.9	2.9	6.4	39.5	0.8351	0.0825	17.46	16.61
3	Biochar 60 ULV: 40 PET	46.8	3.2	2.7	5.8	41.5	0.8868	0.0684	16.88	16.18
4	Biochar 50 ULV: 50 PET	46.2	2.8	2.3	5.1	43.6	0.9437	0.0606	16.47	15.86
5	Biochar 40 ULV: 60 PET	45.7	2.3	1.9	4.5	45.6	0.9978	0.0503	16.03	15.53
6	Biochar 30 ULV: 70 PET	44.8	1.8	1.4	4.2	47.8	1.0670	0.0402	15.47	15.08

[a] Calculate by difference

terephthalate. Additionally, this activated energy which is in this reaction has been measured to be around 43.35 kJ mol^{-1} [1].

3.5 *Case Study IV: Study on Microwave-Assisted Co-pyrolysis and Bio-oil of Chlorella Vulgaris with High-Density Polyethylene Under Activated Carbon [8]*

Authors: Chunxiang Chen, Dianzhao Fan, Jian Zha, Qianhao Qi, Xiaodong Huang, Tianyang Zeng, Yingxin Bi

Year: 2022

3.5.1 Materials Used in the Experiment

The low calorific value, approximate analysis, and elemental analysis of C. vulgaris were obtained from Haotian Plastics. HDPE (high-density polyethylene) and C. vulgaris (Chlorella vulgaris microalgae) are used as sources in the experiment. HDPE and C. vulgaris were crushed, powdered, and dried before the experiment. After that, a homogeneous substance smaller than two hundred milometers was attained using

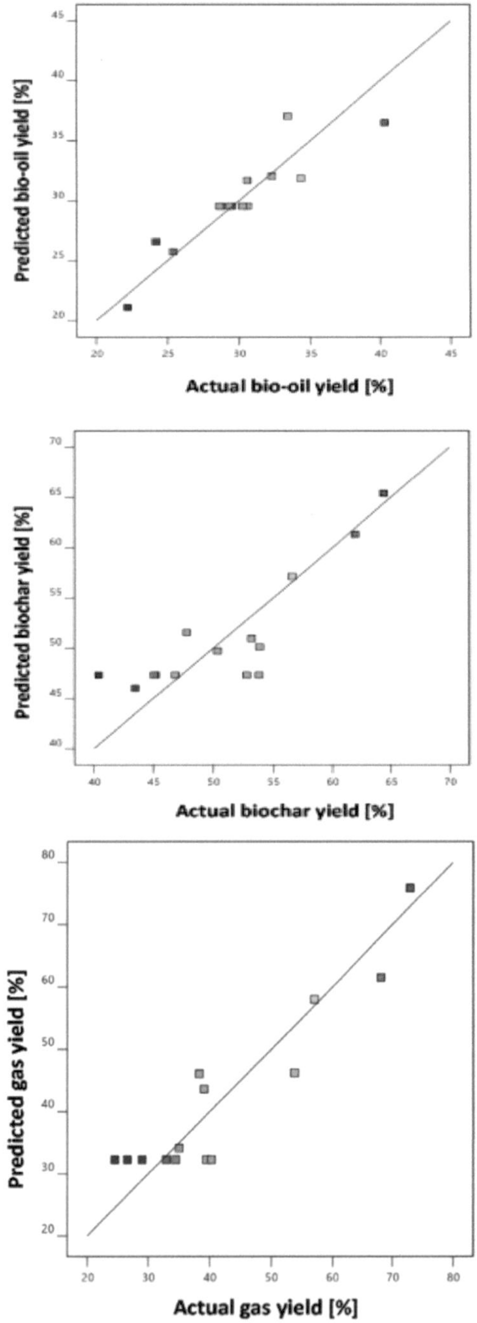

Fig. 21 Real versus expected plots for gas yields, biochar, and bio-oil [1]

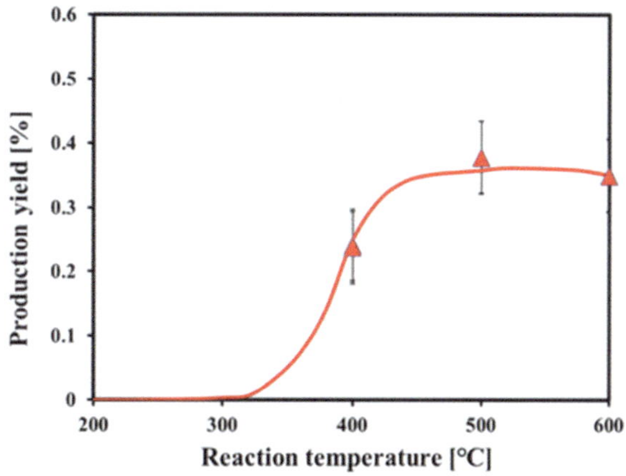

Fig. 22 The bio-oil yield at various temperatures during the co-pyrolysis process of U. Lactuca and PET is shown by the curve-fitting analysis [1]

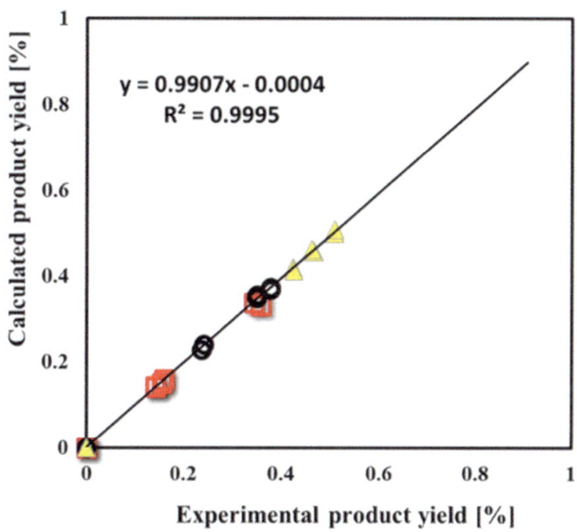

Fig. 23 Plots of parity that compare the bio-oil yields estimated by computations with those obtained from experimental measurements [1]

a molecules filter. A dryer was used to store all the supplies before the experiment. With a size of less than 200 mm and a purity of more than 99%, the catalyst component which is the carbon content which activates the reaction was selected as the microwave absorption. While C. vulgaris and HDPE were drying by the oven machine under one hundred degrees Celsius through twenty-four hours. Furthermore, AC was investigated utilizing X-ray diffraction and scanning electron microscopy, and the results showed that that it possessed a rich porosity structure. The XRD analysis revealed that trace levels of inorganic chemicals, such as SiO2, Al_2O_3, and $CaCO_3$, were present in AC. These compounds might have been left over from the initial raw materials. The pore size, pore volume, and specific surface for the area of AC were also investigated using BET (Brunauer–Emmett–Teller) [8].

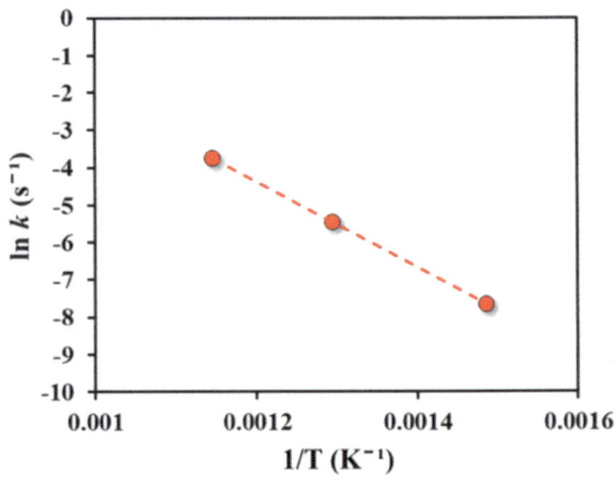

Fig. 24 Arrhenius graphs showing the thermal breakdown of PET and U. lactuca during co-pyrolysis [1]

3.5.2 Experimental Set-Up and Methodology

As shown in Fig. 25, using the microwave helped thermal decomposition apparatus contents of a microwave thermal decomposition for the reaction process, results gathering process, collecting by the condensing process, plus other auxiliary equipment. Using a research-grade microwave device with an assigned repeated of two points forty-five gigahertz and an output rating of eight hundred watts, an investigation was conducted. Weight of the species feedstocks was 30 ± 0.005 g. First, investigations utilizing microwave thermal decomposition were conducted using a control group of 30 g C. vulgaris. Following thorough mixing of C. vulgaris with HDPE in a range of rates (four: one, two: one, one: one, one: two, and one: four) microwave co-thermal decomposition was completed. The reference group was found to have the strongest stimulating effect after analyzing the co-thermal decomposition properties of HDPE and C. Voulgaris mixed at various rates. Following that, different amounts of AC (10, 20, 30, 40, and 50%) were added in order to perform the co-pyrolysis tests for the group with the optimal mixing ratio.700 mL/min of inert gas (N_2) was pumped to the chamber for the reactant through ten minutes prior to the experiment; this flow rate was constant until the test was over to check that the thermal decomposition process was occurring in the place that's empty from O_2 gas. Throughout the experiment, the fixed microwave energy was eight hundred watts. For each experiment category underwent 3 iterations to check that the faults percent reach five percent. The thermal decomposition experiment involved condensing the gas using the spiraling tubes in the container for condensation and releasing the gas which is not condensed using exhaust gas treatment equipment. For the best condensation result, a mixture of water and ice was poured into the condenser. To dehydrate, Na_2SO_4 was added to a solid extractor column. The collected liquid, was then filtered through a microporous membrane. A dark brown bio-oil with a repeated coke-like flavor was the result. For the Gas Chromatography–Mass Spectrometry evaluation, the bio-oil had been diluted with C_6H_5Cl prior to being set in the sealed bottle. The data obtained by the results

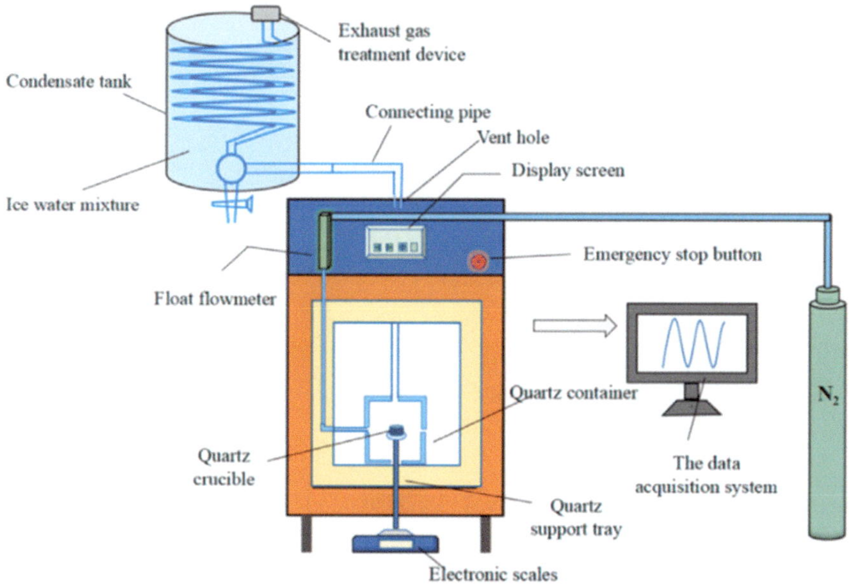

Fig. 25 The system design for microwave-assisted pyrolysis [8]

gathering process was utilized to generate the sample's graph of losing weight. The sample's weight ratio was then calculated using the differential calculation of the weight loss curve [8].

3.5.3 Results and Discussion

The co-thermal decomposition properties parameters are computed based on experimental data and are displayed in Table 8. Equation is used to compute the R_v (10).

$$R_v = {}^{W_t}\!/_{T_s} \tag{10}$$

where W_t stands for total weight loss, T_s for period according to stabilize the weight, and R_v for medium ratio of losing weight [8].

The curves representing heat-based weight loss analysis (TG) and its derivative (DTG) for the microwave-assisted co-thermal decomposition of Chlorella vulgaris and high-density polyethylene at various blend ratios reveal a three-stage degradation process (Fig. 26). The process of evaporating moisture and the release of low volatility are linked to the first phase, which is marked by minimal weight loss. Since the biomass had previously been pre-dried, there was very little mass reduction in this process. The next step is the primary phase of decomposition where the macromolecular components of C undergo a significant weight decrease of around 60%. High

Table 8 Characteristics of C. vulgaris and HDPE co-pyrolysis at varying mixture ratios [8]

Group	R_m[a] (wt%/s)	R_v[b] (wt%/s)	t_m[c] (s)	T_s[d] (s)	W_t[e] (wt%)
PC	0.04782	0.01183	3591	6075	71.857 ± 0.2
C4HP1	0.03803	0.01329	3341	6028	80.141 ± 0.2
C2HP1	0.04111	0.01218	5356	6684	81.379 ± 0.2
C1HP1	0.04580	0.01741	2629	5072	88.265 ± 0.2
C1HP2	0.05222	0.01467	2972	6460	94.774 ± 0.2
C1HP4	0.06261	0.01335	4155	7114	94.969 ± 0.2

[a]R_m—Maximum rate of weight loss
[b]R_v—Average rate of weight loss
[c]t_m—Time corresponding to R_m
[d]T_s—Time corresponding to weight stabilization
[e]W_t—Total weight loss

microwave-induced temperatures cause Vulgaris (such as proteins, lipids, and carbohydrates) and HDPE to thermally shatter. Any remaining material from the previous stage that did not dissolve entirely must be gradually broken down in the last step. The losing weight profiles in the beginning step were largely consistent across all blend ratios according to the large quantity wetting contents and limited microwave absorption; however, there were notable differences in stage which is on the next. Prior to the control category (PC), blend groups C1HP1, C4HP1, and C1HP2 inserted the primary pyrolysis phase, indicating a higher thermal response that is most likely the result of the synergistic interactions between biomass components and HDPE-derived free radicals. The algae claims that while oxygenated compounds from the biomass may facilitate HDPE chain scission, these radicals appear to enhance the cleavage of macromolecules. Furthermore, it is believed that HDPE's hydrogen emissions play a part in secondary pyrolysis processes. The second stage, however, began later in groups C2HP1 and C1HP4, with a drop in weight loss at first, followed by an increase. The melting of HDPE, which envelops portions of C, is responsible for this action. Vulgaris and reduces the effectiveness of pyrolysis by blocking heat transfer and gas diffusion. This physical encapsulation by molten plastic is also said to slow down the kinetics of the biomass's overall decomposition [8].

Figure 27 displays the number of the C component content and content of distributed of three prototype bio-oils made by pyrolyzing Chlorella vulgaris (PC), Chlorella vulgaris with polyethylene which distinguished a high value of density and activated carbon (C1HP1-40%AC), and Chlorella vulgaris with polyethylene which distinguished a high-density value (C1HP1). Oxygenous compounds, particularly amines (34.21%), were abundant in the PC-derived biofuel according to the large protein and lipid of the algae. However, co-pyrolysis with HDPE suppressed amines, ketones, and acids/esters while significantly raising the quantity of alcohol (35.58%) and hydrocarbons (37.88%). This implies that HDPE-generated free radicals promoted hydrocarbon synresearch and deoxygenation. Alcohol and ketones dropped to 14.6% and 0.42%, respectively, but the hydrocarbon content increased to

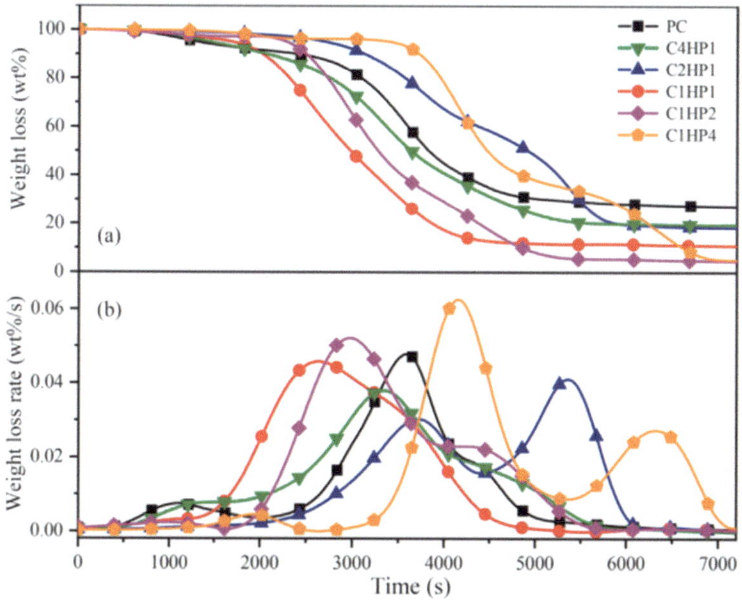

Fig. 26 C. vulgaris and HDPE weight loss curves (**a**) and weight loss rate curves (**b**) at varying mixing ratios [8]

48.88% with the addition of activated carbon. This suggests that the catalytic function of activated carbon promotes primary cracking and deoxygenation. According to carbon number analysis, the PC group contained a high proportion of nitrogenous compounds in the C7–C12 range; however, the addition of HDPE altered the distribution to favor long-chain alcohols and hydrocarbons, particularly the C13–C18 fractions. Activated carbon enhanced the synresearch of long-chain alkanes and decreased oxygenated components. HDPE significantly encouraged denitrification and the conversion of oxygen-containing groups to hydroxyls, whereas activated carbon stimulated further deoxygenation by converting alcohols and ketones into hydrocarbons. All things considered, the combined action of HDPE and activated carbon significantly improved the number of hydrocarbons and compositional excellence of the resulting bio-oil [8].

Figure 28 displays the distribution profiles of nitrogen and oxygen-containing components for three example bio-oil samples: (1) Chlorella vulgaris (PC)-only bio-oil; (2) a 1:1 mixture of C. vulgaris and HDPE (C1HP1); and (3) the same mixture with 40% activated carbon (C1HP1 + AC). The algae-only bio-oil included large levels of both oxygenated and nitrogenous compounds because pyrolysis thermally cracks, decarbonylates, and decarboxylates lipids and proteins. The addition of HDPE significantly altered the bio-oil's composition. To either increase the release of nitrogen as NH_3 and HCN or promote nitrogen fixing in the char, the hydrogen atoms released from HDPE reacted with nitrogen radicals created during protein decomposition.

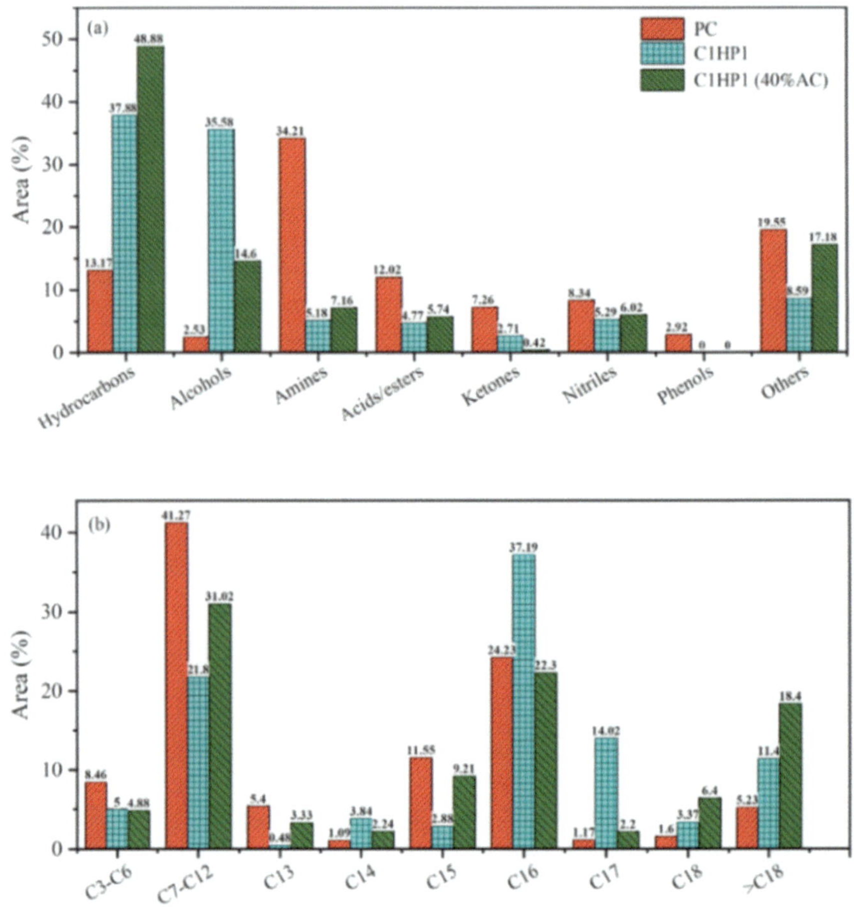

Fig. 27 PC, C1HP1, and C1HP1 (40%) AC are three common bio-oils: The primary components' distribution (**a**); the carbon number distribution (**b**) [8]

The simultaneous reduction of carbon-oxygen double bonds to hydroxyl groups by hydrogen atoms led to an increase in the synresearch of alcohol. As a result, the C1HP1 blends of the N_2 contents component declined by seventy-five-point seventy five percent and its contents of O_2 component by forty sex point forty two percent when made a comparison to the PC group. Alcohol accounted for 35.58% of the total bio-oil. Oxygen-containing compounds were further decreased by forty-five-point sixty-two percent made the difference compared to Chlorella vulgaris pyrolysis and thirty seven point ninety eight percent compared to C1HP1 when 40% activated carbon was added to the C1HP1 + AC group. Because of its high surface area and porosity, activated carbon had a longer vapor residence time, which facilitated deoxygenation reactions. There was less hydrogen available to stabilize nitrogen radicals because of the facilitation of the dehydration between hydrogen atoms

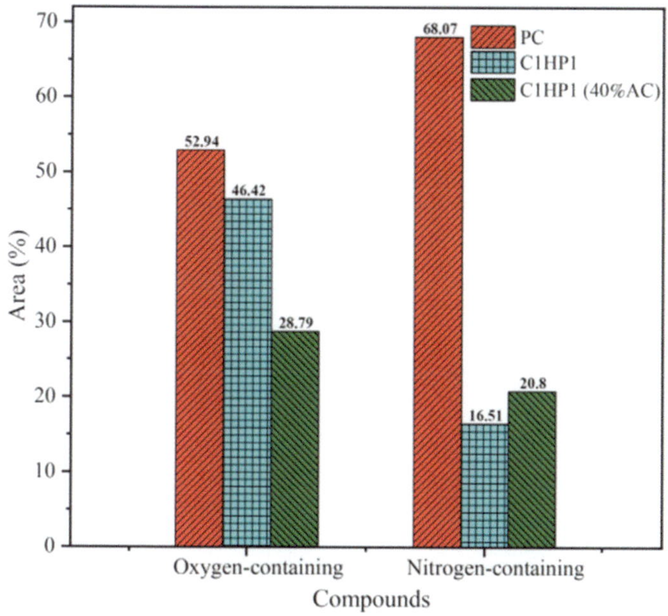

Fig. 28 The distribution of molecules that contain O/N in three common bio-oils (PC, C1HP1, and C1HP1 (40% AC)) [8]

and hydroxyl groups. As a result, the C1HP1 + AC group's nitrogen-containing molecules slightly rose to 20.80%, a 4.29% rise over C1HP1. Furthermore, a notable reduction in the concentration of ketone and alcohol was observed, indicating that activated carbon effectively enhances oxygen deprivation during co-pyrolysis while considerably limiting denitrogenating [8].

Figure 29 displays the proposed reaction routes for chemical formation during the co-thermal decomposition of the algae which is Chlorella vulgaris with the polyethylene which is distinguished with a high density, which are based on previous studies and new findings. The three primary biochemical constituents of C. vulgaris amino acids, triglycerides, and carbohydrates are associated with proteins, lipids, and polysaccharides. When pyrolyzed alone, these macromolecules are widely broken apart, generating many nitrogen and oxygen-containing free radicals. Following their reaction with hydrocarbon fragments, these radicals produce amides (such as hexadecane and palmi oleamide), acids, esters, ketones, amines, nitriles, and alcohols. The addition of HDPE significantly alters the chemistry of pyrolysis. When HDPE breaks down, many hydrogen and methyl radicals are created, which promotes hydrogen transfer, cracking, and oligomerization. Hydrogen radicals reduce the quantity of nitrogen in the bio-oil by stabilizing nitrogen radicals (such –C≡N and $–NH_2$) and converting them into volatile NH_3 and HCN. Like this, radicals that include oxygen are changed into hydroxyl groups, which then join forces with lengthy alkyl chains to form high length links of components which had alcohols contents for example

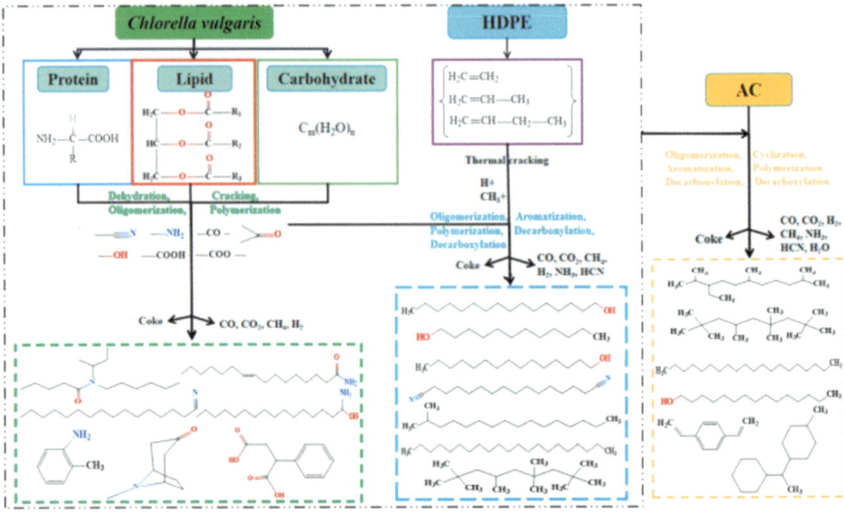

Fig. 29 The proposed microwave co-pyrolysis reaction route between HDPE and C. vulgaris [8]

$C_{12}H_{26}O$ and $C_{16}H_{34}O$. When 40% activated carbon (AC) is added, the C1HP1 blend changes even further. The cyclization, aromatization, decarboxylation, decarbonylating, and dihydroxylation processes that AC catalyzes produce a bio-oil that is primarily composed of cyclic and aromatic hydrocarbons and has a significant drop in alcohol concentration. Hydroxyl groups and other oxygenates are released as CO, H_2O, or CO_2 during deoxygenation. HDPE and AC generally greatly develop the effectiveness of deoxygenation and denitrogenating, which raises the quality of bio-oil. Further investigation into the proposed reaction pathways is required to validate specific mechanisms [8].

4 Summary

This paper analyzes four selected experimental case studies that investigate the co-thermal decomposition of algae and the polymeric materials (plastic)waste for bio-oil yields. While all experiments focus on the chemical of the thermal transformation of biomass (algae) and plastic blends, each study employs different reactor types, feedstocks, catalysts, and operational conditions, thereby offering diverse perspectives on process optimization. To provide a consolidated overview and facilitate comparison, a comprehensive table (Table 9) is presented below. This table summarizes key experimental variables, including feedstock, method employed, particle size, reactor type, product yield, processing time, optimal temperature, catalyst used, higher heating

Table 9 Comparison to the case studies results

Case study	I	II	III	IV
Method	Experimental	Experimental	Experimental	Experimental
Feedstock	Algae (Nannochloropsis sp) + (LDPE)	Algae (Ulva intestinalis) + LDPE	Algae (Ulva Lactuca) + PET	Algae (Chlorella vulgaris) + HDPE
Particle size (mm)	Algae (Nannochloropsis sp) < 0.12, LDPE < 0.38	Algae (Ulva intestinalis) and LDPE ≤ 3 mm	Algae (Ulva Lactuca: 0.25 mm PET: 0.5 mm	(Algae (Chlorella vulgaris) + HDPE) < 0.2 mm
Type of reactor	Fixed-bed reactor	Microwave vacuum pyrolysis reactor	Batch reactor	Microwave-assisted pyrolysis reactor
Maximum Bio-oil yield%	Bio-oil: ~ 82 wt%	Bio-oil: ~ 72wt%	Bio-oil: 37.91wt%	Bio-oil: 47.5 wt%
Time (min)	30	30	47.5	118.57
Ideal temperature (°C)	600	500	400–600	500
Catalyst used	–	Silicon carbide (Sic) (microwave absorbent)	–	Activated carbon (AC)
HHV (MJ/Kg)	–	36.6	15.47	35.23
Condensation system	Ice-water and liquid nitrogen	Dichloromethane solvent wash	–	Ice-water + anhydrous sodium sulfate (Na_2SO_4)
References	[13]	[17]	[1]	[8]

value (HHV), power consumption, and the Condensation system used. These comparative insights form the basis for evaluating synergistic effects, energy efficiency, and product quality across different pyrolysis configurations.

5 Results and Discussion

5.1 Overview

The Results and Discussion chapter provides a thorough comparison of four chosen experimental case studies that examine the co-pyrolysis of plastic waste and algae biomass to produce bio-oil. Table 9 lists the experimental settings, feedstock properties, and outputs obtained from each experiment. In each instance, different types of algae, including Chlorella vulgaris, Ulva intestinalis, Ulva Lactuca, and Nannochloropsis sp, were used as the main biomass sources and co-processed with thermoplastics such as LDPE, PET, and HDPE. Particle sizes were meticulously controlled throughout the investigations to guarantee experimental uniformity and promote effective thermal breakdown. The LDPE and algae in Case Study 1 were pyrolysis the reaction in the reactor which is a bed that is fixed after being ground to less than 0.12 mm for the LDPE and less than 0.38 mm for the algae. A microwave vacuum pyrolysis reactor was used in Case Study 2 to treat feedstocks with particle sizes less than 3 mm. Using a batch reactor system, Case Study 3's particle sizes were 0.25 mm for algae and 0.5 mm for PET. With the use of a microwave-assisted pyrolysis reactor, the fourth investigation produced the tiniest particle size (< 0.2 mm).

5.2 Discussion of Experimental Results

The effect of the variables process (parameters) on conversion efficiency was highlighted by the considerable differentiation in the bio-oil yield among the four tests. Case Study 1 yielded the highest output, at almost 82%, while Case Study 3 yielded the lowest, at 37.91%. In addition, Case Study 2 showed a high yield of almost 72%, whereas Case Study 4 produced 47.5 weight percent of bio-oil. These differences imply that, in addition to feedstock and reactor type, catalysts, heating techniques, and reaction time are important operational factors that affect product yield. Determining the quality and effectiveness of the pyrolysis process also depended heavily on the reaction time and operating temperatures. With processing periods of 30 min, Case Studies 1 and 2 were comparatively brief, but Case Studies 3 and 4 took 47.5 and 118.57 min, respectively. Case Study 1 operated at the maximum temperature (600 °C), which may have contributed to its improved yield. The optimal temperature range for pyrolysis was 400 to 600 °C. Case Study 1 operated at the highest temperature, 600°C, which may account for its higher yield. In Case Study 3, the temperature range was wider (400–600 °C), indicating greater fluctuating thermal behavior. Condensation methods and catalyst utilization were equally important for process results. Although Case Studies 1 and 3 did not use catalysts, Case Study 2 used silicon carbide (SiC) as a microwave absorber, and Case Study 4 used activated carbon (AC), which probably helped to save energy and develop the efficiency of biofuel. Case Study 2 reported the highest heating value (HHV) of bio-oil is thirty-six point six megajoules per kilogram, closely followed by Case Study 4 (35.23 MJ/kg), which suggests higher fuel quality. Case Studies 1 and 4 used ice-water condensation systems, with Case Study 4 also using anhydrous sodium sulfate (Na_2SO_4) to minimize moisture. Case Study 2 used dichloromethane solvent washing to recover bio-oil.

5.3 Economic Feasibility

According to the economic analysis, co-pyrolyzing plastic waste and algae biomass to produce biofuel shows encouraging prospects for widespread and affordable deployment. The co-pyrolysis of Ulva intestinalis and low-density polyethylene (LDPE) demonstrated robust economic indicators in a pilot-scale system that was simulated using Aspen Plus. A total capital cost (TPC) of about US$ 399,147.8 was projected for a processing capacity of 32,114 tons per year, while the operating cost, which includes labor, maintenance, feedstock, power, and silicon carbide (SiC), was estimated to be US$ 22,795.1 per hour. The use of a vacuum environment rather than nitrogen to maintain inert conditions was found to cut operating expenses by 27%, despite the comparatively high energy cost [11].

Based on Aspen Plus simulation, Table 10 shows the anticipated capital and operational costs for producing crude bio-oil by co-pyrolyzing algae (Ulva intestinalis) with 75% plastic waste (LDPE). The biochar production was 2.8 times higher, showing good process efficiency, whereas the bio-oil yields from the simulation were equivalent to actual results. The estimated annual processing capacity of a pilot-scale plant running at 11.4 tons per hour is 32,114 tons, with a total capital cost (TPC) of US$ 399,147.8. The overall cost of operations, including feedstock, electricity, SiC, maintenance, and labor, is US$ 22,795.1/h; labor alone is anticipated to cost US$ 11,974.4/h. Using a vacuum atmosphere instead of nitrogen gas lowers manufacturing costs by 27%, despite the high cost of power. Bio-oil, charcoal, and syngas are expected to generate US$ 31,160.5 every hour for the system. The anticipated cost of producing crude bio-oil is thirty point nine cents per liter, which is eighty one point seven percent less than the market price of one dollar and sixty-nine cents per liter and less than the cost of producing biodiesel, which is US$ 0.53–2.04/L. Co-pyrolysis's full biomass conversion, as opposed to biodiesel processes that depend on lipid extraction, is the reason for this economic advantage [11].

Overall, according to the current analysis, the proposed 20-year lifetime plant would generate 87.78 million US dollars in total revenue and 23.17 million US dollars in net profit. Studies on modified economical and practical viability are presented in this research, which can be used to expand this idea with additional feedstocks under ideal circumstances. Even though the economic outcomes are initial, they offer encouraging and practical signs that the vacuum co-pyrolysis of algae and plastic trash will be implemented successfully. Waste management agencies, the recycling industry, those who make decisions, and other stakeholders focused in the shift to a circular economy may find these findings helpful [11].

Table 10 Assessment of the techno-economic viability of microwave-assisted vacuum co-pyrolysis of Ulva intestinalis algae combined with seventy-five percent waste plastic [11]

Items	Value
I. Capital cost	
1. Plant capacity (ton/h)	11.4
2. Total equipment cost (US$)[a]	315,000.0
3. FCI (US$)	347,085.0
4. SWC (US$)	52,062.8
5. TPC (US$)	399,147.8
II. Operating cost	
1. Feedstock (US$/h)	570.3
2. SIC (US$/h)[b]	1197.0
3. Electricity (US$/h)	1070.4
4. Maintenance (US$/h)	7983.0
5. Labour (US$/h)	11,974.4
Total operating cost (US$/h)	22,795.1
III. Sales and profit	
1. Bio-char (US$/h)	2487.3
2. Bio-oil (US$/h)	28,266.9
3. Syngas (US$/h)	406.2
4. Estimated bio-oil production (L/h)[c]	8726.2
5. Estimated production cost of bio-oil (US$/L)	0.309
Total sales (US$/h)	31,160.5
IV. Total investment (million US$)	64.61
V. Total profit (million US$)	87.78
VI. Net profit (million US$)	23.17

[a] Calculated as the summation of the cost of dryers (US$ 100,000), microwaves (US$ 200,000), condensation system (US$ 2000), adapters (US$ 5000), pipes (US$ 3000), and vacuum pump (US$ 5000)

[b] Based on feedstock: SIC ratio of 20:3 at average market price of US$ 700 ton^{-1} for commercial SIC

[c] At bio-oil production rate of 7522 kg h^{-1} and density of 0.82862 kg m^{-3}

6 Conclusion and Future Work

In conclusion, the co-pyrolysis process involving plastic waste and algae biomass has been thoroughly evaluated in this research, highlighting its potential as a sustainable and profitable method of producing biofuel. The study has shown the significant impact of process parameters on product distribution, energy content, and overall conversion efficiency by methodically examining important operational variables, including reactor selection, thermal regimes, catalytic enhancement, and particle size optimization. The investigation's findings highlight the revolutionary possibility of converting waste streams both biodegradable and non-biodegradable into useful energy sources. This study makes an important participation to the development of renewable energy technology by changing the view of algae and plastic trash from environmental pollutants to strategic power resources. The knowledge gained here provides a solid basis for additional investigation and business improvement, highlighting the function of co-thermal decomposition in the larger structure of the sustainable economy and sustainable energy transitions.

References

1. Farobie O, Amrullah A, Fatriasari W, Nandiyanto ABD, Ernawati L, Karnjanakom S, Lee SH, Selvasembian R, Azelee NIW, Aziz M (2024) Co-pyrolysis of plastic waste and macroalgae Ulva lactuca, a sustainable valorization approach towards the production of bio-oil and biochar. Results Eng 24:103098. https://doi.org/10.1016/j.rineng.2024.103098
2. Tan IJY, Loy ACM, Chin BLF, Cheah KW, Teng SY, How BS, Alhazmi H, Leong WD, Lim HY, Lam MK, Lam SS (2024) Co-pyrolysis of Chlorella vulgaris with plastic wastes: thermal degradation, kinetics and progressive depth swarm-evolution (PDSE) neural network-based optimization. Green Technol Sustain 2(2):100077. https://doi.org/10.1016/j.grets.2024.100077
3. Fan D, Yang R, Chen C, Qiu S, He S, Shi H (2024) Microwave catalytic co-pyrolysis of microalgae and high-density polyethylene over activated carbon supported bimetallic: characteristics and bio-oil analysis. Renew Energy 235:121323. https://doi.org/10.1016/j.renene.2024.121323
4. Hou C, Zheng X, Song Y, Shangguan D, Zhao J, Zhou X, Zhang Y, Shen Z (2024) Influence of polyethylene chain structure on microwave-assisted co-pyrolysis of polyethylene and microalgae: nitrogen reduction and oxygen increase effects in bio-oil. Fuel 374:132463. https://doi.org/10.1016/j.fuel.2024.132463
5. Ormanci EG, Ozcakir G, Karaduman A (2023) Co-pyrolysis of waste polystyrene foam and microalgae at low temperatures. Academic Editor: Bipro Dhar. https://doi.org/10.3390/ecp2023-14608
6. Chen C, Zhao J, Wei Y, Huang X, Lu W, Fan D, Bi Y, Qiu H (2022) Influence of graphite/alumina on co-pyrolysis of Chlorella vulgaris and polypropylene for producing bio-oil. Energy 265:126362. https://doi.org/10.1016/j.energy.2022.126362
7. Jiang L, Zhou Z, Xiang H, Yang Y, Tian H, Wang J (2021) Characteristics and synergistic effects of co-pyrolysis of microalgae with polypropylene. Fuel 314:122765. https://doi.org/10.1016/j.fuel.2021.122765
8. Chen C, Fan D, Zhao J, Qi Q, Huang X, Zeng T, Bi Y (2022) Study on microwave-assisted co-pyrolysis and bio-oil of Chlorella vulgaris with high-density polyethylene under activated carbon. Energy 247:123508. https://doi.org/10.1016/j.energy.2022.123508
9. Chen C, Fan D, Ling H, Huang X, Yang G, Cai D, Zhao J, Bi Y (2022) Microwave catalytic co-pyrolysis of Chlorella vulgaris and high-density polyethylene over activated carbon supported monometallic: characteristics and bio-oil analysis. Biores Technol 363:127881. https://doi.org/10.1016/j.biortech.2022.127881
10. Chen R, Zhang S, Yang X, Li G, Zhou H, Li Q, Zhang Y (2021) Thermal behaviour and kinetic study of co-pyrolysis of microalgae with different plastics. Waste Manage 126:331–339. https://doi.org/10.1016/j.wasman.2021.03.001
11. Abomohra AE, Sheikh HM, El-Naggar AH, Wang Q (2021) Microwave vacuum co-pyrolysis of waste plastic and seaweed for enhanced crude bio-oil recovery: experimental and feasibility study towards industrialization. Renew Sustain Energy Rev 149:111335. https://doi.org/10.1016/j.rser.2021.111335

12. Xu S, Cao B, Uzoejinwa BB, Odey EA, Wang S, Shang H, Li C, Hu Y, Wang Q, Nwakaire JN (2020) Synergistic effects of catalytic co-pyrolysis of macroalgae with waste plastics. Process Saf Environ Prot 137:34–48. https://doi.org/10.1016/j.psep.2020.02.001
13. Tang Z, Chen W, Chen Y, Yang H, Chen H (2018) Co-pyrolysis of microalgae and plastic: characteristics and interaction effects. Biores Technol 274:145–152. https://doi.org/10.1016/j.biortech.2018.11.083
14. Cao B, Sun Y, Guo J, Wang S, Yuan J, Esakkimuthu S, Uzoejinwa BB, Yuan C, Abomohra AE, Qian L, Liu L, Li B, He Z, Wang Q (2019) Synergistic effects of co-pyrolysis of macroalgae and polyvinyl chloride on bio-oil/bio-char properties and transferring regularity of chlorine. Fuel 246:319–329. https://doi.org/10.1016/j.fuel.2019.02.037
15. Dai M, Xu H, Yu Z, Fang S, Chen L, Gu W, Ma X (2018) Microwave-assisted fast co-pyrolysis behaviors and products between microalgae and polyvinyl chloride. Appl Therm Eng 136:9–15. https://doi.org/10.1016/j.applthermaleng.2018.02.102
16. Qi P, Chang G, Wang H, Zhang X, Guo Q (2018) Production of aromatic hydrocarbons by catalytic co-pyrolysis of microalgae and polypropylene using HZSM-5. J Anal Appl Pyrol 136:178–185. https://doi.org/10.1016/j.jaap.2018.10.007
17. Abomohra AE, Sheikh HM, El-Naggar AH, Wang Q (2021) Microwave vacuum co-pyrolysis of waste plastic and seaweeds for enhanced crude bio-oil recovery: experimental and feasibility study towards industrialization. Renew Sustain Energy Rev 149:111335. https://doi.org/10.1016/j.rser.2021.111335

Production of Biofuels from Almond Hull Wastes

Aisha Albalushi and Seyed Mojtaba Sadrameli

Abstract Research on biofuels made from agricultural leftovers has accelerated due to the growing need for sustainable energy on a worldwide scale. Through a variety of thermochemical and biochemical conversion techniques, this research investigates the potential of almond shell wastes as a feasible feedstock for the generation of biofuel. Particle size and residence time are important factors in product dispersion, and in the first instance, pyrolysis of almond hulls at 450 °C. The second research produced high-value bio-liquid with little carbon loss by employing hydrothermal hydrogenation at 250 °C with Ru/CNF catalyst. In the first instance, pyrolysis of almond hulls at 450 °C produced an ideal bio-oil yield of around 45 weight percent; the distribution of the product was greatly influenced by particle size and residence time. In the second investigation, bio-liquids with a 60% conversion efficiency and no carbon loss were produced by hydrothermal hydrogenation at 250 °C using a Ru/CNF catalyst. Smaller particle sizes and longer processing times improved the antioxidant effectiveness of hull extracts, with 89% inhibitory efficiency in corrosion tests, according to the third study, which concentrated on extracting bioactive components. Up to 68.2 mg GAE/g DW in total phenolic were obtained in the fourth example, which employed ultrasound-assisted extraction for phenolic compounds. The comparison shows that the most important factors in optimizing the production of biofuel or bio-products are temperature, particle size, residence time, and catalyst selection. These results highlight the adaptability of almond hulls in energy recovery and their compatibility with the objectives of the sustainable bio-economy. This study encourages the advancement of scalable conversion methods for the value adding of almond hulls.

Keywords Biofuels · Almond hulls · Biomass conversion · Pyrolysis · Thermochemical processes · Agriculture waste · Renewable energy · Sustainable feedstock · Bio-oil · Hydrothermal carbonization · Lignocellulose biomass · Green energy · Environmental sustainability · Energy recovery · Waste valorization

A. Albalushi · S. M. Sadrameli (✉)
Department of Engineering, German University of Technology in Oman, Muscat, Oman
e-mail: seyed.sadrameli@gutech.edu.om

A. Azizi (ed.), *Mechatronic Innovations in Sustainable Energy*, Emerging Trends in Mechatronics, https://doi.org/10.1007/978-981-95-5863-6_5

1 Introduction

1.1 History and Inspiration

The world's energy industry is changing dramatically from fossil fuels to energy sources that are environment friendly. Energy derived from non-depleting source is referred to as renewable energy. They often have a minimal or zero carbon impact and are self-replenishing and natural. Wind, sunlight, bioenergy (fuel made from burning the organic materials), and hydroelectricity, including tidal energy, are a few examples of renewable energy sources [1]. Researches and governments are looking for alternative energy solutions due to the depletion of fossil fuel supplies and rising environmental pollutions as a result of climate change. Because they are inexhaustible and have the potential to lower greenhouse gas emission, biofuels which have become an appealing substitute is receiving a lot of attention. Almond hulls are an underused agricultural waste product that has a lot of potential for generating biofuel among other biomass sources [2]. The crop of almond trees is presently evolving and growing globally. One of the most well-known nut trees in the world, almonds typically thrive in arid regions, which may include times when there is a dearth of water. The international nut and dried fruit council reports that over the past 10 years, almond production has been rising. In the 2019–2020 season, it reached a maximum value of over 1.36 million metric tons, which is 26% higher than the previous 10 years' average and 7% higher than the previous season. A significant amount of hulls waste is produced by the extensive cultivation of almonds in various countries like US, Spain, Italy, and Australia [3]. With about 80% of the world's almond output, the USA continues to lead the field. Australia and Spain come in second and third, with 8% and 6% of the worldwide production, respectively. Nut biomass from hulls and shells is abundant, cheap, and accessible, which makes it appealing to valorize. Especially, almonds are the most prevalent tree nut crop in the world. It should be mentioned that a significant number of byproducts are produced concurrently with the manufacture. Therefore, more details on the biochemical makeup and usefulness of these wastes are essential for developing innovative alternatives for their valorization, which will support the "zero waste economy" while simultaneously lessening its effects.

Almond hulls, which are usually thrown away or utilized for low-value products including animal feed, can be a substantial source of feedstock for the production of biofuel. In addition to producing an alternative energy source, turning the biomass from almond hull into biofuel. It helps with waste management and preservation of the environment [4].

1.2 Biofuel's Significance for Renewable Energy

Compared to fossil fuels, biofuels provide a more ecologically sound sustainable energy source since they are made from biological resources like plants, algal blooms, and waste biomass [5]. They fall into three categories including lingo-cellulosic biomass, agricultural leftover and third generation come from non-food biomass. Third generation products come from algae and other cutting-edge bio-based sources. Particularly, biofuels from this method have drawn attention because they offer an environmental-friendly energy sources and can help to resolve the food versus fuel controversy [6]. The need for food and agriculture land is growing as the world's population rises. The need for biofuel production is heightened by environmental and energy sustainability, and it is anticipated to rise in the upcoming years. This brings up the contentious issues of food security versus biofuel. Numerous studies have demonstrated that biofuels are the primary driver of rising food cots, yet some have argued that the development of biofuel has not substantially changed food prices.

The second-generation biofuel category incorporates almond shells biomass which provide a feasible feedstock without compromising with food production. Almond hulls and other agricultural wastes can be used to cut carbon emissions and lessen reliance on traditional fossil fuels. Biofuels are a workable approach for attaining energy sustainability since they can also be included into current fuel infrastructures.

1.3 Using Biomass from Agriculture as a Feedstock

Organic materials that result from agriculture operations, such as crop remainders, fruits and nut's contaminants, and forestry waste generating agricultural biomass. Due to their high cellulose, hemicellulose, and lignin content, these materials can be utilized as biofuel [7]. Widely accessible as well as renewable agricultural biomass may be processed by various technologies. Biofuels generated by these technologies include bioethanol, biogas, and biodiesel.

Fermentable sugars, cellulose, and other organic components that can transformed into biofuels are abundant in almond husk biomass. This is a byproduct of almond business. Despite being widely available almond hulls, they have historically been underused; the majority are either used as low-value feed or wind up in landfills [8]. This project is to support waste valorization and the production of renewable energy by investigating their potential as feedstock for biofuels.

1.4 Chemical Composition of Almond Hulls and Potential for Pyrolysis

Cellulose, hemicellulose, lignin, and other organic substances make up the majority of composition of almond hulls. They are a great option for producing biofuel since they typically contain 30–40% cellulose, 20–30% hemicellulose, and 10–15% lignin [9]. They also contain a lot of essential oils, polyphenols, and fermentable sugars, which can increase the production of biofuel. These organic components facilitate pyrolysis fermentation, and hydrolysis, among other biofuel conversion processes.

The United States accounts for around 80% of the world's annual production of almonds, which is anticipated to reach around 3 million metric tons [3]. With an estimated 1.5 million metric tons of almond hulls available annually worldwide, pyrolysis offers significant potential for producing biofuel. Pyrolysis of almond hulls can produce up to 60% bio-oil, 25% syngas, and 15% biochar, depending on the process parameter [10].

1.5 Research's Targets

This research's main goal is to determine whether it is feasible to turn the biomass from almond hulls into biofuel or not. The particular goals of the research can be summarized as:

1. Analyzing the chemical makeup of biomass from almond hulls to see whether it is suitable for generating biofuel or not. Conversion of almond hulls to biofuel requires various chemical techniques and processes which are analyzed in this report.
2. Assessing various means of pretreatment to increase the amount of biofuel produced from almond hulls.
3. Investigating conversion methods for the production of biofuel, including thermochemical and biochemical processes.
4. Evaluating the environmental effects and economic feasibility of utilizing biomass from almond hulls as a feedstock for biofuels.
5. Contrasting almond hulls potential for biofuel with those of other widely utilized biomass sources like wood waste from forestry, agricultural crop, sugarcane bagasse, manure, algae, and solid waste.

1.6 Structure of Research

The research work presented has been divided into four main chapters. Section 1 gives basic overview of research involving the heading's background information, importance of biofuels and research's objectives and structure of research. Section 2 is all about literature review. This chapter includes history or pre-existing research on biofuels, biomass feedstock, technologies involved in conversion process and previous studies linked to almond hull utilization for fuel formation. Section 3 is about case studies and their results. This chapter focuses on experimental procedures, chemical techniques, and analytical technologies used in case studies. Results are also included in this chapter. Section 4 is included with title comparison and conclusion. Comparison between various methods like pyrolysis used in case studies and final results are presented in conclusion form. It further evaluates the economic and environmental ramifications and talks about how well different biofuels' conversion techniques performed. It also emphasizes the study's possible uses, and summarizes and offers suggestions for more research.

This extensive framework guarantees a careful investigation of the potential of biomass from almond hulls for the generation of biofuel, advancing the fields of sustainability and renewable energy more broadly.

2 Literature Review

2.1 Introduction

Biofuels, a new business containing global environmental energy goals that can be possible with almond shell waste. This chapter presents a comprehensive review of recent research articles that address a variety of subjects linked to the production of biofuel. Some of these subjects include feedstock characterization, conversion techniques, and economic feasibility mechanisms. The chapter emphasizes the effectiveness and environmental impact of advancements in thermochemical conversion, hydrolysis with enzyme actions, and anaerobic digestion of almond hulls waste. Research on life-cycle assessments of almond hulls, pretreatment methods, and optimizing the production of biofuels are also examined. Through an overview of prior research, this literature study explores innovation approaches to enhance biofuel production while recognizing significant challenges such as cost-effectiveness and technological limitations. At the end of this chapter, gaps in the current data are also analyzed to establish the foundation of further research. This thorough evaluation will support the creation of efficient and sustainable methods for converting waste from almond hulls into biofuels, which will help in fulfillment of energy demand of nation. Salgado et al. [11] look on the chemical composition and potential applications of almond husk biomass. The extraction of bioactive compounds and the use of biomass to produce biofuel are two methods of valorization. This study highlights how almond hulls are suitable for a variety of biotechnological applications due to the presence of high amount of lignin, carbohydrates, and bioactive compounds. Analytical techniques were used to assess the almond hulls that held considerable potential as a feedstock for high-value products like biofuels. The study emphasizes sustainable tactics, such as environmentally friendly extraction techniques. These enhance the utilization of this agricultural waste into useful working. Almond hulls may be efficiently processed to liberate important compounds and utilized as a sustainable energy source.

The authors stress in their conclusion the importance of almond husk as a potential raw material for the production of biofuel and the extraction of important bioactive chemicals. According to the study's findings, it is feasible to optimize the use of almond hulls, reduce agriculture waste, and promote green energy sources by utilizing state-of-the-art sustainable energy technology. While acknowledging that cost-effectiveness and process efficiency are among the barriers to board adoption. This paper argues that more advancements in bio-refining techniques can boost economic viability. This study highlights the potential of almond hulls as a renewable resource, which advances the boarder goal of developing sustainable and ecologically acceptable alternatives of fossil fuels from society.

Offeman et al. [12] compare the quality of carbohydrates in almond hulls from different varieties and Californian production locations to investigate the viability of almond hulls as a feedstock for the production of biofuel. Since California produces 1.65 million tons of almond hulls annually, which is 80% of the world's almonds, it is crucial to know how much fermentable sugar is in almond hulls for biofuel applications. The study focused on fermentable carbohydrates including sucrose, fructose, and glucose and examined six almond varieties from seven countries. According to the findings, fermentable sugars accounted for between 25 and 33% of the dry weight of the hulls. Notably, the nonpareil type, which accounts for around 39% of total output, has an average of 271 kg of fermentable sugars per ton of hulls. The study concludes that almond hulls, particularly the nonpareil kind, offer a lot of potential as a feedstock for the production of biofuel due to their high fermentable sugar content. Theoretical ethanol output ranged from 130 to 176 L per ton of hulls, depending on the cultivar. The study highlights challenges with the economical use of wet, discarded hulls after fermentation. It also emphasizes the importance of finding cost-effective applications for these wastes for the overall feasibility of using almond hulls in the production of biofuel.

According to the Almond Board of California [13] a number of uses including animal feed, as raw materials for bio-products, and as sources to produce bioenergy. It highlights the chemical composition of these biomass residues and their potential for usage in renewable energy applications, such as the anaerobic digestion process that produces biogas and the subsequent conversion. The study also examines the economic and environmental benefits of integrating almond waste into the state's bio-economy to reduce waste and promote sustainability.

The study concludes by highlighting the substantial potential of almond shells and hulls as useful materials in California's bio-economy. These agricultural leftovers may be converted into renewable bio-products by using cutting-edge and sustainable technology, which will support both economic expansion and environmental preservations. To improve resource sustainability and efficiency in the almond business, the research recommends further investigation and application of such valorization techniques.

Kiani et al. [14] examine the effective use of Ultrasound-Assisted Extraction (UAE) to recover Phenolic Compounds (PCs) from Almond Husk Waste (AHW). The goal of the study is to increase Total Phenolic Content (TPC) by optimizing extraction parameters such as solvent-to-AHW ratio, extraction duration, and sonication amplitude. Along with describing the extracted phenolic compounds, the study assesses their antibacterial qualities against infections and probiotic microbe interactions. The findings demonstrate that UAE is a successful technique for eliminating PCs from AWH. This can be used in food preservation and medical applications.

When processed utilizing ultrasound-assisted extraction, the author concludes by highlighting the potential of almond husk waste as a useful source of phenolic chemicals. In addition to increasing the production of phenolic compounds, the improved extraction procedure guarantees the extract's bioactivity, especially their antibacterial efficiency. By valuing agricultural byproducts, this strategy supports the practices of waste management. This advances environmentally friendly technology in both food and pharmaceutical sectors.

Remon et al. [15] examined the co-processing of Almond Hulls (AH) and FFP2 Face Masks (FM) utilizing hydrothermal liquefaction (HTL) in saltwater as a sustainable reaction medium. The study assesses how the reaction medium and feedstock composition affect the synresearch of biofuel at temperature of 350 °C and pressure of 170° bar across reaction durations varying from 20 to 180 min. To evaluate their effects on the yield and characteristics of bio-crude and hydrochar, the study investigates different binary combinations of AH and FM as well as various deionized water and seawater mixes.

Conclusion section shows that a viable method for producing biofuel is the synergistic co-valorization of almond hulls and disposable face masks in saltwater under hydrothermal conditions. The study emphasizes how saltwater may be used as sustainable reaction medium to lessen the need for freshwater resources. According to results, co-processing these waste products can result in the effective production of high energy content hydrochar and bio-crude, which will help to create adaptable, season-independent bio-refineries. This strategy not only solves the problems with disposing plastic and agricultural waste, but it also provides a means to produce energy sustainability and in line with the ideas of the circular economy.

Boulika et al. [16] examine the potential of almond hulls (AH) and almond shells (AS) as natural sources of antioxidants and corrosion inhibitors. To get extracts from AH and AS, the researchers used Soxhlet extraction using methanol and ethanol. Comparing AH extracts to AS extracts, phytochemical studies showed that the former had a more varied composition of secondary metabolites. Total polyphenols, flavonoids, and condensed tannins were measured in the research; AH extracts showed the highest amounts. DPPH and FRAP tests were used to measure antioxidant activities, and the results showed that AH extracts had better antioxidant qualities. Furthermore, mild steel having 1 molar HCl medium was used to test the extract's effectiveness in preventing corrosion. The findings showed that the effectiveness of inhibition rose with the concentration of extract, reaching maximum efficiencies of 86.46% for AS and 88.37% for AH at 200 ppm and 25 °C temperature. However, when the temperature rose, the inhibitory efficiency dropped.

The author concludes by highlighting the possibility of turning lignocellulose waste from the almond business into valuable chemicals by removing bioactive substances that act as powerful corrosion inhibitors and antioxidants. Applications needing natural antioxidants may find AH extracts suitable because of their greater phytochemical concentration and antioxidant activity. Both the AH and AS extracts' noteworthy corrosion inhibition efficacy suggests that they might be used in industrial settings as environmental benign substitutes for synthetic corrosion inhibitors. This study supports the development of green technology and sustainable waste management into demonstrating the viability of turning agricultural waste into useful items.

Chen et al. [17] examine the creation of biodegradable packing materials using groundnut shell cellulosic fibers in combination with leftover almond shell and hull material. By offering a sustainable substitute that makes use of agricultural waste, the study seeks to allay environmental worries related to plastic packaging. Almond hulls and shells are cleaned, dried, and powdered as part of the process, and then they are combined with the groundnut shell cellulosic fibers. After that, the bio-composite is transformed into packing materials, and its qualities are assessed for possible uses.

The study concludes by showing that it is feasible to blend groundnut shell cellulosic fibers with leftover almond shell and hull material to create biodegradable packaging materials. This method not only provides a long-term solution for surrounding problems caused by plastic trash, but it also gives agricultural wastes that are usually thrown away. According to the research, these bio-composites have the potential to be effective substitutes for traditional packaging materials, which would help the packaging sector become more sustainable and reduce waste level.

Offeman et al. [12] examine the feasibility of producing bioethanol from almond hulls, a byproduct of processing almonds, using a renewable resource. Almond hulls, which make up between 25 and 33% of their dry weight, are abundant in fermentable carbohydrates including glucose, fructose, and sucrose. Almond hulls' high sugar content makes them a viable option for producing bioethanol, which is in line with the rising need for renewable energy sources. To maximize the choice of feedstock for the manufacturing of bioethanol, the study looks at the variations in sugar content among several almond species and geographical areas in California.

The study concludes that because of their high fermentable sugar content, almond hulls have the potential to be good feedstock for the manufacturing of bioethanol. Targeted feedstock selection is crucial to maximizing bioethanol outputs, as evidenced by the observed differences in sugar concentrations across several almond types and geographical regions. The almond business can increase the value of its byproducts and support a more sustainable and circular economy by turning almond husks into bioethanol. This method tackles the waste management issues related to processing almonds in addition to providing an alternate energy source.

McCaffrey et al. [18] investigate the possibility of using almond processing residues especially including shells and hulls that can be used for gasification operations that produce bioenergy. Characterizing the syngas generated and evaluating the operational difficulties related to air and steam gasification techniques are included as main objectives of this study. The amount of energy and chemical composition of resultant syngas under various gasifying agents were assessed by experiments carried out utilizing a laboratory-scale fluidizing bed gasifier. Syngas composition during this process was among the important criteria examined.

The conclusion emphasizes how almond biomass, especially the shells and hulls, may be used as a practical feedstock for gasification-based bioenergy applications. The study emphasizes how crucial it is to use the right gasifying medium to maximize syngas quality and reduce operational problems such as bed agglomeration. According to the results, steam gasification could be superior than air gasification in meaning of reactor stability and syngas energy content. With the use of agricultural leftovers, these insights aid in the creation of sustainable bioenergy solutions.

Vasic et al. [19] examine the thorough analysis of the enzymatic hydrolysis method for producing bioethanol from different lignocellulosic biomasses. The author presents lignocellulosic biomass as a viable substitute for fossil fuels and highlights the growing need for renewable energy sources. They go over the cellulose, hemicellulose, and lignin structural elements of lignocellulosic materials and how they affect the effectiveness of enzymatic hydrolysis. The study emphasizes various pretreatment techniques intended to increase the accessibility of enzymes to polysaccharides, thereby increasing the efficiency of hydrolysis. The authors also look at different enzyme techniques and how new, efficient enzymes might help reduce the cost of the hydrolysis process. To evaluate the possibility for effective bioethanol production, the study also investigates the use of lignocellulosic feedstocks, such as agricultural waste, wood residues, and marine algae.

The promise of lignin-rich biomass as a practical feedstock for the manufacture of sustainable bioethanol is highlighted in the conclusion. They point out that the selection of pretreatment techniques and the creation of efficient enzyme combinations have a major impact on the effectiveness of enzymatic hydrolysis. To increase sugar yields and lower production costs, the authors recommend more study into streamlining these procedures. They also stress how crucial it is to investigate a variety of lignocellulosic sources, especially underused resources like marine algae. In this way, we can increase the durability of the manufacture of bioethanol and diversify feedstock possibilities. According to the assessment, developments in pretreatment techniques and enzyme technology are essential for the economic feasibility of lignocellulosic bioethanol, supporting the global transition to renewable energy sources.

In this study, Sillero et al. [20] proposed a thorough Life-Cycle Assessment (LCA) in order to assess the environmental effects of six different bio-refinery scenarios that aim to turn almond shell into high-value products. The cellulose conversion pathways and delignification techniques differ for each circumstance. The study evaluates environmental performance across many impact categories, such as photochemical oxidation and abiotic depletion, using experimental data processed by Semipro software.

With a high relative effect of 74% recorded in one scenario, the results show that the enzyme hydrolysis stage has the greatest environmental impact, independent of the delignification technique used. Furthermore, it was discovered that organosolve delignification procedure had more significant adverse impacts than alkaline treatments, especially in the areas of abiotic deficiency and photochemical oxidation. The study finds that auto-hydrolysis stage is the best bio-refinery pathway for valorizing almond shells. This process produces oligosaccharides, lignin, and nanocrystals among goods with value.

The study by Idowu [21] explores the value addition of leftover cassava peel and almond hull utilizing microwave-assisted processing methods with an emphasis on producing fermentable sugars, bio-oil, and micro-fibrillated cellulose (MFCs).. It investigates how well microwave-assisted pyrolysis breaks down almond husk biomass to produce syngas, charcoal, and bio-oil. To improve the quantity and quality of bio-oil, the process parameters such as temperature, processing time, and microwave power are adjusted. To evaluate the bio-oil's potential as a chemical feedstock or renewable energy source, its chemical makeup is examined.

The results show that turning almond husks into bio-oil via microwave-assisted pyrolysis is a successful process, which gives notable yields under ideal circumstances. The diverse blend of hydrocarbons in the bio-oil suggests that it may be used as an intermediate for chemical synresearch or as a sustainable fuel. Rapid heating and effective energy transfer are two benefits of microwave-assisted pyrolysis that the study emphasizes and which raise the process's general effectiveness. These findings provide credence to the viability of using leftover almond hulls to produce sustainable bio-oil, supporting efforts for utilization of waste and renewable energy.

In the context of California's bio-economy, the paper proposed by Almond board of California [22] examines the several uses of almond biomass, especially hulls and shells. It draws attention to the high levels of cellulose, lignin, and hemicellulose found in almond hulls, which are relevant to the methods used to produce biofuel. The viability of using almond hulls in anaerobic digestion systems to generate biogas which may then be converted to compressed natural gas (CNG) for onsite use. It is covered in the document. The high moisture content of hulls and the expense of drying them for effective digestion are other issues covered in this research.

The result indicates that although almond hulls include the components needed for anaerobic digestion-based biogas production, practical issues such as moisture content and related drying expenses must be resolved to maximize the process. In order to support California's renewable energy goals, the research highlights the significance of creating affordable pretreatment techniques to improve the feasibility of almond hulls as an environmentally friendly substrate for biogas generation.

In order to improve the production of biofuel, Swami et al. [23] in their study explored the potential of ionic liquids as environmentally friendly solvents for the processing of lignocellulosic biomass. By using ionic liquids to break down the complex and hard polymeric structure of lignocellulosic biomass, the research aims to overcome its resistance and increase its susceptibility to enzymatic hydrolysis. The paper offers a thorough analysis of several pretreatment techniques, highlighting the special qualities of ionic liquids like their capacity to be recycled and reused that set them apart from more conventional pretreatment techniques. Additionally examined are the critical factors controlling the biomass dissolving process and the potentially dangerous characteristics of ionic liquids.

According to the results, ionic liquids have a lot of potential as environmentally friendly solvents for processing lignocellulosic biomass, which may be used for almond hulls. Ionic liquids are appealing substitutes for conventional pretreatment techniques due to their special qualities, which include their capacity to break down the lignocellulosic matrix as well as their capacity to recycle. To fully utilize the possibilities in biomass pretreatment processes, however, issues like the potentially harmful characteristics of some ionic liquids and the requirement for economical recycling techniques must be resolved. In order to promote sustainable biofuel production, future research should concentrate on creating eco-friendly, efficient, and effective bio-energy.

This paper proposed by Dele-Afolabi et al. [24] does a bibliometric analysis in order to examine research trends and advancements in the valuation of shell-based crop residue biomass for the generation of biofuel with the circular bio-economy. Using data from the World Wide Web of knowledge and VOSviewer for academic mapping, the analysis covers the time frame of 1997–2023. An analysis of 1333 research publications throughout time showed significant changes in the focus of study, from gasification to enzyme-driven hydrolysis and lignocellulosic biomass combustion, and from pyrolysis and biomass power. The study emphasizes how important pretreatment procedures like pyrolysis, torrefaction, and combustion are to increase the yield of biofuel from shell-based agricultural byproducts such as almond hulls.

The results highlight the substantial potential of agricultural wastes derived from shells, including almond hulls, as sustainable feedstocks for the manufacturing of biofuel. A rising interest in pretreatment process optimizing to increase biofuel production is reflected in the evolution of research focus during the period under analysis. In order to effectively use the potential of these biomass resources and support the development of the circular bio-economy, the study highlights the significance of ongoing research and development in this field.

In this study, Monroe [25] investigated the use of almond hulls, a plentiful agricultural waste product, as a substrate to produce protein-rich fungal biomass appropriate for animal feed additives. Finding the best fungal strains and fermentation conditions to increase the protein content of almond hulls is the main goal of the study. Several filamentous fungi were screened using solid-state fermentation (SSF), and myceliophthora thermophila proved to be the most successful strain, with a maximum crude protein concentration of 18.10% in the fermentation residue. The ideal parameters for fermentation were a solid loading 6.90% (w/w), a fermentation length of 96 h, and an almond hull particle size of 75–250 μm. The fermentation residue that resulted showed promise as a protein-rich addition for chicken feed due to its full amino acid profile and high in vitro digestibility.

Through SSF, Monroe's research shows that it is feasible to turn almond hulls into protein-enriched fungal biomass, providing a sustainable method of repurposing this agricultural waste. The study emphasizes how crucial it is to choose the right fungal strains and adjust fermentation conditions to optimize digestibility and protein output. Although the fermentation byproduct has potential as a supplement for chicken feed, its high fiber concentration would prevent direct use in aquatic diets, requiring further processing to extract proteins. This study emphasizes how fungal bioconversion may be used to turn low-value biomass into lucrative products, supporting waste management and sustainable farming methods.

El Achaby et al. [26] examined the structural morphology and chemical makeup of almond hulls and shells. The goal of the study is to assess these byproducts' potential for a range of uses, such as the manufacturing of bio-fuel. According to the results, almond hulls have a high concentration of fermentable sugars that may be used to produce bioethanol, with 60.24% extractives and 37.67% holocellulose.

Almond shells, on the other hand, are more suited for combustion-based fuels like charcoal because they contain more lignin. The study highlights the special qualities of almond byproducts that can be used for certain biofuel applications by contrasting these findings with those of other agricultural leftovers.

El Achaby concludes that because of their high fermentable sugar content, almond byproducts, especially hulls, have a great deal of potential for the manufacture of biofuel. The study highlights how crucial it is to choose the right conversion technologies depending on the unique chemical makeup of each byproduct. For example, almond hulls are excellent candidates for fermentation-based bioethanol production due to their high extraction and holocellulose content. On the other hand, lignin-rich almond shells are more suitable for biochar production by thermochemical conversion techniques like pyrolysis. To maximize biofuel outputs and advance sustainable waste management techniques, this comparative research emphasizes the necessity of customized valorization strategies for various agricultural leftovers.

Remon et al. [27] examined the viability of employing hydrothermal treatment (HTT) to construct a bio-refinery concept based on almond hulls. Variations in temperature (200–300 °C), pressure (100–180 bar), time (20–180 min), and solid loading (5 to 25 weight percent) were carefully investigated in relation to the conversion of almond hulls into four main products: gas (2–13%), bio-oil (2–12%), aqueous phase (4–69%), and hydrochar (17–89%). A combination of hydrogen, carbon dioxide, carbon monoxide, and methane made up the gaseous product, which had a lower heating value between 1 and 13 MJ/m^3. Alkane, aldehydes, ketones, phenol, furans, benzenes, and nitrogen compounds were all present in the bio-oil which had a heating value of 21–31 MJ/kg. The aqueous phase included important molecules such as saccharides and tiny oxygenated compounds, whereas the hydrochar showed up as an energy dense substance with an HHV of 21–31 MJ/kg.

According to the study's findings, almond hulls might be efficiently treated hydrothermally in a bio-refinery to provide a number of useful chemicals. The process was optimized to convert up to 43% of the original feedstock into energy-rich bio-oil and hydrochar at 256 °C, 100 bar, 5 weight percent solid loading, and 157 min. Almond hulls have the potential to be sustainable feedstock value-added chemicals and biofuels, according to this study.

2.2 *Yields of the Biofuels Produced in the Review*

The analysis of producing biofuel from leftover almond hulls shows encouraging results using a range of conversion techniques. Research shows that depending on pretreatment methods and fermentation efficiency, bioethanol production yields might vary from 20 to 35% (w/w). Transesterification of extracted almond husk oils has been shown to provide biodiesel yields with conversion efficiencies ranging from 85 to 95%. Furthermore, the yields of syngas and biochar from pyrolysis vary from 20 to 40% and 30 to 50%, respectively. These findings show that almond hulls have the potential to be a sustainable feedstock biofuel.

The examined literature emphasizes major developments in the use of almond hulls for the generation of biofuel, with a focus on several conversion methods such as anaerobic digestion, thermochemical processing, and enzymatic hydrolysis. Because of their excellent chemical characteristics and carbohydrate rich content, studies have shown that almond hulls have the potential to be a sustainable feedstock. Nonetheless, a number of gaps in the literature indicate that more investigation is required to completely comprehend the viability and optimization of producing ethanol from this biomass source.

The dearth of thorough investigations comparing various biofuel conversion methods under consistent experiment circumstances is one of the main gaps found. Finding the most effective and saleable strategy is difficult since the majority of research concentrates on distinct approaches without conducting a direct comparison. Furthermore, although the chemical characterization of almond hulls has been the subject of several research, few have investigated the possibility of large-scale manufacturing, long-term sustainability, and economic viability. The practical implementation of these findings is limited by the lack of pilot-scale or industrial-scale investigations.

The dearth of studies of the life-cycle assessment and environmental effects of producing biofuel from almond hulls is another gap. There is little mention of carbon emissions, waste management, and other ecological implications, despite the fact that several researches focus on energy output and conversion efficiency. Concerns regarding commercialization and market acceptance are also raised by the under-studies' state of industrial integration and governmental framework. More research is needed to determine how almond hull biofuel fit into the current energy infrastructure and how competitive they are with other bioenergy feedstock by filling these gaps via multidisciplinary research and pilot-scale applications.

3 Case Studies and Results

3.1 Introduction

Biofuel production from biomass has drawn a lot of interest as a sustainable substitute for fossil fuels. Because of their high lignocellulosic content, almond hulls and shells are a suitable feedstock among other biomass sources. One method that works well for turning almond shells into biofuels is pyrolysis, a thermochemical process that breaks down organic material at high temperatures without the presence of oxygen. The pyrolysis process's efficiency and yield are influenced by a number of critical variables, including temperature, residence time, heating rate, and particle size.

A thorough analysis of case studies concentrating on the pyrolysis of almond hulls for the generation of biofuel is provided in this chapter. The chosen studies shed light on experimental procedures, process condition optimization, and the effects of various factors on the yields of bio-oil, biochar, and syngas. Understanding how temperature, catalyst usage, and the reactor types affect product dispersion is given special attention. These studies' findings which include tabular and graphical data are examined to identify patterns and variances in biofuel production. This chapter seeks to advance knowledge of the viability and effectiveness of almond shell pyrolysis as a feasible biofuel production process by assessing these findings.

Case Study 1

Title: "Comprehensive Investigation of Almond Shells Pyrolysis Using Advanced Techniques"

Gupta, A., Patel, P., and Shah (2024).

3.2 Material Used in Experiment

Gupta et al. [28] used almond shells as the main biomass feedstock in their investigation on the pyrolysis of almond shells. To guarantee uniformity in the experimental setting, the almond shells were obtained from nearby agricultural fields and went through a number of preparation procedures. To get rid of any contaminants including dust, grime, and leftover almond kernels, the gathered shells were first cleansed. To lower the moisture content and improve the effectiveness of pyrolysis, the shells were cleaned and then dried for 24 h at 105 °C in a forced air oven.

The dried almond shells were crushed and sieved into various size fractions, usually ranging from 0.5 mm to 2 mm, in order to examine the effect of particle size on pyrolysis efficiency. Prior to research, the processed biomass was kept in an airtight container to stop moisture absorption.

The pyrolysis products were characterized using a number of chemical reagents in addition to the biomass, some of these chemicals regents are given in Table 1. Bio-oil was extracted and purified using solvents like hexane and dichloromethane, and samples were prepared using deionized water. Standard gases, such as helium and nitrogen, were utilized to make the pyrolysis reactor inert for gas analysis. In certain experimental runs, analytical-grade catalysts including zeolite and zinc oxide (ZnO) were also used to examine their impact on the composition and yields of biofuel. The dependability and repeatability of the pyrolysis tests carried out in this investigation were guaranteed by these carefully chosen materials. An organized summary of the supplies and substances used in this experiment is given in this table.

3.3 Methods and Experimental System Used

Figure 1 shows a fixed bed reactor system, which guarantees exact control over operating parameters, used for the pyrolysis studies. A stainless steel tube reactor housed inside an electric furnace with temperature control was part of the system

Table 1 Case study; materials and chemical used [28]

Category	Materials/Chemicals	Chemical structures	Purpose
Biomass feedstock	Almond shells	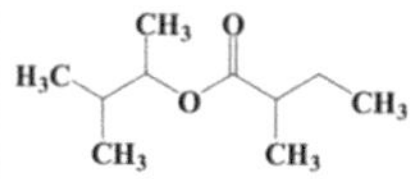	Raw material
Catalysts	Zinc oxide (ZnO)	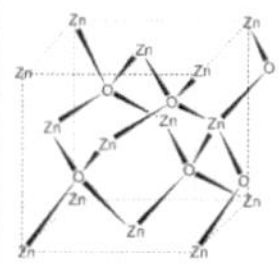	Impact of catalysis on the composition and production
	Zeolite	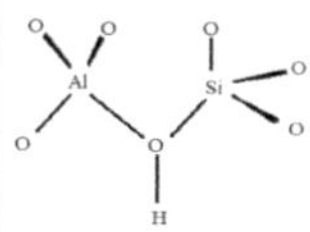	Potential to improve the efficiency of pyrolysis
Gases	Nitrogen	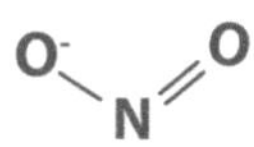	Inert carrier gas
	Helium	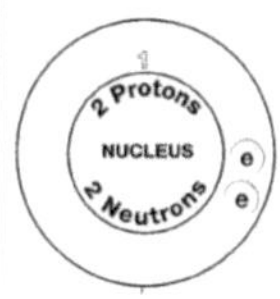	Used as a carrier gas for gas chromatography
Solvents	Hexane	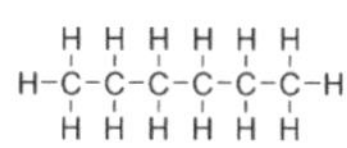	Used to extract and purify bio-oil
	Dichloromethane	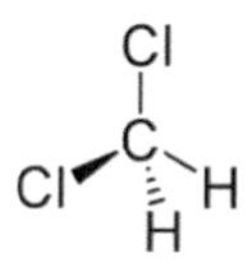	A solvent for removing the components of bio-oil
Chemical for analysis	Acetone	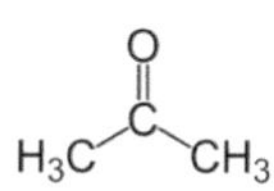	Preparing the sample and cleaning
	Deionized water	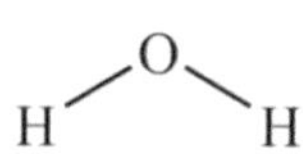	Used in analytical and sample preparation processes

(continued)

Table 1 (continued)

Category	Materials/Chemicals	Chemical structures	Purpose
Analytical standards and reagents	Analytical standards		Used in chromatographic studies for calibration and quantification
	Potassium bromide	K^+ Br^- KBr	Used in the production of pellets using FTIR analysis

[29]. To keep the environment inert and stop unintended oxidation processes, nitrogen gas was delivered at a steady flow rate. To investigate the impact on product yields, the reactor's temperature was adjusted between 300 and 700 °C. The reactor was filled with biomass samples, each weighing around 20 g, and the heating rate was regulated at 10 °C/min. To guarantee full pyrolysis, the residence period was kept at the desired temperature for 30 min. A fixed bed reactor system is shown in Fig. 1.

To calculate yield, the solid residence (biochar) from pyrolysis was collected and weighed. A succession of condensers chilled by an ice water combination was used to collect condensable vapors, which led to the collection of bio-oil. Gas sample bags were used to collect non-condensable gases for further examination. Gas Chromatography Mass Spectrometry (GC–MS) was used to identify and measure the chemical components of the bio-oil as shown in Fig. 2.

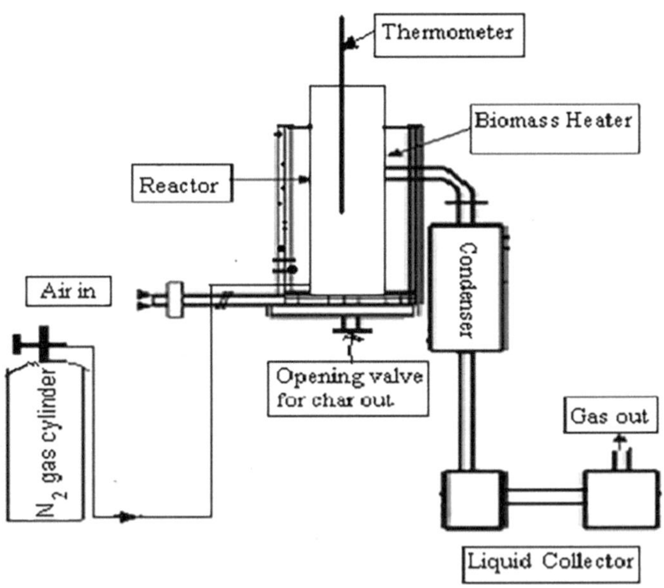

Fig. 1 Fixed bed reactor pyrolysis setup [29]

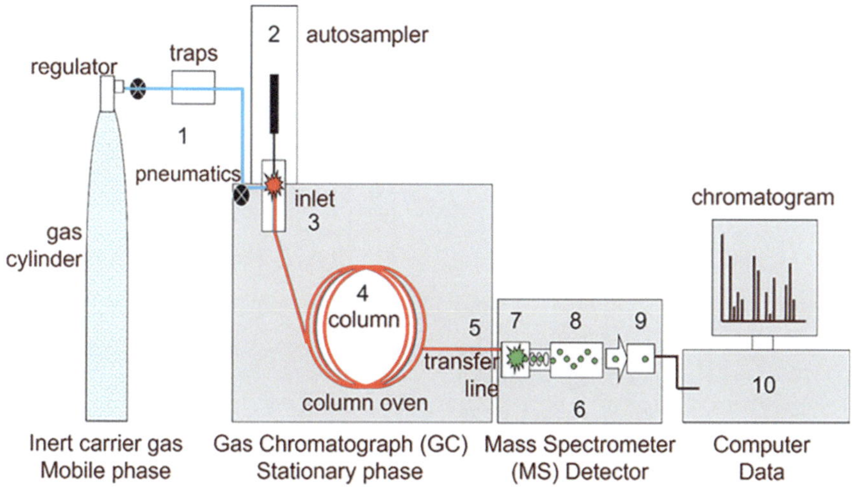

Fig. 2 Bio-oil treatment with gas chromatography mass spectrometry [30]

Brunauer–Emmett–Teller (BET) analysis was used to examine the surface area and pore structure of the biochar, while CHNS (carbon, hydrogen, nitrogen, and sulfur) analysis was used to ascertain its elemental makeup. Gas chromatography with a Thermal Conductivity Detector (GC TCD) was used to evaluate the gaseous products and quantify elements such as CO_2, CO, CH_4, and H_2 [30]. The behavior of almond shells' heat deterioration was investigated using thermogravimetric (TGA). In a nitrogen environment, samples were heated at several speeds as 5, 10, 15, and 20 °C/min from ambient temperature to 800 °C. The activation energy and reaction mechanisms controlling the pyrolysis process were revealed by applying the Coats-Redfern technique to compute kinetic parameters using the TGA data.

Through the use of zinc oxide and zeolite as catalysts, catalytic pyrolysis experiments (shown in Fig. 3) were carried out to improve the quality of bio-oil and change the distribution of the product. Prior to pyrolysis, the catalysts were physically mixed with almond shell particles in 1:10 ratio, and their impact on the composition of pyrolysis products was regularly assessed.

The impacts of temperature, particle size, heating rate, and catalyst present on the yield and composition of biochar, bio-oil, and syngas were evaluated by analyzing the experimental data. The pyrolysis settings were optimized for optimal bio-oil output using Reponses Surface Methodology (RSM), taking into account the interaction impacts of the factors under study.

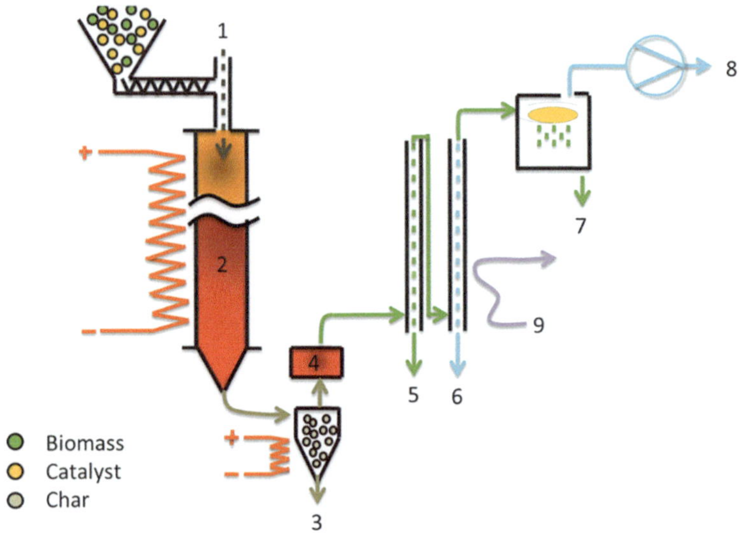

Fig. 3 Zinc oxide and zeolite as catalysts, catalytic pyrolysis experiments. 1. Carrier gas, 2. Pyrolysis reactor, 3. Solids recovery, 4. Hot particle filter elements, 5.6.7. Liquid collection, 8. Gas to analysis unit, 9. Coolant [31]

3.4 Results

The yield and content of pyrolysis products produced from almond shells are strongly influenced by temperature. Pyrolysis tests between 300 and 700 °C showed clear patterns in the formation of bio-oil, charcoal, and syngas. Due to partial thermal decomposition of biomass, the output of biochar was very high at lower temperatures (between 300 and 400 °C) (around 35% at 300 °C). However, because solid carbon decomposed more readily into volatiles higher temperatures, the production of biochar rapidly decreased, reaching around 15% at 700 °C [28]. Effect of temperature on pyrolysis yield of almond shells is given in Table 2.

Figure 4 supports the choice of ideal pyrolysis settings for certain biofuel applications by clearly demonstrating the shift in product distribution with increasing temperature.

Table 2 Effect of temperature on biofuel yield [28]

Temperature (°C)	Bio-oil yield (%)	Biochar yield (%)	Syngas yield (%)
300	25	35	20
400	35	28	25
500	46	22	30
600	38	18	38
700	30	15	40

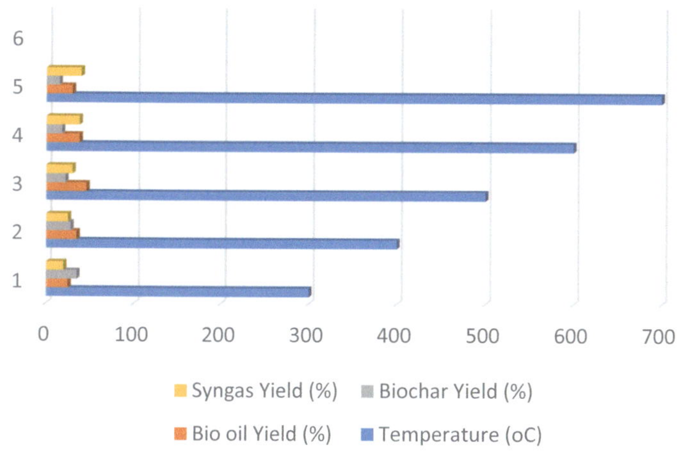

Fig. 4 Effect of temperature on biofuel yield [28]

The yield content of pyrolysis products is mostly determined by residence time and particle size. Smaller particle sizes improve heat transfer efficiency, which speeds up pyrolysis processes and increases the output of bio-oil. An excessively high particle size causes uneven heat penetration, which leads to partial thermal decomposition and a rise in the production of biochar. Researches show that whereas bigger particles (above 2mm) tend to retain more solid residue because of slower internal heating, smaller particles (less than 1mm) encourage the synresearch of bio-oil [28]. This is due to the fact that smaller particles offer a greater surface area-to-volume ratio, which facilitates effective volatile release and heat absorption.

Product distribution also greatly impacted by residence time, which is the amount of time biomass vapors stay in the reactor. Because they avoid further cracking events that turn liquid intermediates into syngas, shorter residence periods (< 2 s) are ideal for optimizing bio-oil output [28]. However, because extended temperature exposure causes bigger organic molecules to break down into gaseous components, longer residence periods (over 5 s) encourage the creation of syngas times and promote the development of coke, which lowers overall efficiency of biofuel. Observations of particle size and residence time are represented in Table 3.

Graphical presentation of these effects is provided in Fig. 5.

Thus, whereas bigger particles and longer residence encourage the generation of char and syngas, the ideal combination of smaller particle size (< 1 mm) and moderate residence time (2–3 s) improves the synresearch of bio-oil [28]. These results are essential for maximizing the production of biofuel from almond shells by adjusting the pyrolysis settings.

Heat transmission happens slowly at low heating rates, giving secondary processes like carbonization and polymerization enough time to improve the synresearch of biochar. As a consequence, less bio-oil is produced and a greater yield of biochar (~35%) is obtained. Furthermore, a slower rate of heating makes it easier to manage

Table 3 Effect of particle size and residence time on biofuel yield [28]

Particle size (mm)	Residence time (s)	Bio-oil yield (%)	Biochar yield (%)	Syngas yield (%)
< 0.5	1	50	18	32
< 0.5	3	48	20	32
< 0.5	5	42	23	35
0.5–1.0	1	48	19	33
0.5–1.0	3	45	22	33
0.5–1.0	5	40	25	36
1.0–2.0	1	40	22	37
1.0–2.0	3	38	25	38
1.0–2.0	5	32	30	37
> 2.0	1	35	28	38
> 2.0	3	30	32	38
> 2.0	5	25	35	40

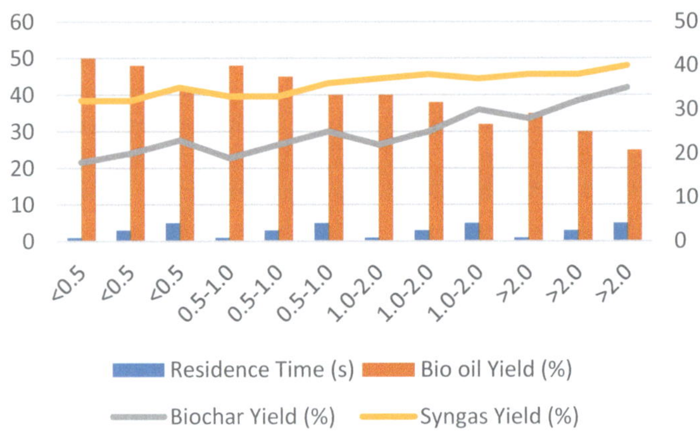

Fig. 5 Effect of particle size and residence time on biofuel yield [28]

the release of volatile components, which promotes more uniform breakdown. On the other hand, rapid temperature increases at high heating rates encourage quick volatilization, which increases the generation of bio-oil (~ 50%). Rapid heating enhances the release of volatile materials and reduces char formation by limiting further cracking reactions. Excessive heating rates, however, may cause gasification effects, which would increase the output of syngas (~ 40%) at the expense of bio-oil. Effect of heating rate is shown in Table 4 and graphical form is shown in Fig. 6.

According to these findings, the ideal heating rate for producing bio-oil is between 20 and 50 °C/min, which balances the conflicting pyrolysis processes.

Table 4 Effect of heating rate on biofuel yield [28]

Heating rate (°C/min)	Bio-oil yield (%)	Biochar yield (%)	Syngas yield (%)
5	35	40	25
10	40	35	25
20	45	30	25
50	50	25	25
100	42	20	38

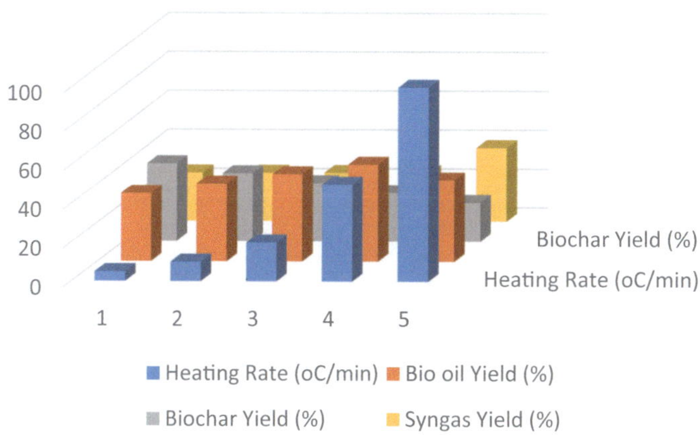

Fig. 6 Effect of heating rate on biofuel yield [28]

Case Study 2

Title: "Pyrolysis of Almond Shells: Kinetic Analysis and Bioenergy Potential"

Martinez, J.D., Ramos, J., and Gomez, J. (2021).

3.5 *Material Used in Experiment*

The kinetic behavior and bioenergy potential of pyrolysis of almond shells were examined in the work by Martinez et al. [32]. Almond shells, gathered from nearby processing firms, served as the experimental's main raw material. Before being subjected to pyrolysis, these shells were completely dried to eliminate any remaining moisture. Contaminants including dirt, oils, and other organic residues that can impede heat breakdown were removed from the shells as part of the study. The researches employed proximal and ultimate analytical techniques to examine the almond shells' chemical makeup. While the ultimate analysis identified the elemental composition, including the amount of carbon, hydrogen, oxygen, nitrogen, and sulfur, the proximate analysis examined the amount of moisture, volatile matter, fixed carbon, and ash.

A thermogravimetric analyzer (TGA) was used for kinetic analysis in order to examine the biomass's breakdown at various heating speeds. To increase the production of bio-oil and boost the effectiveness of the pyrolysis process, the study also used a variety of catalysts, including zeolites and metal oxides (Fe_2O_3 and CaO). Furthermore, in order to minimize undesirable oxidation processes during pyrolysis, an inert atmosphere was created using carrier gases such as nitrogen. For further characterization using gas chromatography mass spectrometry (GC–MS), Fourier Transform Infrared Spectroscopy (FTIR), and elemental analysis, the researchers additionally gathered samples of bio-oil, biochar, and syngas. A detailed list of all used material, chemical, and reagent used in this case study is given in Table 5.

The components utilized in the pyrolysis experiment for almond shells, such as the biomass, feedstock, catalysts, analytical tools, and end pyrolysis products, are all listed in detail in this table.

Table 5 List of materials, techniques, and chemicals used [32]

Experimental material used	Working	Descriptions
Almond shells	Pyrolysis main biomass feedstock	Gathered from local processing industries
Nitrogen gas	Pyrolysis gas carrier	Necessary for inert atmosphere and avoid oxidation
Zeolite	Catalyst in pyrolysis process	Increase bio-oil productivity along with improved sensitivity of product
Iron oxide (Fe_2O_3)	Decomposition catalyst	Promotes secondary cracking process
Calcium oxide (CaO)	Tar reduction catalyst	Improving the quality of final product
Thermogravimetric analyzer (TGA)	Kinetic analysis used for biomass decomposition	Weight loss measurement at different heating rates
Fourier transform infrared spectroscopy (FTIR)	Act as chemical characterization	Functional group's identification in bio-oil and biochar
Gas chromatography mass spectrometry (GC–MS)	Analysis technique for bio-oil composition	Chemical compounds determiner present in bio-oil
Elemental analyzer	Analyzer for biochar and bio-oil	Used for measurement of C, H, O, N, and S
Muffle furnace	Exterminator for ash content	Used quantify fixed carbon and burn leftover biochar
Moisture analyzer	Measurement of moisture content	Establish the almond shells' initial water content
Bio-oil, biochar, and syngas	Products of pyrolysis process	Gathered and examined for potential energy

3.6 Methods and Experimental System Used

Almond shells were purchased from nearby processing facilities, washed to get rid of contaminations, and then dried for 24 h at 105 °C to get rid of any remaining moisture [33]. Prior to pyrolysis, the dry shells were crushed and sieved to achieve consistent particle sizes between 0.2 and 2 mm. For the purpose of avoiding oxidation and undesirable secondary reactions, the pyrolysis studies were conducted in a fixed bed reactor with an inert nitrogen environment as shown in Fig. 7.

In order to examine the kinetic behavior of biomass breakdown, the working temperature was adjusted between 300 and 700 °C, with heating rates at 5, 10, and 20 °C/min. To ascertain its impact on product distribution, the biomass's residence period within the reactor was varied from 5 to 30 min. Syngas was evacuated, biochar was gathered for further examination, and a condensation unit was part of the reactor system to extract bio-oil.

The behavior of almond shells' heat deterioration was investigated using a thermogravimetric analyzer (TGA). By examining weight loss data at various temperatures and heating rates, the Redfern technique was used to ascertain the kinetic characteristics, such as activation energy and reaction order. Several analytical methods were used to characterize the pyrolysis products that were gathered. Gas Chromatography Mass Spectrometry (GC–MS) was used to identify the main chemical components in bio-oil and determine its composition as shown in Fig. 8. Biochar was analyzed using Fourier Transform Infrared Spectroscopy (FTIR), which revealed information on its elemental makeup and functional groups. The amount of carbon, hydrogen, oxygen, nitrogen, and sulfur in the solid residue was also measured using elemental analysis. CO, CO_2, H_2, and CH_4 concentrations in syngas were measured using a gas analyzer.

The thorough assessment of the effects of temperature, heating rate, particle size, and residence time on the composition and production of biofuels was made possible

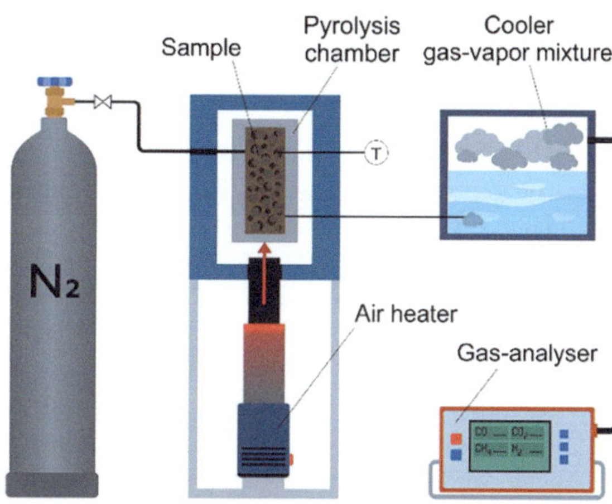

Fig. 7 Fixed bed reactor with N_2 gas experimental setup [33]

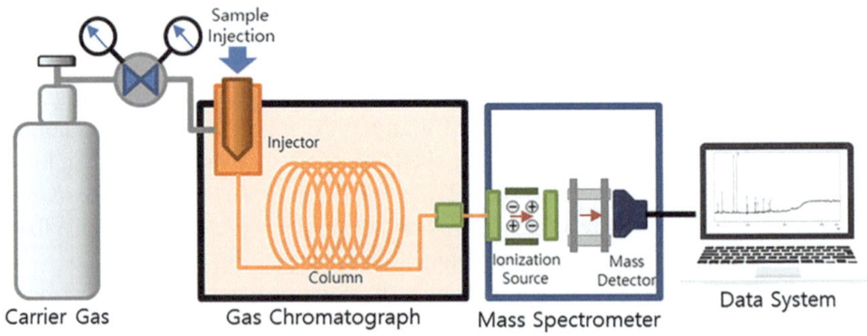

Fig. 8 Gas chromatography mass spectrometry (GC–MS) arrangement [34]

by this experimental configuration. The study enhanced almond shell's potential as a sustainable bioenergy source by offering insightful information on how to best optimize pyrolysis conditions to enhance biofuel generation.

3.7 Results

The yield of biofuels produced during pyrolysis process is significantly influenced by temperature. Product's distribution changes dramatically with increasing temperature. Because of partial thermal breakdown, the production of biochar is larger at lower temperatures (300–400 °C), but the yield of bio-oil is still relatively low. Due to improved thermal cracking of biomass components, bio-oil production peaks as the temperature rises to an ideal range (500–600 °C). Secondary processes like gasification and deoxygenation cause the yield of bio-oil to decrease and output of syngas to rise over 600 °C. The study found that while higher temperatures promoted the synresearch of syngas, bio-oil output peaked at about 550 °C [32]. This pattern emphasizes how crucial it is to adjust the pyrolysis temperature in order to efficiently balance the production and composition of bio-oil. Tabular form of this effect is described in Table 6. Graphical representation of this effect is shown in Fig. 9.

Because particle size affects mass diffusion and heat transfer, it has a significant impact on the pyrolysis process. The study found that the ideal particle size for maximizing bio-oil production was between 0.5 and 1.0 mm. In addition to this range, biochar production is boosted while too much small particles cause secondary cracking which produce syngas [32]. Smaller particles provide a larger surface area for heat absorption, which speeds up decomposition and increase bio-oil yield. This effect is illustrated through tabular (Table 7) and graphical form (Fig. 10).

The duration of residence has an impact on the degree of secondary reactions and thermal deterioration. While longer residence durations boost syngas production through greater decomposition, shorter residence times promote bio-oil generation

Table 6 Effect of temperature on biofuel yield [32]

Temperature (°C)	Bio-oil yield (%)	Biochar yield (%)	Syngas yield (%)
300	45	30	25
350	42	35	23
400	38	40	22
450	34	45	21
500	30	48	22
550	27	49	24
600	25	42	33
650	22	38	40
700	20	35	45

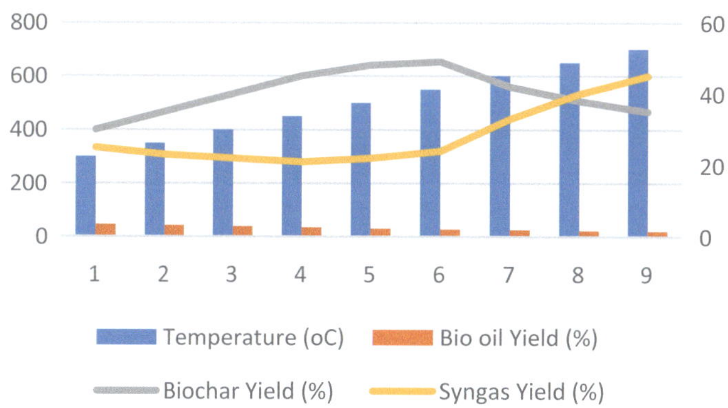

Fig. 9 Effect of temperature on biofuel yield [32]

Table 7 Effect of particle size on biofuel yield [32]

Particle size (mm)	Bio-oil yield (%)	Biochar yield (%)	Syngas yield (%)
< 0.2	25	35	40
0.2–0.5	22	42	36
0.5–1.0	20	48	32
1.0–2.0	28	40	32
> 2.0	35	32	33

by reducing secondary cracking. According to the study, 10–15 min was the ideal residence period to maximize the output of bio-oil. Beyond this, too much heat resulted in more gas production and less liquid output. During pyrolysis, the distribution of the product is greatly influenced by the rate of heating. By encouraging quick thermal breakdown and reducing char formation, higher heating rates (> 20°C/min) enhance the generation of bio-oil [32]. Prolonged pyrolysis at lower heating rates (< 5°C/

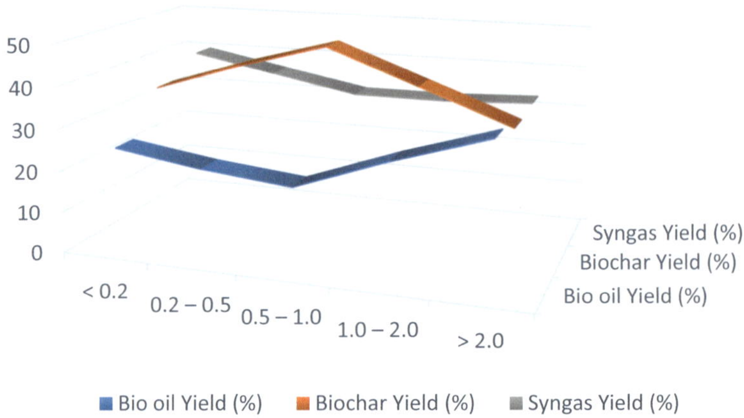

Fig. 10 Effect of particle size on biofuel yield [32]

min) produces more charcoal and excessive carbonization. The ideal heating rate for optimizing the synresearch of bio-oil was determined by the study to be between 15 and 20 °C/min. Tabular (Table 8) and graphical representation (Fig. 11) is shown.

The prospective use of almond shells as a renewable source of biofuel is highlighted by this study. The development of renewable energy may be supported by

Table 8 Effect of heating rate and residence time on biofuel yield [32]

Heating rate (°C/min)	Residence time (min)	Bio-oil yield (%)	Biochar yield (%)	Syngas yield (%)
5	5	40	32	28
5	10	38	35	27
5	15	35	33	32
10	5	35	40	25
10	10	32	42	26
10	15	30	40	30
15	5	30	45	25
15	10	28	48	24
15	15	26	47	27
20	5	28	44	28
20	10	26	47	27
20	15	24	45	31
25	5	25	40	35
25	10	22	38	40
25	15	20	35	45

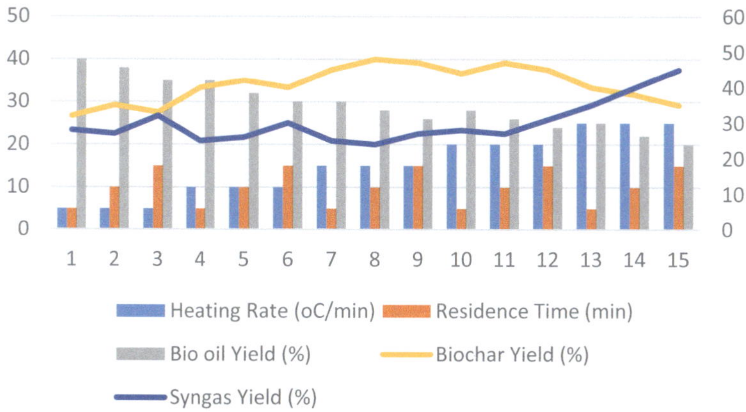

Fig. 11 Effect of heating rate and residence time [32]

maximizing bio-oil outputs through the optimization of pyrolysis settings. Catalyst impacts and scaling up procedures should be investigated in future research to improve biofuel efficiency and economic feasibility in the field of renewable energy.

Case Study 3

Title: "Hydrothermal Carbonization Coupled with Fast Pyrolysis of Almond Shells; Valorization and Production of Valuable Chemicals"

Fernandez, A., Lopez, G., and Olazar, M. (2023).

3.8 Material Used in Experiment

Almond shells were used as the main biomass feedstock for hydrothermal carbonization (HTC) as shown in Fig. 12 and rapid pyrolysis in the work by Fernandez et al. [35]. A local agricultural provider provided the almond shells, guaranteeing a representative sample of lignocellulosic waste. In order to get a constant moisture content, the raw biomass was washed to get rid of dirt and foreign objects before being dried for 24 h at 105 °C. In order to acquire various particle size fractions for controlled testing, the dried almond shells were then ground and sieved. Deionized water was employed as the reaction medium for the hydrothermal carbonization process, which allowed the biomass components to be broken down thermally. HTC used a stainless steel batch reactor that ran under autogenous pressure at temperatures between 180 and 250 °C [36].

Hydrochar, a solid carbon-rich substance produced by the method, was subsequently exposed to rapid pyrolysis. During the rapid pyrolysis step, catalysts such as HZSMS-5 and γ-Al_2O_3 were used to improve the selective synresearch of important

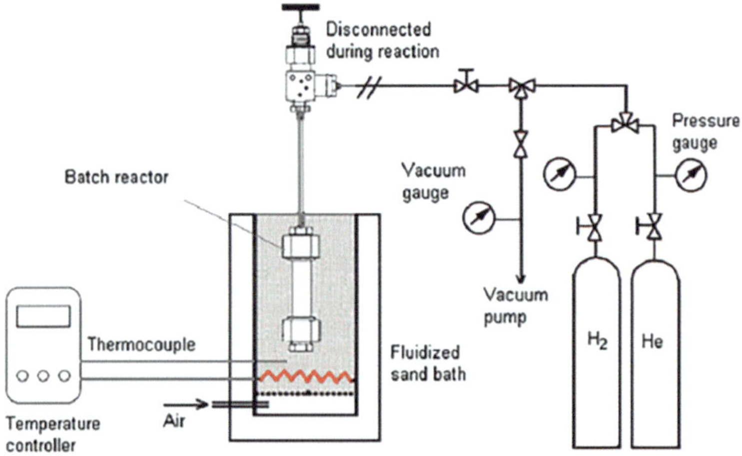

Fig. 12 HTC stainless steel batch reactor experimental setup [36]

compounds [35]. Complex organic molecules were broken down into simpler hydrocarbons with the help of the catalysts. In order to avoid oxidation during pyrolysis, nitrogen gas was also employed as an inert carrier gas.

Liquid bio-oil fractions were extracted and characterized using a variety of solvents, such as ethanol and acetone. For chemical analysis and pH corrections, analytical reagents including potassium hydroxide (KOH) and hydrochloric acid (HCl) were employed. All of these materials, chemicals, and reagents are listed in Table 9. These resources worked together to guarantee a methodical assessment of the pyrolysis and HTC processes for efficient biomass valorization.

3.9 Methods and Experimental System Used

The study used rapid pyrolysis and hydrothermal carbonization (HTC) to turn almond shells into useful chemicals and biofuel [36]. Almond shells were first pretreated in the experiment by being washed, dried for 24 h at 105 °C, then ground, and sieved to produce a consistent particle size of 0.5–2 mm. Before being processed further, the ready biomass was kept in a controlled atmosphere to avoid absorbing moisture. Under autogenous pressure, hydrothermal carbonization was conducted for 3–6 h at temperatures between 180 and 250 °C in a high-pressure reactor. To start thermal decomposition, the biomass was combined with deionized water in a 1:10 biomass-to-water ratio and heated. The flow diagram of whole process is shown in Fig. 13. This method yielded gaseous byproducts, mostly CO_2, and hydrochar as a solid product. Following cooling, it is filtered out, carefully cleaned, and dried to eliminate any remaining moisture before being subjected to pyrolysis.

Table 9 Detailed list of material and chemical used [35]

Category	Items	Descriptions
Materials	Almond shells	Primary feedstock raw material
	Deionized water	Medium for DTC process
	Nitrogen gas	Inert gas carrier for pyrolysis process
Techniques	Hydrothermal carbonization (HTC)	Production of hydrochar under 180–250 °C
	Fast pyrolysis	Conversion of biomass into bio-oil at 450–600 °C
	Catalytic pyrolysis	Improvement of bio-oil composition
	Milling and sieving	Particle size controller
	Drying	Removal of moisture content
	Extraction and fractionation	Separation and purification of bio-oil components
	Gas chromatography mass spectrometry (GC–MS)	Analyzer of bio-oil compositions
	Thermogravimetric analysis (TGA)	Examination of thermal stability of biomass
Chemicals	Ethanol (C_2H_5OH)	Act as solvent
	Acetone (C_3H_6O)	Fractionation of liquid used in pyrolysis
	Potassium hydroxide (KOH)	Set pH and characterization
	Hydrochloric acid (HCl)	For chemical treatment and examination
Catalysts	HZSMS-5	For selective hydrochar cracking process
	γ - Al_2O_3	Thermal decomposition is enhanced along with bio-oil yield

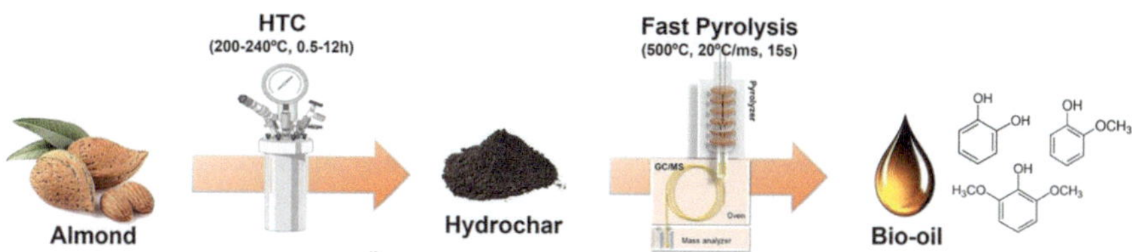

Fig. 13 Hydrothermal carbonization (HTC) process flow diagram [36]

Using fluidized bed reactor, fast pyrolysis was carried out at temperatures ranging from 450 to 600 °C while nitrogen was used as an inert carrier gas to stop oxidation. To examine its impact on the yield of biofuel, the residence duration was kept between 2 and 10 s, while the heating rate was adjusted between 10 and 50 °C/min. Biochar, pyrolysis gases, and bio-oil were the end products of the process. The solid biochar was removed for further examination, and the bio-oil was obtained by condensation and solvent extraction using ethanol and acetone.

Using HZSMS-5 and γ-Al_2O_3 catalytic pyrolysis was carried out to improve the quality of bio-oil by encouraging the selective breaking of big hydrocarbons to smaller ones. The composition of the bio-oil was examined using thermogravimetric analysis (TGA) to evaluate the thermal stability of the resultant biofuels, Fourier Transform Infrared Spectroscopy (FTIR) to identify functional groups, and Gas Chromatography Mass Spectrometry (GC–MS) to identify key chemical composition.

3.10 Results

The biomass was more carbonized when the HTC temperature was raised from 180 to 250 °C. This produced hydrochar with greater carbon content but lower yield because of the accelerated breakdown. Larger pyrolysis temperatures promote the generation of bio-oil rich in phenolic compounds during the rapid pyrolysis phase, which was carried out at temperatures ranging from 450 to 600 °C [35]. However, excessive temperatures resulted in larger gas yields at the expense of liquid product. Longer residence periods during pyrolysis also enhanced secondary cracking events, which further converted bio-oil into gaseous products and decreased the total liquid output, according to the research. Tabular (Table 10) and graphical presentation (Fig. 14) of this effect are shown.

By accelerating the thermal breakdown of biomass components and reducing secondary reactions that may otherwise turn volatiles into char or gas, raising the heating rate from 10 to 50 °C/min resulted in a significant increase in the production of bio-oil [35]. However, very high heating rates might reduce the output of bio-oil by causing incomplete heat transfer inside bigger particles. The yield of bio-oil was also significantly influenced by particle size. Higher bio-oil yields and homogeneous pyrolysis were promoted by the effective heat transmission made possible by smaller particle sizes (0.5–1 mm). On the other hand, because of limits in heats transport, bigger particles (> 2 mm) showed lower bio-oil yields, resulting in incomplete pyrolysis and increased char formation. By ensuring efficient heat penetration,

Table 10 Effect of temperature and residence time [35]

HTC temperature (°C)	Pyrolysis temperature (°C)	Bio-oil yield (%)	Biochar yield (%)	Gas yield (%)	Residence time (s)
180	450	52	35	13	2
200	500	50	32	18	4
220	550	47	30	23	6
250	600	42	27	31	8
250	650	38	25	37	10

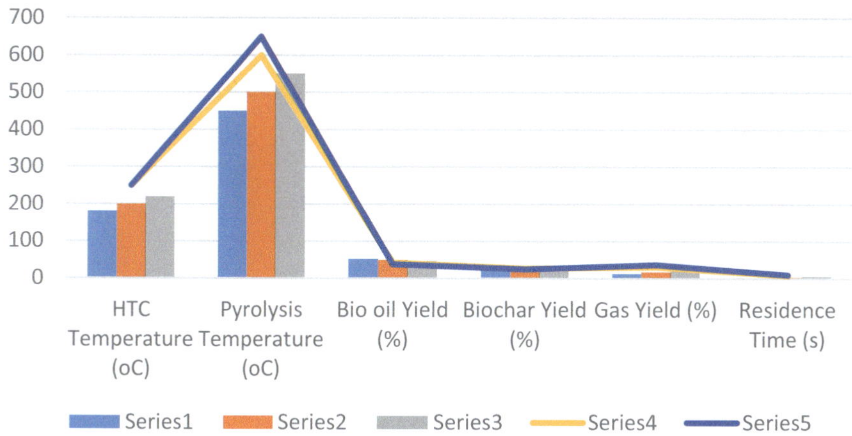

Fig. 14 Effect of temperature and residence time [35]

Table 11 Effect of particle size and heating rate [35]

Heating rate (°C/min)	Particle size (mm)	Bio-oil yield (%)
10	0.5–1	45
10	2	38
30	0.5–1	52
30	2	44
50	0.5–1	55
50	2	47

particle size optimization improves the pyrolysis process's efficiency as shown in Table 11. Graphical display of this effect is mentioned in Fig. 15.

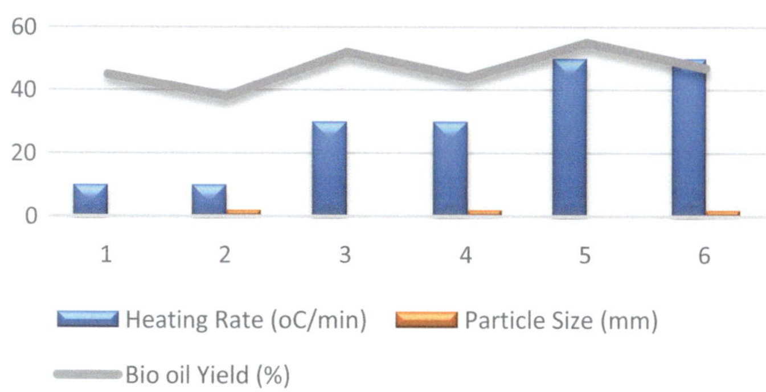

Fig. 15 Effect of particle size and heating rate [35]

Table 11 shows how, when pyrolyzing almond shells, smaller particle sizes and greater heating rates work together to increase the output of bio-oil. To optimize the efficiency of bio-oil production, these factors must be carefully optimized.

4 Summary

The synresearch of bio-oil by pyrolysis is highly dependent on important process variables, including residence time, temperature, particle size, and heating rate. Optimizing yields requires an understanding of their consequences. Based on three case studies, Table 12 compares these factors and highlights their ideal circumstance (Fig. 16).

To get the best results, it is essential to choose the right reactor and regulate variables like temperature, heating rate, residence time, and particle size.

5 Comparison and Conclusion

5.1 Introduction

The outcomes of research investigations on the pyrolysis of almond shells for the generation of biofuel are compared in this chapter. This section evaluates the main findings and shows how different experimental methodologies affect biofuel yield, distribution of products, and efficiency. The goal of the debate is to determine the best practices for maximizing the production of bio-oil by examining the effects of reactor types, pretreatment techniques, and process conditions. The cost-effectiveness of producing biofuel from almond shells is also evaluated by economic study that takes into account variables including feedstock availability, processing expenses, and possible market value. This chapter summarizes the general findings from these investigations, highlighting the viability of using almond shells as a source of bioenergy and looking at possible advancements for further study.

5.2 Case Studies: Discussion of Results

Under ideal circumstances, pyrolysis of almond shells can provide significant bio-oil yields, according to a research by Gupta et al. (2024). According to their results, temperature has a significant impact on the dispersion of the product; the largest amount of liquid biofuels is produced in an ideal range of 500–550 °C. Their kinetic study also showed that the activation energy changed with the pace of heating, which affected how almond shells broke down. By lowering the oxygen level and raising

Table 12 Comparison between case studies [28, 32, 35]

Parameters	Gupta et al. (2024)	Martinez et al. (2021)	Fernandez et al. (2023)	Ideal conditions
Reactor type	Fixed bed reactor	TGA and lab-scale pyrolysis reactor	Fluidized bed reactor	Both can be used for bio-oil generation
Feedstock	Waste flowers	Almond hulls	Almond shells	Anyone can be used
Catalyst	None reported	Zeolite based	Ru/CNF	Depend on experimental process
Temperature (°C)	The output of bio-oil rose at higher temperatures (500–600 °C) while more gas was formed at higher temperature (> 700 °C)	550 °C provided the most bio-oil, lower temperatures resulted in more char	At 500 °C, the best results were obtained, at this temperature, thermal cracking decreased liquid production	The ideal temperature range for optimizing bio-oil yield is 500–550 °C
Particle size (mm)	The production of bio-oil was increased by the improved heat transmission caused by small particles (~ 0.5 mm). Incomplete pyrolysis resulted from large particles (> 2 mm)	Higher bio-oil output and improved breakdown were the outcomes of finer particles (~ 0.5 to 1 mm)	Particle sizes under 1mm produced the best results, higher sizes enhanced the output of char	The ideal range for effective pyrolysis is 0.5–1 mm
Heating rate (°C/min)	Liquid production was enhanced by higher heating rates (30–50 °C/min), however secondary cracking occurred at 50 °C/min	The optimal pace for bio-oil output was 30 °C/min, lower rates resulted in more char formation	At 40 °C/min, the best bio-oil yields were recorded, too high rates encouraged the creation of gas	The ideal range for optimizing bio-oil yield is 30 to 40 °C/min
Residence time (s)	Vapor residence period was optimized at 1 to 2 s, greater durations (> 3 s) resulted in secondary cracking	Prolonged residence periods resulted in increased gas generation, while shorter residence times (~ 1.5 s) promoted bio-oil output	In order to prevent excessive deterioration, the best results were achieved at 2 s residence time	The optimal residence period for the generation of bio-oil is 1.5–2 s
Bio-oil yield (%)	55–65%	50–58%	60–68%	–
Biochar yield (%)	15–25%	20–30%	12–22%	–
Gas yield (%)	10–20%	12–22%	15–25%	–

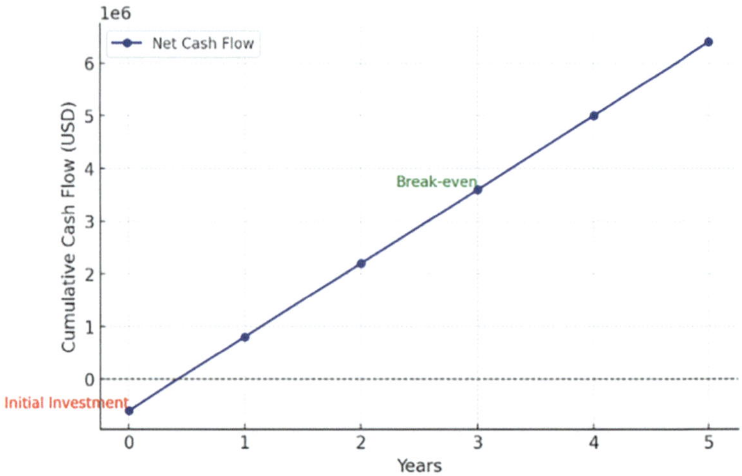

Fig. 16 Net cash flow, Gupta et al. (2024) pyrolysis process [28]

the calorific value, the use of catalysts improved the quality of bio-oil and made it more appropriate for use in industrial settings.

Martinez et al. (2021), on the other hand, concentrated on the thermogravimetric and kinetic features of pyrolysis of almond shells. Their study identified three primary phases of decomposition and offered a thorough assessment of thermal degradation trends. The study underlined how crucial residence time and particle size are to maximizing the yields of biochar and bio-oil. It was discovered that longer residence durations led to more char formation, whereas a faster heating rate favored the synresearch of bio-oil. Better control over the pyrolysis process for energy generation was made possible by their kinetic model, which provided insightful information about reaction processes.

Fernandez et al. (2023) investigated how quick pyrolysis and hydrothermal carbonization (HTC) interact with almond shells. Their results showed that by lowering the moisture and volatile content, HTC pretreatment greatly increased the production and stability of bio-oil. The study showed that improved energy recovery in the pyrolysis process that followed was caused by lower temperatures in the HTC step. Additionally, the study's catalytic method increased the bio-oil's proportion of important compounds, making it more suitable for bio-refining. According to the study's findings, combining HTC and pyrolysis may be a viable way to transform biomass into valuable chemicals and fuels.

When taken as whole, these case studies after a thorough grasp of how various pyrolysis methods and process variables affect the amount and caliber of biofuels produced from almond shells. The main results from these three case studies on the pyrolysis of almond shells are summarized in Table 13.

Table 13 Case studies: comparison [28, 32, 35]

Parameter	Gupta et al. (2024)	Martinez et al. (2021)	Fernandez et al. (2023)
Pyrolysis type	Catalytic pyrolysis	Thermogravimetric and kinetic pyrolysis	Hydrothermal carbonization (HTC) with fast pyrolysis
Reactor type	Fixed bed reactor	Thermogravimetric analyzer (TGA)	Fluidized bed reactor
Temperature range (°C)	400–600	300–700	HTC: 180–250, pyrolysis: 400–600
Optimal temperature (°C)	500–550	550	HTC: 220, pyrolysis: 500
Heating rate (°C/min)	10–50	5–40	20–60
Residence time (min)	10–30min	5–20 min	15–40
Catalysts used	Zeolites, Al_2O_3	No catalyst	Ni-based catalyst
Particle size (mm)	0.2–2.0	0.1–2.5	0.3–2.0
Bio-oil yield (%)	55–65	50–58	60–68
Biochar yield (%)	15–25	20–30	12–22
Gas yield (%)	10–20	12–22	15–25
Key findings	Enhanced fuel qualities and a high bio-oil production with catalytic cracking	Thorough kinetic modeling, ideal char and bio-oil conditions	Enhanced bio-oil stability and quality due to HTC pretreatment improved synresearch of chemicals
Economic feasibility	Moderate cost, catalyst regeneration required	Low-cost method, suitable for kinetic analysis	High cost due to HTC pretreatment, better fuel stability

5.3 *Economic Evaluation of Pyrolysis of Almond Shells for Biofuel*

A number of variables, including reactor design, yield efficiency, processing costs, catalyst prices, and overall profitability, affect how economically feasible it is to produce biofuel from almond shells. These case studies may be analyzed to gain knowledge on revenue potential, capital investment, and cost-effectiveness.

Catalytic pyrolysis, which necessitates an initial investment in catalysts like zeolites and alumina-based materials, was used in the work by Gupta et al. (2024) in a fixed bed reactor. Catalysts enhance the output and quality of bio-oil, but their cost and regeneration needs raise operating costs. Energy prices are also influenced by the heating rate and responses time, which makes the process somewhat costly but long-term feasible because of improved fuel qualities.

Using Thermogravimetric Analyzer (TGA) as their primary tool, Martinex et al. (2021) concentrated on thermogravimetric and kinetic analysis. Despite being essential for optimization, TGA experiments are usually employed in small-scale research rather than large-scale industrial settings. Although the poor bio-oil output may restrict its economic viability in comparison to catalytic pyrolysis, the lack of catalysts saves expenses. However, the process gains from less capital expenditure and complexity.

Fernandez et al. (2023) used a combination of rapid pyrolysis and hydrothermal carbonization (HTC), which greatly enhance the quality of bio-oil but necessitates a large initial outlay. HTC pretreatment necessitates more energy-intensive conditions, more water use, and extra equipment. Notwithstanding these expenses, the resulting bio-oil is more suited for commercialization in high-value applications, such as bio-based chemicals, due to its higher stability and chemical makeup.

5.4 *Biofuel Yield and Market Feasibility*

Each method's yield efficiency is crucial to its viability from an economic standpoint. According to Gupta et al. (2024), the catalytic method produced a comparatively high bio-oil yield of 55 to 65%. Because a larger yield of bio-oil results in more marketable product production, this improves the total economic return. On the other hand, long-term profitability is impacted by catalyst regeneration and deactivation costs. Economic analysis for pyrolysis process is provided in Table 14.

Martinez et al. (2021) reduced material costs by achieving a reasonable bio-oil yield of 50 to 58% without the use of catalysts. But because of the comparatively lower yield, more feedstock is needed to make the same quantity of gasoline, which might raise the cost of raw materials (Table 15 and Fig. 17).

Due to HTC pretreatment, Fernandez et al. (2023) showed the best bio-oil yield (60–68%). HTC lowers upgrade costs and improves product stability despite raising process costs. The method is commercially feasible despite its hefty initial investment because the high-value biofuels and chemicals it produces may command premium pricing (Table 16 and Fig. 18).

5.5 *Energy Use and Financial Success*

Another important economic element is energy use. Energy costs increased as a result of the regulated heating and catalyst regeneration needed for the catalytic process in Gupta et al. (2024). Because Martinez et al. (2021) used the least amount of energy, it was the most economical way to produce basic bio-oil. Yet, because of the extra HTC phase, Fernandez et al. (2023) had the greatest energy needs; yet, the improved quality of the bio-oil made the costs worthwhile.

Table 14 Economic analysis of Gupta et al. (2024) pyrolysis process [28]

Parameter	Estimated values	Remarks
Capital investment	\$500,000–\$700,000	Fixed bed reactor settings with catalytic pyrolysis system
Feedstock cost	\$50–\$80 /ton	Almond shells
Catalyst cost	\$200–\$400/ton	Zeolite-based catalysts
Energy consumption	1.5–2.2 kWh/kg	Heating and reactor operations
Bio-oil yield	55–65%	Based on optimized pyrolysis conditions
Bio-oil production cost	\$0.90–\$1.30/liter	Includes raw material, energy, and operational costs
Market price of bio-oil	1,500,000–2,000,0000 L	Assumed plant capacity
Annual bio-oil production	1,500,000–2,000,000 L	Assumed plant capacity
Annual revenue	\$2,250,000–4,000,000	Based on bio-oil sale price
Annual operating cost	\$1,200,000–2,000,000	Includes raw materials, labor, and maintenance
Net profit	\$1,050,000–\$2,000,000	Revenue, operating cost
Return on investment (ROI)	20–30%	Based on profit relative to capital investment
Payback period	3–5 years	Estimated to recover capital investment

Table 15 Economic analysis of Martinez et al. (2021) pyrolysis process [32]

Parameter	Estimated values	Remarks
Initial investment	\$500,000	Include equipment and setup
Operating cost per year	\$120,000/year	Includes labor and maintenance
Biomass processing rate	5 tons/day	Based on experimental setup
Bio-oil yield	45%	Mass yield from biomass
Annual bio-oil production	750 tons/year	Based on yield and input
Bio-oil selling price	\$600/year	Market value estimate
Annual revenue	\$450,000/year	Bio-oil sales
Payback period	3.5 years	Estimated return period
Net present value (NPV)	\$200,000	After 5 years of operations
Internal rate of return (IRR)	18%	Profitability measure
Breakeven point	3 years	Time to cover costs

6 Conclusion

The feasibility of pyrolysis as an effective conversion method is demonstrated by the study on the synresearch of biofuels from almond shell wastes. Three case studies were compared, and it was shown that temperature, particle size, heating rate, and

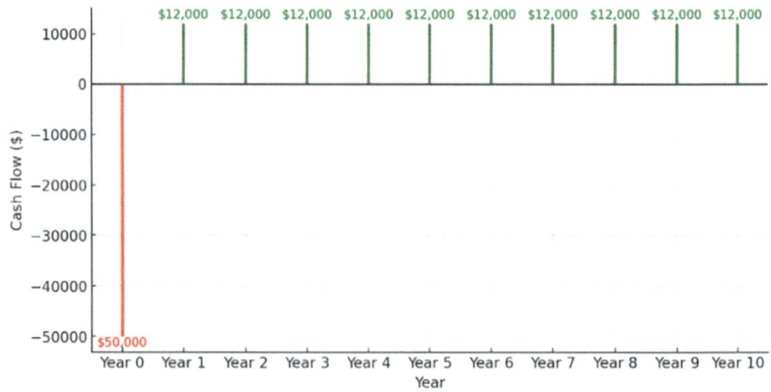

Fig. 17 Cash flow diagram, Martinez et al. (2021) pyrolysis process [32]

Table 16 Economic analysis of Fernandez et al. (2023) pyrolysis process [35]

Parameter	Estimated values	Remarks
Initial investment	$1.5 million	Include reactor setup, catalyst cost, and infrastructure
Feedstock cost	$50 per ton	Cost of almond hulls as raw material
Processing capacity	5000 tons/year	Annual feedstock processing capability
Operational cost	$300,000 per year	Includes labor, maintenance, and utilities
Catalyst cost	$80,000 per year	Cost of catalysts used in hydrothermal carbonization
Bio-oil yield	45% by weight	Average yield based on experimental results
Biochar yield	30% by weight	Secondary product for additional revenue
Syngas yield	25% by weight	Used for energy recovery
Selling price of bio-oil	$1.2 per liter	Market price of produced bio-oil
Selling price of biochar	$200 per ton	Used as a soil amendment or activated carbon
Annual revenue	$3.2 million	Estimated revenue from bio-oil and biochar sales
Net profit margin	25%	After deducting operational expenses
Payback period	~ 5 years	Time required to recover initial investment
Return on investment (ROI)	20% per annum	Profitability assessment of the project

residence time all had a substantial impact on the production and composition of biofuel. Economic evaluations showed that pyrolysis might be profitable, with ideal circumstances increasing process effectiveness. All things considered, almond hulls offer a sustainable biofuel feedstock that advances waste valorization and renewable

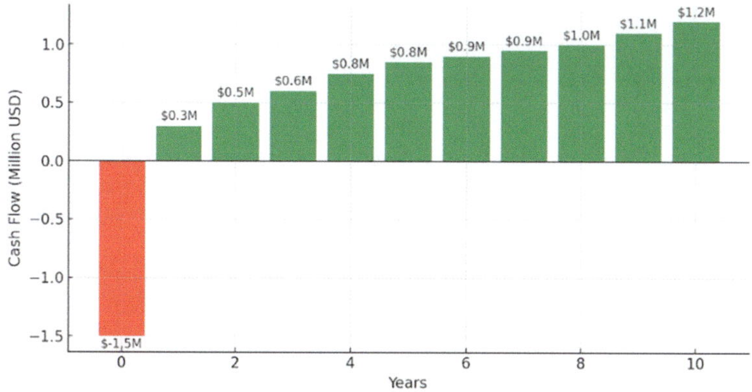

Fig. 18 Cash flow diagram, Fernandez et al. (2023) pyrolysis process [35]

energy. Enhancing catalyst efficiency and scaling up procedures for commercial applications in the bio-energy industry should be the main goals of future research.

References

1. International Energy Agency (2020) Renewable 2020: analysis and forecast to 2020. IEA Publications
2. Demirbas A (2017) Importance of bioenergy and biofuels for environmental sustainability. Energy Convers Manage 39(5):343–353
3. USDA (2021) Almond industry report. United States Department of Agriculture
4. Gonzalez-Garcia S, Moreira MT, Feijoo G (2018) Environmental performance of lignocellulosic bioethanol production from agricultural residues. Renew Sustain Energy Rev 98:107–117
5. Sims RE, Mabee W, Saddler JN, Taylor M (2010) An overview of second-generation biofuel technologies. BioSource Technol 101(6):1570–1580
6. Naik SN, Goud VV, Rout PK, Dalai AK (2010) Production of first and second-generation biofuels: a comprehensive review. Renew Sustain Energy Rev 14(2):578–597
7. Zabed H, Sahu JN, Suely A, Boyce AN, Faruq G (2017) Bioethanol production from renewable biomass. Bio Resources 12(1):31–70
8. Zhang Y, Xu L, Wang J (2020) Utilization of almond hull biomass for bioenergy production. J Renew Energy 45:214–226
9. Miranda I et al (2019) Chemical characterization of almond hull biomass for biofuel applications. Biores Technol 293:122–129
10. Mohan D et al (2006) Pyrolysis of wood/biomass for bio-oil: a critical review. Energy Fuels 20(3):848–889
11. Salgado-Ramos M, Martí-Quijal FJ, Huertas-Alonso AJ, Sánchez-Verdú MP, Cravotto G, Moreno A, Barba FJ (2022) Sequential extraction of almond hull biomass with pulsed electric field (PEF) and supercritical CO_2 for the recovery of lipids, carbohydrates, and antioxidants. SSRN Electron J
12. Offeman RD, Holtman KM, Covello KM, Orts WJ (2014) Almond hulls as a biofuels feedstock: Variations in carbohydrates by variety and location in California. Ind Crops Prod 54:109–114

13. Almond Board of California (2016) 2016 Almond Almanac. Available at: https://www.almonds.com/why-almonds/almond-living-magazine/introducing-2016-annual-almanac-report. Accessed: 12 Mar 2025
14. Kiani A, Torabi P, Mousavi ZE (2024) Green recovery of phenolic compounds from almond hull waste using ultrasound-assisted extraction: phenolics characterization and antimicrobial investigation. J Food Sci Technol 61:1930–1942
15. Remón J, Latorre-Viu J, Matharu AS, Pinilla JL, Suelves I (2022) Direct conversion of almond waste into value-added liquids using carbon-neutral catalysts: hydrothermal hydrogenation of almond hulls over a Ru/CNF catalyst. Sci Total Environ 826:154044
16. Boulika H, Hajji Nabih M, Hmada A, Dkhireche N, Idrissi Kandri N, Zerouale A (2024) Almond hulls and shells extract characterized as a high-performance eco-friendly antioxidant and corrosion inhibitor. Waste and Biomass Valorization, 16 October
17. Chen P, Cheng Y, Deng S, Lin X, Huang G, Ruan R (2010) Utilization of almond residues. Int J Agric Biol Eng 3(4):1–18. Available at: https://www.ijabe.org/index.php/ijabe/article/viewFile/348/pdf. Accessed: 15 Mar 2025
18. McCaffrey Z, Thy P, Long M, Oliveira M, Wang L, Torres L, Aktas T, Chiou B-S, Orts W, Jenkins BM (2019) Air and steam gasification of almond biomass. Front Energy Res 7:84
19. Vasić K, Knez Ž, Leitgeb M (2021) Bioethanol production by enzymatic hydrolysis from different lignocellulosic sources. Molecules 26(3):753
20. Sillero L, Morales A, Fernández-Marín R, Hernández-Ramos F, Dávila I, Erdocia X, Labidi J (2021) Life cycle assessment of various bio refinery approaches for the valorization of almond shells. Sustain Prod Consum 28:749–759
21. Idowu O (2020) Waste cassava peel and almond hull valorizations via microwave processing
22. Almond Board of California (2017) Biomass: building a California bioeconomy with hulls and shells
23. Swami S, Suthar S, Singh R, Thakur AK, Gupta LR, Sikarwar VS (2024) Potential of ionic liquids as emerging green solvent for the pretreatment of lignocellulosic biomass. Environ Sci Pollut Res 31:12871–12891
24. Dele-Afolabi TT, Kamil M, Al-Gheethi A (2023) Shell-based agricultural wastes biomass valorization for biofuel production: a bibliometric analysis. Environ Sci Pollut ResEnviron Sci Pollut Res
25. Monroe H (2020) Conversion of almond hulls into protein-enriched animal feed via solid-state fermentation. University of California e Scholarship Repository
26. El Achaby M, El Moussout H, El Khoukhi M, Aboulkas A (2024) Physico-chemical properties and valorization perspectives of almond by-products: a comparative study. Biomass Convers Bio Refinery
27. Remón J, Latorre-Viu J, Matharu AS, Pinilla JL, Suelves I (2021) Analysis and optimization of a novel 'almond-refinery' concept: simultaneous production of biofuels and value-added chemicals by hydrothermal treatment of almond hulls. Sci Total Environ 765:142671
28. Gupta A, Patel P, Shah K (2024) Comprehensive investigation of almond shells pyrolysis using advanced techniques. Renew Energy 220:1105–1118
29. Available at https://www.researchgate.net/figure/Schematic-diagram-of-fixed-bed-pyrolysis-system_fig2_272148397
30. Available at https://www.technologynetworks.com/analysis/articles/gc-ms-principle-instrument-and-analyses-and-gc-msms-362513
31. Imran A, Bramer EA, Seshan K, Brem G (2016) Catalytic flash pyrolysis of biomass using different types of zeolite and online vapor fractionation. Energies 9(9):721. Available at: https://doi.org/10.3390/en9090721
32. Martínez JD, Ramos J, Gómez J (2021) Pyrolysis of almond (Prunus amygdalus) shells: Kinetic analysis and bioenergy potential. Biores Technol 330:125–135
33. Gorshkov A, Berezikov N, Kaltaev A, Yankovsky S, Slyusarsky K, Tabakaev R, Larionov K (2021) Analysis of the physicochemical characteristics of biochar obtained by slow pyrolysis of nut shells in a nitrogen atmosphere. Energies 14(23):8075. Available at: https://doi.org/10.3390/en14238075

34. Available at https://oizom.com/glossary/gcms/
35. Fernández A, López G, Olazar M (2023) Hydrothermal carbonization coupled with fast pyrolysis of almond shells: valorization and production of valuable chemicals. Waste Manage 152:90–102
36. Alcazar-Ruiz A, Villardon A, Dorado F, Sanchez-Silva L (2023) Hydrothermal carbonization coupled with fast pyrolysis of almond shells: valorization and production of valuable chemicals. Waste Manage 152:90–102. Available at: https://doi.org/10.1016/j.wasman.2023.01.012

Applications of a Phase Change Materials (PCMs) in Solar Collector Systems

Mallak Salim Al Habsi and Seyed Mojtaba Sadrameli

Abstract This thesis investigates the integration of phase change materials (PCMs) for thermal energy storage in solar collector systems. Addressing solar energy intermittency, the research explores innovative PCM applications across different solar collector configurations. Three experimental case studies are analyzed, including conventional evacuated tube heat-pipe systems and direct-absorption collectors featuring shape-stabilized PCMs (SSPCMs). The experimental investigations evaluate the system's charging and discharging processes under varied solar irradiance conditions and the implementation of heat transfer enhancement techniques, such as porous media and finned configurations. Results from Case Study I show that integrating PCM with a thermal battery and porous media results in a system efficiency of 50% ± 9.3%, with the system heating over 120 L of water per day to 38 °C under sunny conditions. In Case Study II, the use of SSPCM in a direct-absorption collector improved thermal efficiency from 66 to 82%, achieving an optimal balance of charging and discharging performance at moderate water flow rates. Case Study III, comparing finned and un-finned configurations, revealed that the un-finned system achieved higher daily efficiency, while the finned configuration provided more uniform heat distribution and prolonged heat release. These findings suggest that optimal PCM integration, coupled with appropriate system modifications, offers a cost-effective, sustainable solution for domestic and industrial heating applications. The work contributes to understanding latent heat energy storage mechanisms and highlights the potential for future PCM-based solar thermal systems.

Keywords Phase change material (PCM) · Solar collector systems · Thermal energy storage · Direct absorption · Evacuated tube solar collector · Latent heat storage · Porous media · Heat transfer enhancement · Renewable energy · Sustainable technology

---

M. S. Al Habsi · S. M. Sadrameli (✉)
Department of Engineering, German University of Technology in Oman, Muscat, Oman
e-mail: seyed.sadrameli@gutech.edu.om

A. Azizi (ed.), *Mechatronic Innovations in Sustainable Energy*, Emerging Trends in Mechatronics, https://doi.org/10.1007/978-981-95-5863-6_6

1 Introduction

1.1 History of Solar Energy Collector Technology

According to historians, the use of solar energy can be traced back to ancient civilizations, as evidence of its use can be found in Greek and Roman architecture [1]. However, modern solar collector technology emerged in the 1780s with Swiss scientist Horace-Benedict de Saussure first developing the solar collector, an insulated box covered with glass that allowed solar heat to be trapped [2]. During the nineteenth century, steam engines powered with solar energy by Auguste Mouchout in 1866, alongside a solar water heating system developed by Charles Tellier in 1885, marked a period of progress which noted. Clarence Kemp introduced the first commercially available solar water heating system termed, "Climax," in 1891. William J. Bailey designed the thermosyphon system in 1908, which incorporated separated collector and storage units permitting more efficient heating to be stored. The invention of flat-plate collectors by Hottel and Whillier in the 1950s, followed by the evacuated tube collectors in the 1970s, were major developments in the mid-twentieth century [3]. Following the oil crisis in 1973, there was a boom in the rate of solar technology development. In recent decades, further advancements such as selective coatings have been incorporated into renewable energy systems to improve efficiency [4].

1.2 Solar Collectors as a Sustainable Energy Source

Solar collectors are one of the crucial technologies that drive the change to sustainable energy systems. As opposed to traditional fossil fuel technologies, solar collectors gather renewable solar radiation without inflicting further harm to the environment. Their operation does not result in direct emissions of greenhouse gases, which makes them a part of efforts to address climate change.

The advantages of sustainability go beyond environmental aspects. The benefits of solar collectors include:

- Reduction in total primary energy use.
- Energy self-sufficiency and security.
- Possibility for decentralized energy „production".
- Long service life (20–30 years).
- Economically viable because of decreasing costs.

Solar thermal systems have provided high energy conversion efficiencies (usually 40–70% for modern collectors), surpassing the photovoltaic systems in the direct conversion of solar energy to useful energy [5]. This makes them particularly useful in areas where heat is needed, rather than electricity, such as in domestic hot water, space heating, and a variety of industrial thermal processes.

1.3 Solar Collector Challenges

While beneficial, solar collectors also have cautions on their practical and technical functionalities, including:

1. Intermittency: Solar energy is available at different times of the day and year, and its utilization depends on the weather and place [6].
2. Low energy density: Energy capture requires large collection areas, which results in decreased energy density.
3. Temperature limitations: Varying maximum operating temperatures exist across different collector design types

    Flat-plate collectors: 30–80 °C.
    Evacuated tube collectors: 50–200 °C.
    Concentrating collectors: 100–400 °C (decreased for specialized designs).

4. Thermal losses: Efficiency is reduced, especially with higher operating temperatures [4].
5. Integration challenges: Existing systems and infrastructure can pose problems to integration.
6. Economic barriers: There still exists a high initial investment cost, even with decreasing lifetime costs [7].

These challenges stress the need for an efficient thermal energy storage system, enabling the reduction of intermittency and enhancing system efficiency.

1.4 Introduction to PCMs

Phase change materials (PCMs) are substances that absorb or release substantial amounts of heat as they melt or solidify, all while maintaining an almost constant temperature. This latent-heat mechanism allows PCMs to store between five and fourteen times more thermal energy per unit volume than materials relying solely on sensible-heat storage [1]. Table 1 presents the classification and key properties of several common PCMs. PCMs are classified into several groups:

1. Organic PCMs:

    a. Paraffin and Fatty acids
    b. Advantages: Non-corrosive, chemically stable, low degree of supercooling.
    c. Disadvantages: Low thermal conductivity, Flammability, Larger volume changes.

2. Salt hydrates and metallics

    a. Inorganic PCMs:
    b. Advantages: Higher latent heat, higher thermal conductivity, nonflammable.

Table 1 PCM properties and classification [1]

PCM type	Examples	Melting temperature range (°C)	Latent heat Range (kJ/kg)	thermal conductivity (W/m K)	Key advantages	Key limitations
Organic						
Paraffins	*n*-Octadecane, *n*-Eicosane	20–70	200–240	0.2–0.4	Chemical stability, minimal supercooling, non-corrosive	Low thermal conductivity, flammability
Fatty acids	Lauric acid, stearic acid	30–65	150–200	0.15–0.3	Renewable sources, consistent performance	Mild corrosion, odor, cost
Inorganic						
Salt hydrates	$CaCl_2 \cdot 6H_2O$, $Na_2SO_4 \cdot 10H_2O$	30–130	180–300	0.5–1.0	High volumetric storage density, non-flammable, cost-effective	Phase separation, supercooling, corrosion
Metallics	Gallium, aluminum	30–800	70–230	15–40	Very high thermal conductivity	High cost, weight
Eutectic						
Organic-organic	Capric–lauric acid	15–90	120–160	0.15–0.3	Tailorable melting point, no phase separation	Limited data, thermal cycling stability
Organic–inorganic	Urea-$NaNO_3$	30–130	170–220	0.4–0.7	Higher thermal conductivity than organic PCMs	Complex behavior
Inorganic inorganic	$MgCl_2 \cdot 6H_2O$-$CaCl_2 \cdot 6H_2O$	20–130	200–250	0.5–0.8	Higher latent heat, sharper melting point	Supercooling

   c. Disadvantages: Phase Separation, supercooling, potentially corrosive.

3. Eutectic mixtures:

   a. Organic and/or inorganic compounds, jointed combinations.
   b. Advantages: Sharp melting point and customizable transition temperatures.
   c. Disadvantages: Limited commercial availability, not much research data.

An ideal PCM for solar applications should have:

- Desired range melting temperature.
- High latent heat of fusion.
- Good thermal conductivity.
- Long term chemical stability.
- Minimal supercooling.
- Compatible with containment materials.
- Economically and environmentally friendly.

1.5 Advancements in PCM Applications with Solar Collectors

In recent decades, the way renewable energy technologies are incorporated into building materials has evolved dramatically, giving rise to a variety of innovative integration approaches.

1. Systems that incorporate PCM directly into the solar collectors allow you to both capture and store solar heat in one module, reducing thermal losses during transfer and uniting two functions into a single unit from a system design perspective [8]
2. External storage units without PCM enable storage of thermal energy in PCMs connected to the collector fluid [2]
3. Driven by the heating demands of a building, PCMs incorporated into building components with solar collectors for production of space heating and domestic hot water [9]
4. Implementation of PCM-enhanced methods:

   - Using metal fins, matrices, or nanoparticles of materials for increasing thermal conductivity
   - Microencapsulation for better handling and heat transfer
   - Form-stable composites of PCMs for physical stability
   - Multi-layered or cascaded systems of PCMs for larger temperature regions.

Enhancements for these recently include:

- Nano-enhanced PCMs include components of nano thermals, with a result in increased, more than in reported 15–65 percent, in literature [10]
- Smart systems with PCMs: Via control systems-coupled for better prediction capability for Solar coupled charging/discharging.

1.6 Statement of the Problem

Despite Oman's abundant sunshine receiving roughly 5.5–6.0 kWh/m^2 of solar radiation each day (International Renewable Energy Agency [11]) scaling up solar thermal energy remains difficult. To hit its 30 percent renewable-energy target by 2030 [11], the country needs highly efficient solar thermal systems. Unfortunately, the typical water-tank sensible-heat storage approach isn't well suited to Oman's climate: it demands very large tank volumes and suffers substantial heat losses, particularly when power is needed after sunset or on cloudy days [12].

While offering high energy density and near-isothermal behavior, phase change materials (PCMs) present promising solutions; however, their effective integration with solar collectors in Oman poses several challenges:

1. Extreme temperature limits in heat transfer:

The majority of PCMs are characterized by low thermal conductivity (0.2–0.7 W/m K), which results in charging/discharging rates that do not match the operation of the solar collector an issue exacerbated by Oman's extreme irradiance and high sub-45 °C ambient temperatures during summer months [13].

2. Performance decline due to prolonged thermal cycles:

Repeated exposure to high-temperature conditions in a desert climate can severely degrade the reliability of PCMs, as their thermal stability declines over time [14].

3. Dust storms and particulate accumulation:

Oman is known for frequent dust storms that deposit significant particulate matter on solar collectors, reducing heat transfer efficiency. This environmental challenge necessitates different system integration approaches than those typically used in temperate regions [15].

4. Temperature matching regional needs:

Specific engineering challenges arise when aligning the PCM phase change transition temperatures with the operating temperatures of solar collectors and with Oman's particular cooling needs and industrial processes. Reconciling these factors is demanding, given the locality-specific requirements [16].

5. Economic feasibility amid subsidized energy prices:

Although the low prices of PCMs make them economically attractive, their integration must be justified against the backdrop of subsidized standard energy prices in Oman. The added benefits of PCM integration must exceed the cost of subsidized energy for it to be a viable solution [16].

6. Limited freshwater resources:

Oman's restricted freshwater supply requires that solar thermal systems be designed with innovative heat transfer fluid formulations and cleaner PCM integration techniques that use less water [17].

Contending with these issues calls for systematic investigations into PCM–solar collector integration approaches tailored to Oman's climatic, energetic, and financial peculiarities—merging thermal efficiency, system robustness in severe conditions, and cost considerations within the framework of the country's energy transition objectives.

1.7 Objective

This research aims at characterizing the thermophysical properties of appropriate PCMs and the system configuration parameters in conjunction with the performance evaluation of the integrated thermal energy storage systems towards more efficient, sustainable, and economically environmentally friendly solar energy utilization.

1.8 Thesis Layout

This thesis consists of four main chapters. The first chapter introduces the research topic by presenting background information, the study's objectives, and its overall significance. The second chapter is a literature review that examines previous research, discussing theories, methodologies, and key findings related to the subject. The third chapter is devoted to case studies and includes three in-depth analyses of related subjects, detailing the materials used, experimental setups, methodologies, and results of each case. The fourth chapter concludes the thesis with a discussion on future aspects, evaluating the findings from the case studies, and providing insights into the potential for further research along with economic and environmental considerations.

2 Literature Review

The integration of Phase Change Materials (PCMs) with solar collector systems represents a promising approach to addressing the inherent intermittency challenges of solar energy utilization. This chapter presents a comprehensive review of recent research advancements in PCM-solar collector integration, focusing particularly on material characterization, system configurations, performance evaluation metrics, and economic considerations. The literature examined spans experimental investigations, numerical modeling approaches, and practical implementations across various climatic conditions and application domains. Special attention is given to studies exploring organic PCMs, inorganic salt hydrates, metallic PCMs, and eutectic mixtures, as these material categories offer distinct advantages and limitations that significantly impact overall system performance. By systematically analyzing the methodologies, findings, and conclusions of these studies, this review aims to identify the current state of knowledge, emerging trends, and critical research gaps that will inform the objectives and methodological approach of the present investigation.

In their experimental work, Naghavi Sanjani et al. [18] evaluated an evacuated-tube, heat-pipe solar collector fitted with a "thermal battery" of paraffin PCM and a metal porous matrix. By packing the manifold with paraffin, they stored incoming solar heat as latent energy, while the porous metal scaffold boosted the PCM's conductivity. Field tests revealed that on sunny days this PCM battery could heat over 120 L of water to about 38 °C delivering hot water well into the evening whereas on cloudy days supplemental heating was required. The study concluded that embedding a PCM-based thermal battery (reinforced with porous media) directly within the collector not only stabilizes the outlet temperature but also yields efficiencies around 50% ± 9.3% under varying solar conditions.

Senobar et al. [19] experimentally investigated a conventional evacuated tube solar water heater augmented with a small latent heat storage unit filled with a paraffin-based PCM. To improve charging/discharging rates, the PCM was enhanced by embedding a copper metal foam (forming a PCM–MF composite). The results showed that without metal foam the pure PCM was too slow in absorbing and releasing heat, failing to deliver sufficient hot water after sunset. In contrast, the PCM–metal foam composite stored thermal energy 2–3 times faster and delivered water 10–77% hotter (compared to inlet) for about 2 h after solar input stopped. Thus, adding high-conductivity metal fam dramatically improved the system's efficiency and extended the hot water supply duration in the absence of sunlight.

Sadeghi et al. [20] developed a novel tankless solar water heater by integrating shape-stabilized PCM directly into a direct-absorption evacuated tube collector. The system eliminates the bulky storage tank by filling an evacuated tube's absorber with a composite PCM (sodium acetate trihydrate mixed with graphite for stability and conductivity) that absorbs solar heat and gradually releases it to water flowing through the tube. In their experiments, PCM successfully stored excess solar heat and then maintained water outlet temperatures during periods without sunshine, effectively addressing the mismatch between solar availability and hot water demand. The authors report that the PCM-embedded collector provided reliable hot water in a "tankless" configuration, demonstrating the feasibility of using latent heat storage within the collector itself to supply domestic hot water continuously.

Sharma and Gupta [21] examined a flat-plate solar water heater with a eutectic PCM mixture for thermal storage. During experiments, a paraffin-based PCM (60% polyethylene glycol and 40% beeswax) was encapsulated at the back of the absorber plate of the collector. It was observed that the PCM melted at peak sun, storing heat, and later solidified releasing heat to the water. The collector with PCM was ~ 15% more efficient than a similar collector with no PCM and provided hot water for a longer time after sunset. The abstract concludes that a collector with the PEG–beeswax PCM had much superior thermal performance, pointing out that a well-selected eutectic PCM can enhance efficiency and hot water availability during evenings for flat-plate solar heaters.

Rizal et al. [22] examined a parabolic-trough solar water heater enhanced with beeswax PCM to extend its heat-storage capacity. In their design, a curved reflector concentrates sunlight onto the storage tank, inside which a copper tube filled with beeswax PCM is submerged. Field trials in Aceh, Indonesia showed that, without PCM, the system could supply roughly 16 L of water at 40–60 °C, sustaining that temperature for about four hours into the evening. When the beeswax PCM was employed, the same volume of water remained between 40 and 60 °C for approximately five hours an extra hour of usable hot water. Overall, the prototype delivered up to 60 L of warm water per day, demonstrating that the latent-heat storage of beeswax effectively prolongs hot-water availability. The authors concluded that integrating beeswax PCM into a parabolic trough collector significantly improves thermal-storage duration, offering promising potential for domestic-scale applications.

Abokersh et al. [23] experimentally evaluated a direct flow evacuated tube solar collector integrated with phase change material (PCM) to enhance energy storage and improve system efficiency, demonstrating that innovative design modifications such as comparing finned versus un-finned U-tube heat exchangers significantly influence thermal performance. Their study revealed that factors like PCM melting temperature, natural convection effects, and discharge flow rates play pivotal roles in optimizing both the charging and discharging phases of solar water heaters, providing critical insights into achieving higher operational temperatures and reduced heat losses. The research further employed rigorous experimental design and statistical analyses to systematically ascertain the contributions of individual factors, ultimately highlighting the considerable potential of integrating PCM for sustainable and efficient renewable energy systems.

Shareef et al. [24] experimentally evaluated a conventional flat-plate solar water heater retrofitted with a latent heat storage unit in the hot water tank. In their setup, 25 kg of paraffin wax PCM was sealed in cylindrical aluminum capsules and submerged in the storage tank, which also had a helical coil heat exchanger for heat transfer. Tests in Kerbala (Iraq) showed that the PCM-enhanced tank produced outlet water up to 9.5 °C hotter than a similar system without PCM during charging under sunlight. After sunset, the water in the PCM-integrated tank stayed about 28.5 °C above ambient (due to latent heat release), whereas the non-PCM tank's temperature dropped much more quickly. The study demonstrates that adding encapsulated paraffin to the storage tank can effectively store daytime solar heat and keep water warm well into the evening, thereby increasing the useful heat output and average efficiency of the solar water heater.

Natesamurthi et al. [25] examined a compound parabolic concentrating (CPC) evacuated tube solar collector coupled to a dual thermal storage system (sensible plus latent). They added a top layer of commercial paraffin (OM65) PCM occupying the upper quarter of the storage tank, above the water layer, to retain heat and sustain stratification. The experimental setup achieved an energy storage capacity of 21 MJ per day, with the PCM melting and solidifying buffer temperature swings. Importantly, the CPC collector's peak thermal efficiency reached ~ 49% when it was paired with the latent heat storage, significantly higher than when only sensible heat storage was used. This work highlighted that using a PCM in the storage tank of a CPC-based heater not only increases stored energy and keeps the return water cooler (improving collector efficiency) but also boosts the overall system efficiency for domestic hot water production.

Almari et al. [26] developed a domestic evacuated-tube solar water heater fitted with a latent-heat storage tank containing 25 kg of paraffin wax and a helical-coil heat exchanger. By circulating water through the coil, they enabled efficient heat transfer to and from PCM. In charging trials at flow rates ranging from 5 to 13 L/min, they found that upping the flow rate significantly reduced the time required for the paraffin to melt—from 218 min at 5 L/min down to 157 min at 13 L/min and boosted the system's overall thermal efficiency from 28.4 to 40.5%. In evening discharge tests, the PCM reliably released stored heat to maintain outlet temperatures suitable for use. The authors concluded that coupling a latent-heat tank to an evacuated-tube collector is highly effective, and that optimizing the water flow rate markedly enhances both the charging speed and overall performance of the solar heater.

Algarni et al. [27] examined an evacuated-tube collector with a U-pipe immersed in a nano-enhanced PCM, where paraffin wax was doped with 0.33 wt% copper nanoparticles to increase its thermal conductivity. During sunny periods, the PCM absorbed and stored heat, then released it when solar intensity dropped, smoothing out temperature fluctuations at the outlet. Compared to a standard collector without PCM, the nano-PCM version lifted collector efficiency by about 32%. At a modest flow of 0.08 L/min, it also extended the provision of water at roughly 50 °C by an additional two hours after sunset. These results highlight that embedding nanoparticle-enhanced PCM within ETC can significantly raise efficiency and prolong hot water supply in household systems.

Zhao et al. [28] looked at a flat-plate solar water heater equipped with a PCM storage module occupying less than 20% of the tank volume. Thanks to the PCM's high latent-heat capacity, it stored roughly half of all the solar energy captured. In discharge experiments, this PCM tank kept outlet water between 45 and 50 °C for about three hours, whereas a conventional tank's temperature fell by 20 °C in just 1.5 h under the same conditions. Overnight, the PCM-integrated tank lost only 7 °C, compared with a 12 °C drop in the standard tank. Seasonal energy analysis suggested a 34% energy saving versus a normal water tank, and the PCM design allowed for a five-times smaller tank for equivalent performance. Economically, they estimated a payback period of under one year based on operational savings.

Pathak et al. [29] carried out both energetic and exergetic assessments of a heat-pipe evacuated-tube system using palmitic acid as the PCM in the manifold of 25 heat-pipe tubes. Tested at a constant 20 L/h flow on a university rooftop under clear and cloudy skies, the inclusion of PCM increased the total solar energy absorbed, improved overall thermal efficiency, and reduced exergy destruction during hot water production. Their measured outlet temperatures closely matched predictions from a mathematical model, with deviations under 2–6%, validating the model and demonstrating the PCM's stabilizing effect in variable solar conditions. They confirmed that PCM-enhanced heat-pipe ETCs deliver more usable energy and higher exergetic performance than conventional systems.

Alshukri et al. [30] experimentally assessed the impact of adding medical-grade paraffin PCM both inside the evacuated tubes around the heat pipes and within the insulated storage tank. This dual PCM strategy provided "late heat release," melting under daytime solar input and solidifying during periods of high demand or low sunlight to extend hot water availability. Compared to a no-PCM baseline, the fully integrated system achieved a 55.7% efficiency gain; adding PCM only in the tubes yielded a 49.9% improvement, while PCM only in the tank gave 36.5%. Their findings demonstrate that latent-heat storage in both collector and storage tank serves as an effective thermal buffer, significantly enhancing the duration of hot water delivery when solar input is insufficient.

Pawar and Sobhansarbandi [31] investigated a method to enhance a heat-pipe evacuated tube collector by embedding the PCM with a high-conductivity porous matrix. They packed a copper metal foam with paraffin inside the evacuated tube's annulus (around the heat pipe) to form a composite thermal storage. Under peak solar radiation, the PCM–foam composite increased the absorber's peak temperature by about 21 °C compared to a standard tube. Thanks to the improved heat transfer, the modified collector achieved 85.6% daily energy efficiency, far above the ~ 36.9% efficiency of the conventional system without PCM. Over long-term tests, the PCM/foam-enhanced tubes also yielded notable cost savings in hot water production (over 10% savings in 5–10-year projections) due to their higher thermal performance. This study underlines that integrating metal foams or other conduction enhancers with PCMs can dramatically boost a solar water heater's efficiency, making latent storage more effective in practical applications.

Shafieian et al. [32] designed a solar hot water system to meet winter domestic demand by using 25 evacuated heat-pipe tubes with integrated PCM storage. They optimized the number of glass tubes via modeling, then built a prototype with a latent heat storage module. Experimental testing demonstrated that the PCM-enhanced system could supply a greater quantity of hot water and achieved higher efficiency than a conventional system, particularly during peak usage times in the evening. The presence of PCM reduced exergy losses by storing excess energy and releasing it when sunlight was insufficient, thus smoothing out the supply. Importantly, the predicted outlet temperatures from their theoretical model agreed well with the measured values (within ~ 2–6%), confirming the accuracy of their design and the performance gains due to the PCM. This work validates that even in cold seasons, incorporating PCM in an evacuated tube array substantially improves heat capture and utilization in solar water heating.

Ahmed et al. [33] examined an integrated collector–storage solar water heater that used helical-finned heat pipes in the absorber along with a PCM for storage. The experimental system consisted of a flat-plate collector with heat pipes (equipped with spiral fins for better heat transfer) that transferred energy to a PCM-filled storage unit. The results demonstrated "substantial improvements" in thermal performance relative to a traditional water-only storage. In particular, the finned heat pipes effectively charged the PCM during the day, and the PCM's latent heat helped keep the water output warm after sunset. The authors noted that the PCM–heat pipe combination extended the period of hot water availability and increased the system's efficiency by maintaining a higher degree of thermal stratification in the storage tank. This study shows that retrofitting solar water heaters with finned heat pipes and PCMs is a viable strategy to boost energy storage capacity and evening hot water supply.

Li et al. [34] proposed and tested a multi-cell PCM storage tank for solar water heating. Instead of one large volume of PCM, the tank was divided into multiple square cells filled with PCM, interspersed with water. Through experimental discharge tests, they found that this multi-cell PCM tank had significantly better heat-release characteristics than a conventional single-cell (bulk PCM) tank. In particular, the transient heating capacity during hot water draw-off was higher, and more residual heat remained available in the later stages of discharge, due to the enhanced surface area and heat transfer in the multi-cell design. The authors concluded that partitioning the PCM into cells mitigates the slow response of large PCM volumes, thereby improving the usable heat output and overall efficiency of the solar water heater during the delivery of hot water.

Fazilati and Alemrajabi [35]—reported in a recent analysis for context—showed that using paraffin PCM capsules in a solar water heater can increase the stored energy density by up to 39% and improve exergy efficiency by about 16%, compared to water-only storage. Building on such findings, Bonyadi et al. [35] experimentally integrated paraffin-filled spherical capsules into a domestic solar heater's tank and similarly observed higher energy storage and longer delivery of hot water. The PCM-enabled heater could maintain the desired hot water temperature about 25% longer than the conventional system, thanks to better thermal stratification maintained by the PCM's latent heat discharge. These outcomes reinforce that PCM incorporation (even with simple geometries like spheres) meaningfully enhances a solar water heater's performance by reducing heat loss and extending the supply of hot water into the evening.

The review presents a broad range of experimental studies demonstrating that integrating phase change materials into solar collector systems enhances thermal performance by improving heat storage capacity, stabilizing water temperatures, and extending hot water availability into the evening. However, several critical gaps are evident in the literature. Many studies focus on individual systems using different PCM configurations such as pure PCM, PCM combined with metal foam, nano-enhanced PCM, or multi-cell designs yet they lack standardized metrics or direct comparative analyses across various collector types. This makes it challenging to determine the optimum integration method and assess which configuration consistently delivers the best performance.

In addition, most of the current research provides promising short-term experimental results but does not address the long-term stability, durability, and degradation of PCM systems under repeated thermal cycling and extended operational periods. Economic evaluations and lifecycle assessments are also limited, leaving questions about the commercial viability and scalability of these solutions unanswered. Furthermore, while the studies predominantly focus on domestic hot water applications, there is relatively little research on the use of PCM storage in other settings or under a wide range of climatic conditions, including extreme weather scenarios. The integration of advanced control strategies or smart energy management systems to optimize the performance of PCM-enhanced collectors is another area that remains underexplored. Lastly, there is minimal discussion regarding the environmental impact, sustainability, and safety issues such as potential leakage, or the long-term stability of nanoparticles associated with various PCM materials.

Overall, the experimental studies collectively demonstrate that incorporating PCMs into solar collector systems offers significant performance improvements by enhancing thermal storage, stabilizing temperature output, and extending hot water supply. Despite these advancements, the lack of standardized comparative evaluations, long-term durability data, detailed economic assessments, and research on diverse applications and climatic conditions presents a substantial barrier to fully understanding and optimizing PCM-based solar energy storage solutions. Addressing these gaps through systematic, interdisciplinary research will be crucial for refining these systems and accelerating their adoption in both residential and industrial applications.

3 Experimental Case Studies and Results

3.1 Case Study I

Title: **Experimental Investigation on Solar Water Heater Integrated with Thermal Battery Using Phase Change Material and Porous Media** [18]

Authors: Mohammad Sajad Naghavi Sanjani, Mahyar Silakhori, Bee Chin Ang, Hendrik Simon Cornelis Metselaar, Sayed Mohammad Mousavi Gazafroudi and Younes Noorollahi

Year: 2023

This paper discusses the experimental investigation of a solar water heating system integrated with a thermal battery that uses phase change material (PCM) and porous media. The study explores how solar energy can be efficiently captured by evacuated tube heat pipe solar collectors and then stored in a reconfigured manifold—designed as a thermal battery—where the PCM, enhanced by porous media, absorbs, stores, and later releases heat. By operating in both charging and discharging modes, the system aims to provide reliable domestic hot water even under fluctuating solar conditions, while addressing the limitations of conventional solar water heaters [18].

3.1.1 Material Used in Experiment

The experiment's components were chosen to maximize both heat storage and transfer in the solar heater. A paraffin-wax PCM—with a melting range of 53–57 °C provided high latent-heat capacity ideal for low-temperature solar use. To counteract the PCM's low conductivity, a 90%-porous silicon-carbide matrix (≈ 56 W/m·K at 100 °C) was embedded in the storage tank, ensuring more uniform heat distribution. The tank and piping were constructed from mild steel for strength and good thermal conduction. To monitor and control the system's performance, copper–constantan thermocouples tracked temperatures, a Kipp & Zonen pyranometer measured solar irradiance, and a Burkert flowmeter regulated water flow [18].

3.1.2 Experimental Set-Up and Methodology

In the standard ETHPSC-B system, solar heat is picked up by the heat-pipe evaporator and carried via a condenser welded into the manifold to warm water in an insulated storage tank. The new ETHPSC-TM takes that same manifold and turns it into a "thermal battery" by packing it with a phase-change material and a porous medium. The porous matrix speeds heat flow from the condenser into the PCM, offsetting its low conductivity and spreading the warmth evenly. Then, dedicated piping pulls store heat out of the PCM and into incoming cold water, so hot water is available on demand. Figure 1 shows this concept [18].

Figure 2 shows a 3D view of the thermal battery coupled to a bank of ETHPSCs. Each heat-pipe condenser is seated in a socket tube welded directly to the battery and

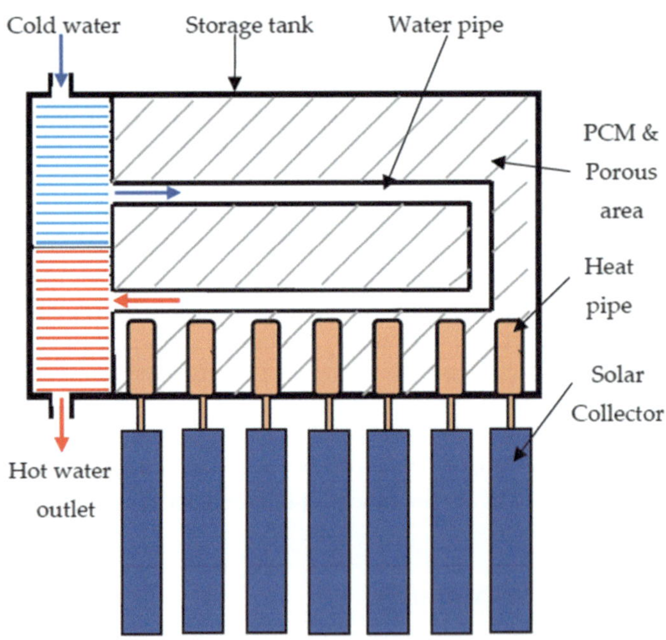

Fig. 1 Diagrammatic representation of the ETHPSC using porous medium and PCM [18]

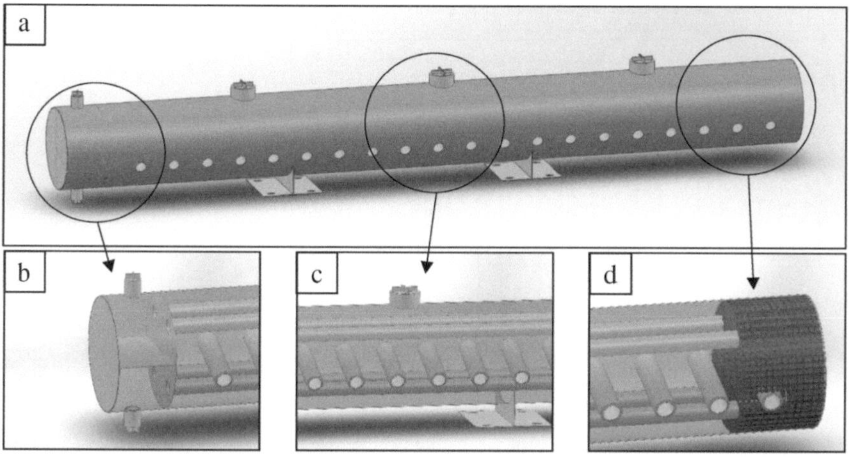

Fig. 2 The thermal battery in three dimensions (**a**), water inlet and outlet sections (**b**), PCM inlet valve and HP solar tube inlet socket (**c**), and porous media within the thermal battery (**d**) [18]

surrounded by a porous matrix, which spreads the incoming heat uniformly through the PCM during charging. When discharging, dedicated piping draws the stored heat out of the PCM and into the cold-water inlet. Small cavities at both the inlet and outlet of these water channels ensure smooth, even flow and keep vapor from reaching the outlet pipe [18].

The ETHPSC-TM setup relies on three key heat-transfer stages: first, the heat-pipe condenser (HPC) captures incident solar radiation from the evacuated-tube collector; next, this energy melts the PCM during the charging phase; and finally, during discharge, heat is drawn off into the cold-water supply, causing the PCM to solidify. Figure 3 presents the overall system schematic, while Fig. 4 shows the thermal-battery assembly and test rig. The full dimensions of the experimental apparatus are listed in Table 2. All tests were carried out at the University of Malaya's Faculty of Engineering (3.1° N, 101.4° E) [18].

System performance depends on five main factors: the type of solar collector, its inclination, the choice of PCM, the characteristics of the porous medium, and the volume of the thermal battery. In Malaysia, optimal collector tilt falls between 10° and 30°, with 20° offering the best orientation for heat-pipe systems to capture solar energy. Selecting a PCM requires weighing its thermodynamic, kinetic, chemical, and economic attributes; for low-temperature solar applications, paraffin wax is ideal, thanks to its melting range of 53–57 °C. Its key thermal properties are listed in Table 3 [18].

The incorporation of porous media in latent heat thermal energy storage (LHTES) systems significantly enhances heat conduction within the PCM, leading to shorter charging and discharging periods. By embedding porous media into PCM, the overall heat transfer process becomes more efficient, accelerating the phase change process due to its superior conduction properties. The specific characteristics of the porous

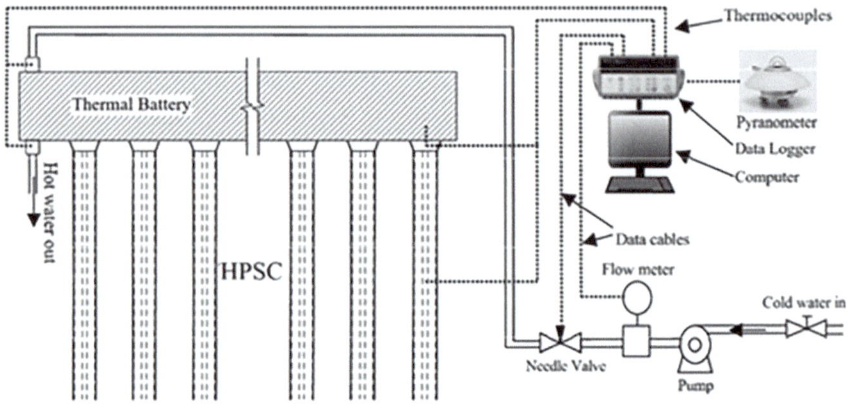

Fig. 3 Diagrammatic representation of the setup [18]

Fig. 4 An image of the experimental setup [18]

media used in this study are detailed in Table 4, and a sample unit of the chosen porous media is depicted in Fig. 5 [18].

The storage tank and connecting pipes are fabricated from mild steel (≈ 45 W/m K), chosen for its strength and thermal conduction. Tank sizing was driven by the local average insolation (≈ 4.5 kWh/m^2·day) and the solar field's specs: twenty collectors (1.77 m^2 total aperture) at ~ 65% absorbance capture about 18.6 MJ of

Table 2 The experimental setup's dimensions [18]

Item	Value (mm)
ETHPSC outer tube diameter	58
ETHPSC inner tube diameter	52
Distance between 2 ETHPSCs	142
HPE length	1675
HP adiabatic length	30
HPC length	70
Thermal battery diameter	200
Thermal battery length	1850
Thermal battery wall thickness	2
Water pipe outer diameter	27
Water pipe length (both U-shape pipes)	~ 7400

Table 3 The paraffin wax's thermal characteristics [18]

Property	Value	Unit
Melting point	53–57(~ 55)	°C
Density solid/liquid at 15/70 °C	990/916	Kg/m^3
Thermal conductivity of solid/liquid	0.349/0.167	W/m K
Specific heat solid/liquid	2.76/2.48	kJ/kg K
Heat storage capacity	164	kJ/kg

Table 4 The silicon carbide porous media's thermal characteristics [18]

Property	Value	Unit
Diameter	200	mm
Thickness	30	mm
PPI	10	–
Porosity	~ 90	%
Thermal conductivity (100 °C)	~ 56	W/(m K)

solar energy daily. In principle, that energy could melt up to 76.6 kg of paraffin wax starting from 25 °C, but once you factor in the 90%porous SiC matrix, pipe and socket volumes, heat losses, irradiance swings, and safety margins, the practical PCM mass is limited to ~ 60% of that ($\approx$ 46 kg). A 0.2 m $\times$ 1.75 m tank, filled with sixty SiC porous elements, accommodates the molten PCM [18].

Cold water is supplied from the municipal grid via a low power, seal less magnetic centrifugal pump; a downstream needle valve fine-tunes the flow. To curb thermal losses, the battery is wrapped in 25 mm of Rockwool and all piping is clad in K-FLEX insulation. Type-T copper–constantan thermocouples monitor temperatures at the inlet, outlet, and a selected heat pipe. A Kipp & Zonen CMP10 pyranometer logs solar irradiance, while a Burkert S030 inline flowmeter tracks water flow; all

Fig. 5 The porous media image [18]

data feed into an Agilent 34970A recorder. With typical local wind speeds of only 1.5–2.0 m/s, convective losses from wind are minimal [18].

3.1.3 Results and Discussion

Two primary tests were conducted on a system comprising 20 evacuated tube heat pipe solar collectors (with a total collector area of 1.77 m^2) that heats tap water in a thermal battery before supplying it to users. Both tests were performed between August and September to evaluate system performance:

1. Mode 1 (Charging/Discharging Mode):
   - Objective: Measure total solar radiation absorbed and quantify the energy transferred to cold water, thereby estimating the system's efficiency.
   - Procedure: The system charges in the morning until around 5 p.m., melting the PCM inside the thermal battery. After 5 p.m., water is drawn at 60 L/h, absorbing the stored heat and delivering hot water until midnight.
2. Mode 2 (Periodic Draw-Off):
   - Objective: Calculate the peak and mean water temperatures across several brief draw-offs in the morning and evening.
   - Procedure: Draw-offs occur four times in the morning and four times in the evening. Each draw-off lasts 5 min at 2 L/min, followed by a 10-min pause, allowing assessment of the system's ability to maintain hot water temperatures in intermittent usage scenarios.

Mode 1 (Charging-Discharging Mode):

In Mode 1, the system runs through two clear stages each day. First, during charging, the ETHPSC alone collects solar energy and passes it through the heat pipes into the thermal battery, melting the PCM and storing heat as both latent and sensible energy—no water flow occurs at this point. Once sunlight wanes and charging can't continue, the discharge stage begins tap water is pumped through the battery's piping, drawing heat from the melted PCM and raising its temperature. During discharge, heat travels predominantly from the heat pipe's adiabatic section into the condenser. The system's behavior under different solar intensities and weather conditions—across four test runs—is detailed in Table 5. [18].

Figure 6 illustrates the results from two low-intensity solar radiation days, labeled as runs #1 and #2. When the discharge phase begins at 17:00, the hot water initially reaches nearly 50 °C, but within approximately 30 min, its temperature drops sharply to around 36–38 °C. Over the following 1–2 h, the hot water temperature further decreases to below 35 °C. Similarly, the cold-water temperature also declines gradually; starting at over 30 °C, it eventually settles around 27–28 °C. Despite the temperature drop during continuous water draw-off, the system maintains a significant thermal differential: about 10 °C for the first 2 h (between 17:00 and 19:00) and at least 5 °C for the subsequent 2 h (from 19:00 to 21:00). This profile indicates that while continuous discharge results in a decline in temperature of hot water, the thermal battery is effective at sustaining a meaningful temperature difference for several hours [18].

Figure 7 displays data for two sunny days (runs #3 and #4) with fluctuating solar radiation due to partial cloudiness. In these runs, the hot water discharge starts with peak temperatures around 55 °C. In run #3, the hot water temperature falls to about 40 °C within 30 min and reaches 35 °C after roughly 2.5 h, with the initial temperature difference exceeding 10 °C and diminishing to around 5 °C, over 2.5 h. In run #4, the hot water cools to 40 °C in about one hour and to 35 °C in roughly three hours. During the first hour, the temperature remains more than 10 °C above the baseline, and it stays over 5 °C higher for the next three hours. Altogether, this shows the thermal battery can deliver adequate hot water for about three to four hours of continuous use, meeting typical household needs [18].

The experimental results show that the efficiency of the thermal battery is defined as the proportion of storable solar energy that is transferred to the hot water varies with solar intensity. On days with lower solar intensity, the efficiency was around 44–47%, while on sunny days it improved to approximately 56%. Overall, the average efficiency of the thermal battery was close to 50%, with an uncertainty of about $\pm$ 9.3% as shown in Table 6. This indicates that the design is reasonably effective at storing and transferring solar energy under varying weather conditions [18].

Table 5 An overview of mode 1 runs [18]

Run No.	Solar intensity category	Weather conditions in morning, noon, afternoon
#1	Low	Cloudy, cloudy, cloudy
#2	Low	Sunny, rainy, rainy
#3	High	Sunny, sunny, sunny
#4	High	Sunny, sunny, sunny

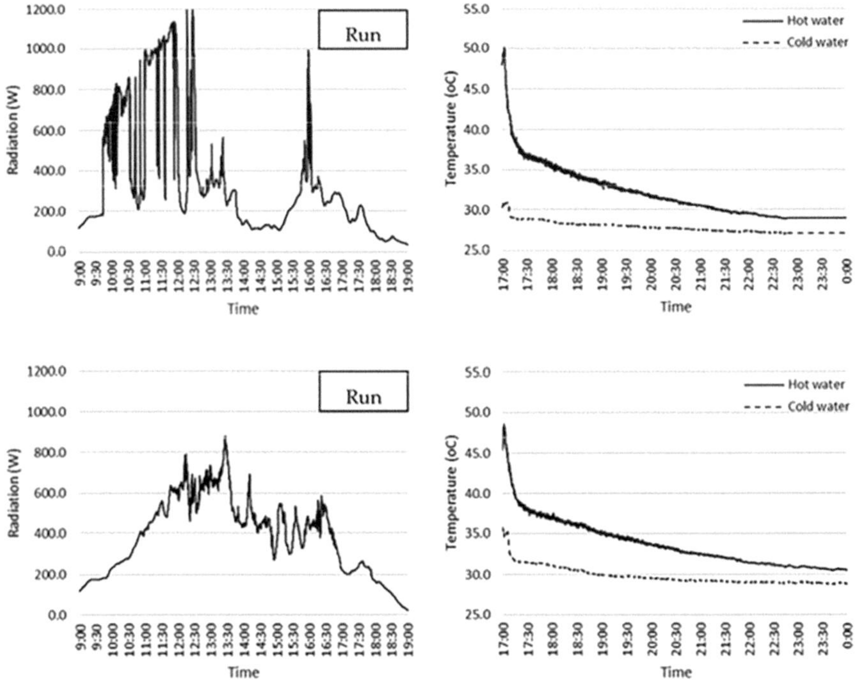

Fig. 6 Low intensity sun radiation is used for runs #1 and #2. (Right) Inlet/outlet hot water temperatures; (Left) Solar radiation [18]

Mode 2 (Periodic Draw-Off):.

Mode 2 employs a batch-wise, concurrent charging-discharging process to assess the system's ability to meet typical domestic hot water requirements under varying climatic conditions. In this mode, the system's performance was tested over four days (runs #5–#8): two days with low solar intensity and two with high solar intensity. As shown in Fig. 8, the solar intensity was recorded throughout these days, while Fig. 9 details the water draw-off profiles for runs #5 and #6. In these experiments, the water was drawn four times in the morning and four times in the evening at a fixed flow rate of 2 L/min. This approach mimics intermittent user demand rather than a continuous discharge pattern. The results indicate that on low solar intensity days (e.g., run #5), the hot water temperature remained mostly below 35 °C during the morning and stayed in a similar low range in the evening, with the thermal battery increasing the water temperature by only about (3–5 °C in the morning) and (5–10 °C in the evening). These findings suggest that under cloudy or rainy conditions, the current thermal battery size and PCM selection are insufficient to consistently reach the required operating temperature of at least 38 °C for domestic use. Therefore, to ensure adequate hot water delivery, the system might need to be scaled up, optimized further in terms of PCM properties, or paired with an auxiliary heater [18].

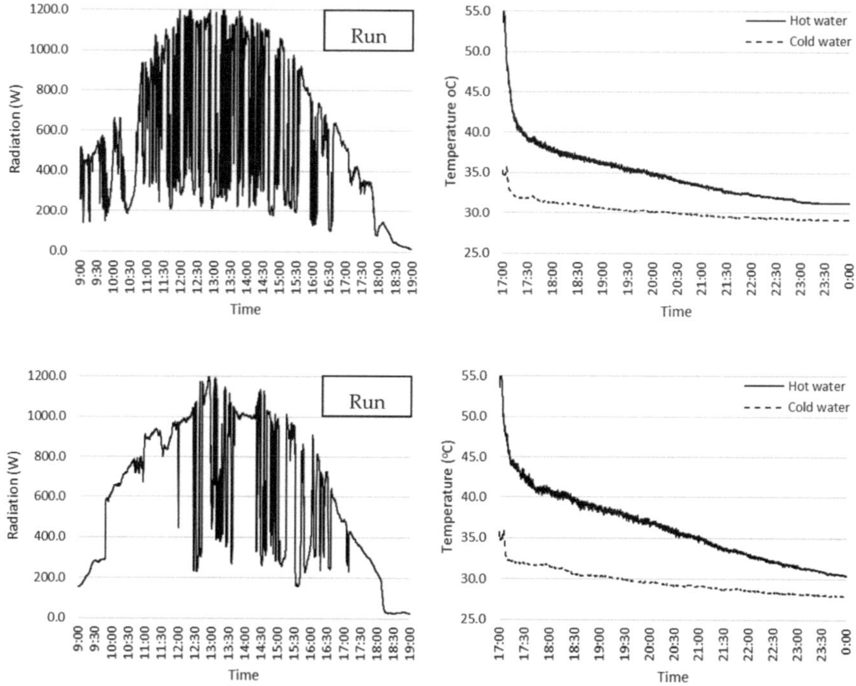

Fig. 7 Solar radiation intensity is high in runs #3 and #4. (Right) Inlet/outlet hot water temperatures; (Left) Solar radiation [18]

Table 6 Totals for energy and efficiency from runs 1–4 [18]

Runs No.	Daily radiation (MJ/m^2)	Daily total solar radiation (MJ)	Storable daily solar radiation (MJ)	Total heat transfer to water (MJ)	Efficiency of thermal battery (%)	Efficiency of system (%)
#1	13.9	24.6	13.3	5.8	44	24
#2	12.3	21.8	11.8	5.5	47	25
#3	18.6	33.0	17.9	7.6	42	23
#4	21.4	37.9	20.5	11.4	56	30

As shown in Fig. 10, runs 7 and 8 took place on days with very high solar irradiance. In those tests, morning draw-off temperatures consistently rose above 38–40 °C, and evening temperatures remained in the 40–45 °C range. Under these bright conditions, the thermal battery boosted the incoming cold water by at least 10 °C in the morning and about 15 °C in the evening, delivering hot water at temperatures well suited for household use. [18].

The "hot-water volume at 38 °C" tells you how much of your heated water can be mixed with cold inlet water and still end up at 38 °C. You work it out by comparing

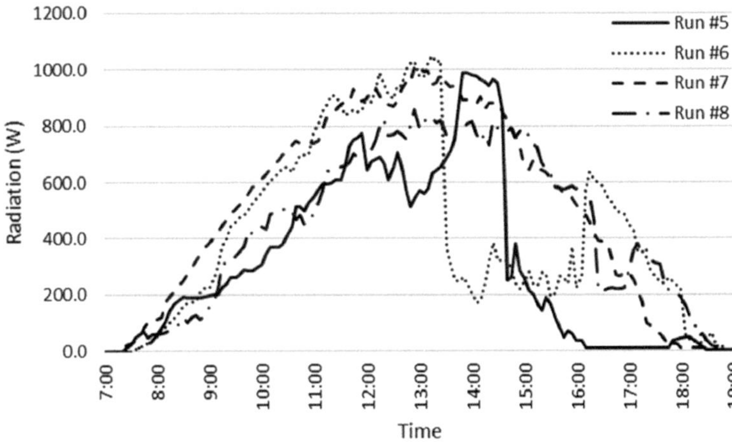

Fig. 8 Four runs of solar radiation in Mode 2 [18]

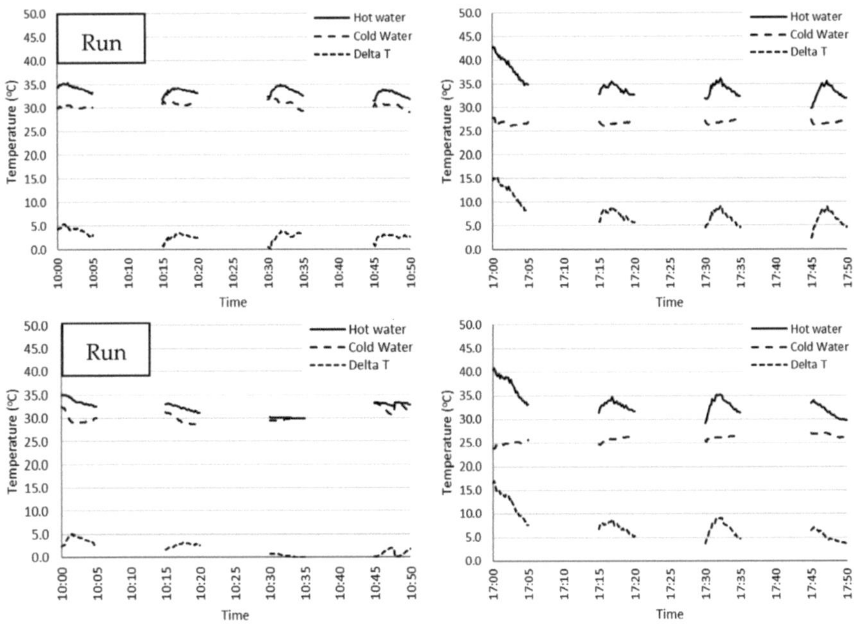

Fig. 9 For days with low levels of sun radiation, run #5 and #6. Morning water draw-off (left). Nighttime water draw-off (Right) [18]

how many degrees your hot water is above 38 °C to how many degrees the cold water is below 38 °C and using that ratio to scale the heated-water volume [18].

Table 7 illustrates that on days with low solar intensity (runs #5–#8), the thermal battery struggles to maintain the necessary temperature for domestic use. Conversely,

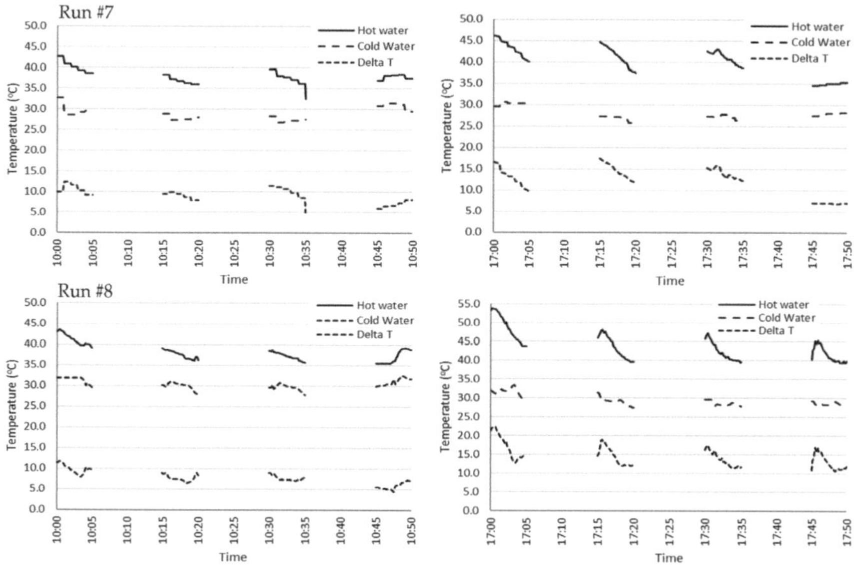

Fig. 10 Runs #7 and #8 on days with significant levels of sun exposure. In the morning, water drew off (left). Evening water draw-off (Right) [18]

under higher solar intensity, the system can reliably deliver a volume of hot water sufficient to meet the typical demand of a 3–5 persons household in Malaysia. If the thermal battery is not completely discharged during the evening, it retains residual heat that can be used the following morning. This means the system can provide hot water in the early hours for domestic use by utilizing the stored heat from the previous day, thereby enhancing energy efficiency and ensuring a more continuous supply of hot water [18].

Table 7 Runs #5–#8: produces hot water at working temperature and receives daily sun radiation [18]

Run No.	Recorded daily solar radiation (Ir/m^2)	Total daily solar radiation (It/day)	Absorbable daily solar radiation (Ia/day)	Hot water volume in Run No. 38 °C (L/day)
#5	13.6	24.1	13.1	0.0
#6	15.9	28.2	15.3	0.0
#7	22.1	39.2	21.2	197.1
#8	20.1	35.5	19.2	130.8

4 Summary

The experimental investigation of the ETHPSC-TM system revealed that the thermal battery, which incorporates a manifold reshaped into a phase change material (PCM) tank with porous media, achieves an efficiency of about 44% on lower solar intensity days and 56% on higher solar intensity days, averaging around 50% with uncertainties. Under sunny conditions, it can heat more than 120 L of water per day to at least 38 °C, making it viable for typical domestic needs. However, on cloudy or rainy days, an auxiliary heater or larger solar collector would be necessary to reach the desired water temperature. In addition to providing immediate hot water upon demand, the system avoids the need for a dedicated hot water storage tank or a water pump, prevents the stratification of hot water, and offers a compact, cost-effective alternative to conventional solar water heaters [18].

4.1 Case Study II

Title: **Experimental Analysis of Shape-Stabilized PCM Applied to a Direct-Absorption Evacuated Tube Solar Collector Exploiting Sodium Acetate Trihydrate and Graphite** [20]

Authors: Gholamabbas Sadeghi, Mohammad Mehrali, Mina Shahi, Gerrit Brem, Amirhoushang Mahmoudi

Year: 2022

This paper discusses the experimental integration of a shape-stabilized phase change material (PCM) into a direct-absorption evacuated tube solar collector, with the aim of eliminating the need for a separate hot water storage tank by storing thermal energy directly in the collector. The authors synthesized a salt-hydrate PCM based on sodium acetate trihydrate with various additive concentrations to create a stable shape-stabilized PCM (SSPCM); an optimal formulation was identified and subjected to multiple charge–discharge cycles to ensure reliable heat storage performance. The SSPCM-filled evacuated tube was then tested by first charging it under stagnation conditions (no water flow) to absorb solar heat and subsequently discharging the stored energy by circulating water through a double U-tube heat exchanger at flow rates of 10, 27, and 40 L per hour (LPH. Key findings indicate that incorporating the PCM within the collector increased the stagnation-mode thermal efficiency from roughly 66–82%, and that raising the water flow rate from 10 to 27 LPH had minimal impact on heat extraction, whereas a high flow of 40 LPH caused a pronounced drop in discharge efficiency. Additionally, a basic economic and environmental analysis showed a payback period of about 6 years for the system and an annual reduction of approximately 5.4 tons of CO_2 emissions due to the improved energy efficiency [20].

4.1.1 Material Used in Experiments

The experiment utilized a shape-stabilized phase change material (SSPCM) as the thermal storage medium for the solar collector, specifically engineered to store heat effectively while preventing leakage during operation. The SSPCM is a composite material comprised of sodium acetate trihydrate (SAT), which serves as the primary phase change material, sodium phosphate monohydrate (SPM) added at 2 wt% as a nucleating agent to promote uniform crystallization, carboxymethylcellulose (CMC) incorporated at 1 wt% to function as a thickening agent and improve mixture stability, and expandable graphite (EG) which enhances both the material's thermal conductivity and its shape stability. As described in Fig. 11a, the SAT is initially melted in an Erlenmeyer flask placed in a water bath on a hot plate under magnetic stirring; subsequently, SPM is added and, after 20 min, CMC is gradually incorporated to prevent agglomeration, before the entire solution is refrigerated for two days to allow it to solidify. Meanwhile, the expandable graphite is separately processed by subjecting it to 900 °C for 30 s in a furnace, converting it into expanded graphite as illustrated in Fig. 11b, c. The solid SAT-SPM-CMC composite is then crushed into a powder and mixed with 4 wt% EG; this powdery mixture is reheated and repeatedly stirred to ensure thorough impregnation of the SAT-SPM-CMC into the pores of the EG [20].

Following this, the SSPCM is packaged and later ground to form a fine powder suitable for use within the solar collector. To optimize the thermal properties of the SSPCM, the preparation process was repeated with variations in CMC and EG concentrations—using 2 and 3 wt% CMC in conjunction with 8 wt% EG—with the resulting samples systematically detailed in Table 8 and visually compared in Fig. 12, which shows formulations with varying levels of CMC and EG [20].

(a) (b) (c)

Fig. 11 **a** Salt hydrate PCM synthesis, **b** crucible-based expandable graphite, and **c** furnace-based expanded graphite [20]

Table 8 Various elements of the SSPCMs created for this study [20]

No. Material	Components			
	SAT (wt%)	SPM (wt%)	CMC (wt%)	EG (wt%)
SSCE (41)	93	2	1	4
SSCE (42)	92	2	2	4
SSCE (43)	91	2	3	4
SSCE (44)	89	2	1	8
SSCE (45)	88	2	2	8
SSCE (46)	87	2	3	8

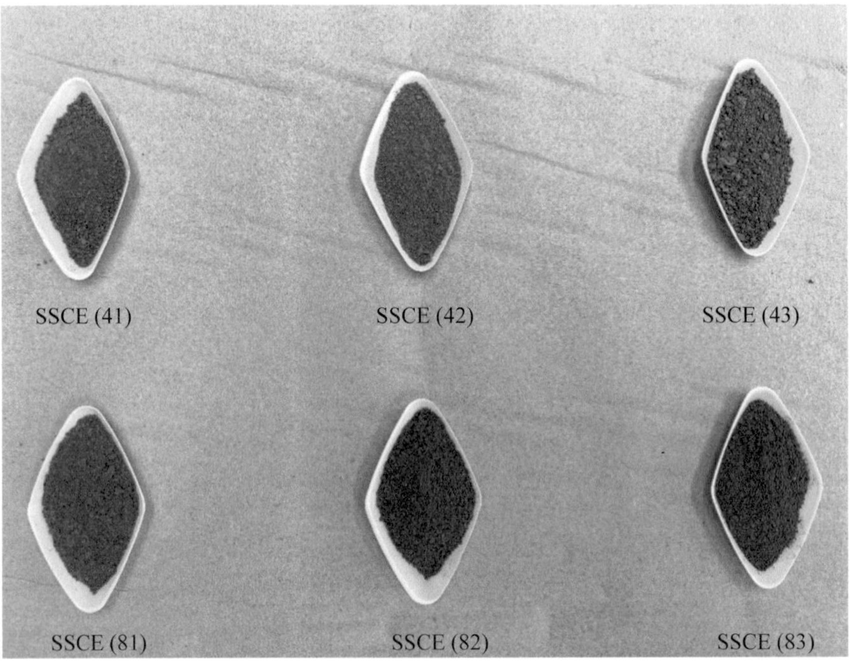

Fig. 12 The synthetic SSPCMs with varying levels of EG and CMC [20]

4.1.2 Experimental Set-Up and Methodology

The experimental set-up and methodology were carefully devised to assess the performance of a direct-absorption evacuated tube solar collector integrated with a shape-stabilized phase change material (SSPCM). The collector was fabricated using borosilicate glass of a defined thickness, and it incorporated a specially designed stainless-steel heat exchanger configured as a double U-pipe to facilitate effective water circulation. Detailed specifications regarding the tube dimensions, construction materials, and overall configuration are provided in Table 9 [20].

Table 9 The physical qualities of the produced DA and commercial -ETSC [20]

Characteristic	Unit	Dimension and substance
Length of the DAET	m	0.5
Inner diameter of the DAET	m	0.048
Outer diameter of the DAET	m	0.075
Number of the DAETs	–	1
Aperture area of the DA-ETSC	m^2	0.0375
Length of the double U-pipe HEX	m	2.2
Inner diameter of the U-pipe HEX	m	0.004
Outer diameter of the U-pipe HEX	m	0.006
Tilted angle of the collector	Degree	45
Working temperature range	Degree C°	20–80
Material of double U-pipe HEX	–	Stainless steel
Material of glass	–	Borosilicate
Material of frame	–	aluminum
Material of insulation	–	Polyethylene foam

In the experimental arrangement, as illustrated in Figs. 13 and 14, the DA-ETSC was assembled within the laboratory environment at the University of Twente and evaluated under controlled conditions using a solar simulator. During the charging phase, the SSPCM housed within the evacuated tube directly absorbed incident solar radiation under stagnant water conditions, thereby eliminating any intervening thermal barrier. The setup was instrumented with multiple thermocouples placed at the top, middle, and bottom segments of the tube to capture the spatial temperature distribution, while additional sensors measured the inlet and outlet water temperatures during the discharging phase [20].

The methodology entailed a systematic charging of the system under standard 1 kW/m^2 solar intensity until the SSPCM reached its designated thermal threshold, followed by a controlled discharging process at flow rates of 10, 27, and 40 L per hour. These flow rate variations were introduced to ascertain their impact on parameters such as heat extraction, discharging efficiency, and thermal losses. Temperature evolution throughout the charging and discharging cycles was continuously recorded, with the acquired data graphically depicted in Fig. 15. To ensure the reliability of the experimental data, each test condition was repeated, allowing for a robust analysis of the collector's performance metrics including useful heat gain and overall efficiency. This integrated approach, combining detailed experimental setup with rigorous data acquisition and repeatability, provided a comprehensive framework to evaluate the potential of the proposed solar collector system [20].

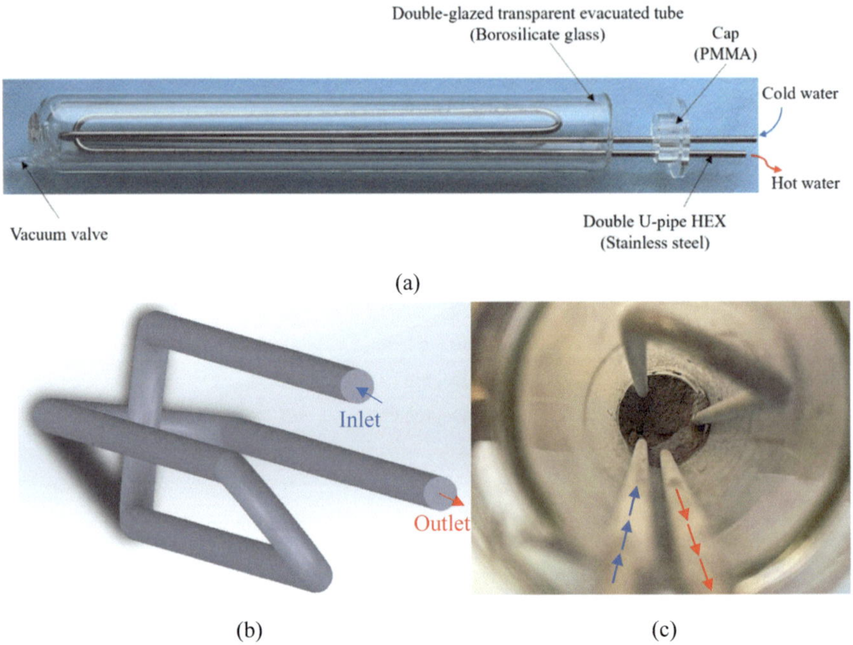

Fig. 13 **a** The HEX inside the DAET is configured; **b** the suggested twin U-pipe HEX is designed in three dimensions; and **c** a view of the tube's opening portion shows how the SSPCM and the HEX within the DA-ETSC are in touch [20]

4.1.3 Results and Discussion

The results of the study demonstrate that the direct absorption evacuated tube solar collector integrated with the SSPCM exhibits promising thermal performance under varying operational conditions. Temperature profiles recorded during the charging phase showed that the SSPCM was effectively heated by the solar simulator under stagnant conditions, as evidenced by the distinct temperature evolution along the tube. During the subsequent discharging phase, experiments conducted at different water flow rates (10, 27, and 40 LPH) revealed that while the collector maintained a high heat absorption efficiency of approximately 82%, the discharging efficiency and heat extraction values varied significantly with the flow rate. These trends are clearly illustrated in Fig. 15 and detailed in Table 10, where the optimal performance was observed at 27 LPH. At this flow rate, the discharging process provided a balanced heat transfer with an average outlet-inlet temperature difference that enabled adequate heat extraction without incurring substantial heat losses [20].

Further analysis, as depicted in Fig. 16, confirms that the useful heat gain of the solar collector remained relatively stable for a period during the discharge at 27 LPH, whereas operating at 10 LPH resulted in prolonged discharge times and 40 LPH led to a rapid decline in heat extraction. Additionally, the study extended these findings to design a multi-tube system, with Fig. 17 providing insights into

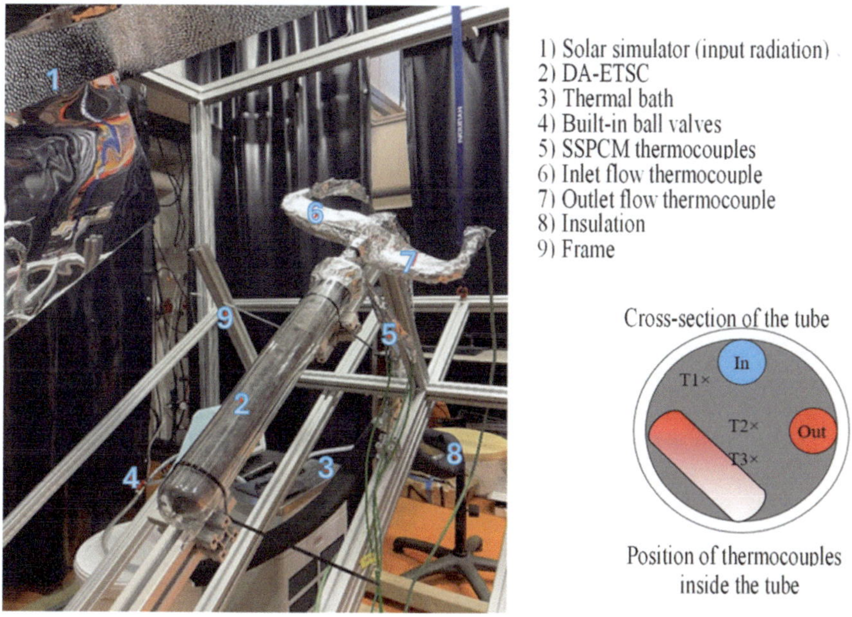

Fig. 14 Experimental setup of the SSPCM-DA-ETSC tested under the solar simulator in the laboratory [20]

the performance generalization across multiple collectors. This approach is validated that the proposed DA-ETSC configuration, when scaled up, can successfully meet the hot water demands of a typical household while minimizing space requirements and achieving enhanced energy efficiency. Overall, the experimental results underscore the technical and practical viability of integrating an SSPCM within a direct-absorption solar collector for efficient thermal energy storage and utilization [20].

5 Summary

The study demonstrates a novel integration of a direct-absorption evacuated tube solar collector with a shape-stabilized phase change material (SSPCM) that effectively enhances thermal energy storage and overall system efficiency. The developed SSPCM, formulated with sodium acetate trihydrate, sodium phosphate monohydrate, carboxymethylcellulose, and expandable graphite, exhibits excellent heat storage capacity and stability, mitigating leakage issues typically encountered in solar thermal systems. Experimental investigations under controlled conditions reveal that the integrated system achieves high heat absorption efficiency, with an optimal balance of charging and discharging performance at a moderate water

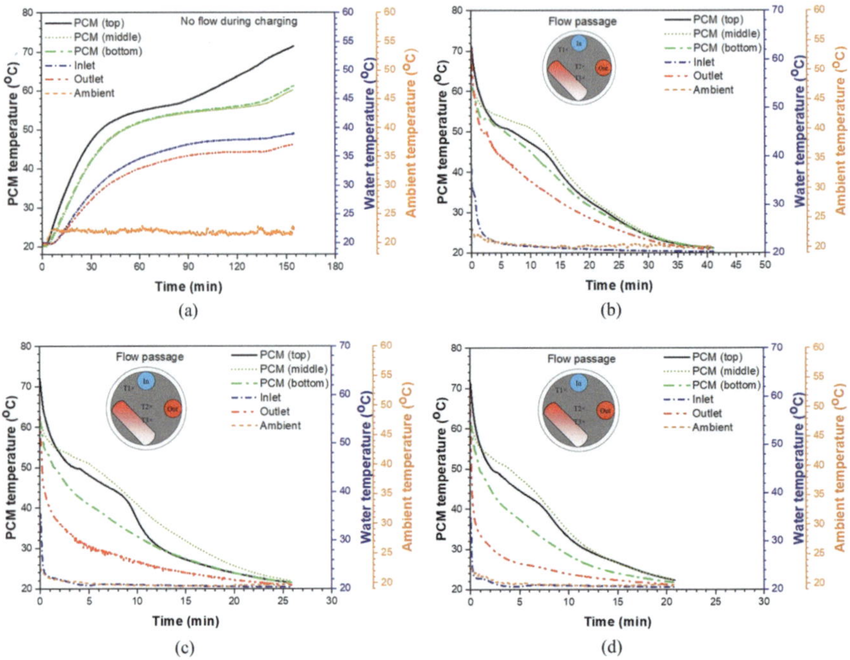

Fig. 15 SSPCM, inlet–outlet water, and ambient temperature variation charts for the following scenarios: **a** charging without flow rate, **b** discharging at 10 LPH flow rate, **c** discharging at 27 LPH flow rate, and **d** discharging at 40 LPH water flow rate[20]

flow rate, thereby ensuring efficient energy extraction and minimized thermal losses. Overall, the findings underscore the technical viability and scalability of the proposed system for sustainable domestic hot water production, offering a promising approach to improve space utilization and reduce energy consumption in solar thermal applications [20].

5.1 Case Study III

Title: **An Experimental Evaluation of Direct Flow Evacuated Tube Solar Collector Integrated with Phase Change Material** [23]

Authors: Mohamed Hany Abokersh, Mohamed El-Morsi, Osama Sharaf, Wael Abdelrahman

Year: 2017

This paper discusses an experimental investigation into integrating phase change material (PCM) within a direct flow evacuated tube solar water heater. The study

Table 10 DA-ETSC thermal analysis with three flow rates during the discharging process and no water flow during the charging process [20]

No.	Parameter	Flow rate during the discharging process (LPH)		
		10	27	40
1	t_{ch} (s)	9650	9650	9650
2	t_{dis} (s)	2462	1550	1206
3	Q_{in} (kJ)	361.87	361.87	361.87
4	Tini, PCM (°C)	20	20	20
5	Tf, PCM (°C)	63.86	64.37	64.54
6	Q_{st} (kJ)	295.41	297.23	296.79
7	Q_u (kJ)	232.02	229.2	187.96
8	$Q_{loss,ch}$ (kJ)	66.46	64.63	65.07
9	$Q_{loss,dis}$ (kJ)	63.39	68.03	108.083
10	F_R	0.79	0.78	0.64
11	$U_{L,ch}$ (W/m^2 K)	4.08	3.96	3.99
12	$U_{L,dis}$ (W/m^2 K)	15.25	26.01	53.47
13	U_L (W/m^2 K)	6.35	7.01	9.49
14	η_{st}	0.78	0.77	0.63
15	η_{dis}	0.64	0.63	0.51
16	η_{ab}	0.82	0.82	0.82
17	η_{ab} regarding water as the working fluid inside an ETSC = 66%			

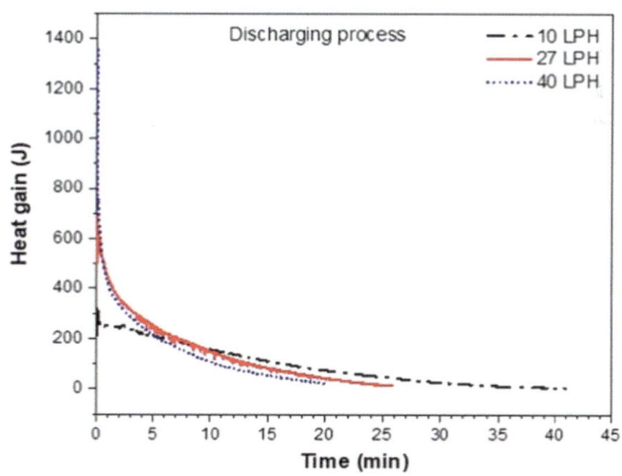

Fig. 16 Qu readings of the DA-ETSC under investigation at water flowrates of 10, 27, and 40 LPH during the experiment's discharge periods [20]

evaluates two configurations: A system with a conventional, un-finned U-tube heat exchanger and one that incorporates external fins; to enhance heat transfer. By embedding paraffin wax inside each evacuated tube, the paper examines how the PCM stores and later releases solar thermal energy. The experimental setup assesses both

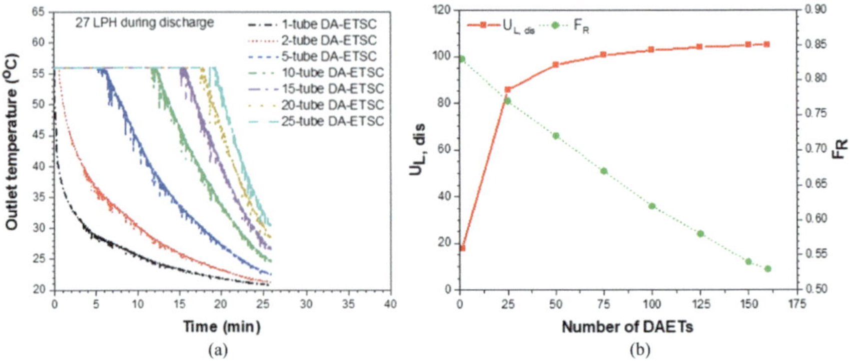

Fig. 17 Qu readings of the DA-ETSC under investigation at water flowrates of 10, 27, and 40 LPH during the experiment's discharge periods [20]

the charging phase (where solar energy is absorbed and stored) and the discharging phase (where stored energy is transferred to the circulating water). Key performance indicators such as system efficiency, heat loss coefficient, and total effective energy discharged are analyzed using methods compliant with recognized standards. The results highlight that while the un-finned system achieves higher daily system efficiency due to elevated PCM temperatures, the finned configuration can prolong energy retention and improve energy discharge under specific operating conditions [23].

5.1.1 Material Used in Experiment

This study employs a direct flow evacuated tube solar water heater system that integrates phase change material (PCM) for thermal energy storage. Each evacuated tube is filled with approximately 0.8 kg of commercial grade paraffin wax, known as ALEX WAX 600, which serves as the core PCM due to its favorable thermal properties. As detailed in Table 11, the paraffin wax exhibits a melting temperature range between 58 and 62 °C, a latent heat of fusion of 189 kJ/kg, and a density of 0.920 kg/m^3 when solid and 0.795 kg/m^3 when liquid. Its specific heat capacity is 2.3 kJ/kg K, and its thermal conductivity is 0.21 W/m K, characteristics that make it especially suitable for capturing and storing solar energy while operating safely up to 90 °C [23].

The system configuration features a U-tube heat exchanger embedded within the evacuated tubes, which facilitates both the charging and discharging phases of the system. In one experimental variant, the U-tube is unmodified, while in the other, aluminum fins are attached externally to the heat exchanger. These fins are introduced to enhance the heat transfer rate by increasing the available surface area, thereby helping to overcome the intrinsic low thermal conductivity of the wax during its

Table 11 Thermophysical characteristics of paraffin wax [23]

Material base	Organic
Melting temperature	58–62 °C
Latent heat of fusion	189 kJ/kg
Density solid (at 20 °C)	0.920 kg/m^3
Density liquid (at 70 °C)	0.795 kg/m^3
Specific heat capacity	2.3 kJ/kg K
Thermal conductivity (Solid and liquid)	0.21 W/m K
Volume expansion	12.5%
Max. operation temperature	90 °C
Oil content	1–2%

phase transition. A schematic diagram illustrating the experimental setup and the placement of fins can be seen in Fig. 18 [23].

Complementing the core materials, the experimental setup includes a suite of precision instrumentation to monitor and control the system parameters. Temperatures within the PCM are recorded via strategically placed negative temperature coefficient (NTC) thermistors along the axial direction of the evacuated tubes, ensuring a comprehensive profile of temperature distribution during both solar charging and subsequent discharging. The system is managed by a data acquisition unit built on an Arduino Mega 2560 platform, which is responsible for both the measurement and control functions. Solar irradiance is measured using a pyranometer, while ambient conditions such as temperature, humidity, wind speed, and direction are continuously monitored by dedicated sensors. Additionally, the water flow rate through the U-tube is carefully regulated and measured using rotameters and water pulse flowmeters, which are essential for maintaining the stability of the system during experimental trials. Together, these materials and instrumentation form an integrated experimental

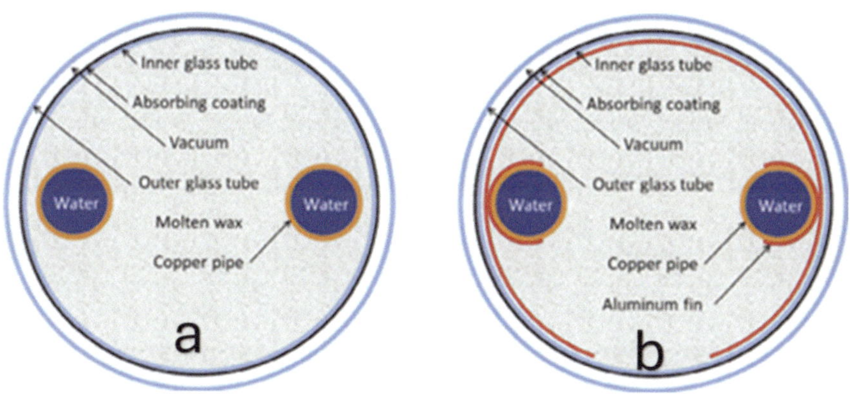

Fig. 18 The integration of PCM in a U-tube HX inside an evacuated tube solar collector is shown schematically in **a** System A, without fin, and **b** System B, with fin [23]

framework that allows for the effective evaluation of thermal performance and energy discharge efficiency in the modified solar water heating system [23].

5.1.2 Experimental Set-Up and Methodology

An experimental rig was developed to assess the charging and discharging performance of a direct-flow evacuated-tube solar water heater enhanced with a phase-change material (PCM). The collector comprises 21 evacuated tubes, each housing a U-tube heat exchanger filled with 0.8 kg of paraffin wax and presents an aperture area of 2.13 m^2. Installed at a 45° tilt and oriented due south, the collector exhibits an optical efficiency of 74.2%. Two distinct circuits charging and discharging govern the system's thermal operation. In the charging circuit, water is maintained at a set temperature in a 32 L storage tank by means of a 3-kW electric heater and temperature controller. The discharging circuit employs a 1 HP centrifugal pump, pressure-relief valves, a pressure regulator, and an expansion tank to deliver heated water to the load. Prior to each charging test, cold water is circulated through the collector at a constant flow rate until thermal uniformity is achieved. Concurrently, a pyranometer records incident solar irradiance, while ambient temperature, relative humidity, and wind speed are logged continuously. Thermistors embedded along each evacuated tube monitor the spatial and temporal evolution of the PCM's solid–liquid phase transition, as illustrated in Fig. 19 [23].

In addition, the experimental methodology follows established standards such as the CNS B7277-12558, which allows for the evaluation of integrated energy storage systems. During the charging phase, a daily system efficiency is determined by comparing the total solar energy incident on the collector with the energy stored in both the PCM and the circulating water. The system's thermal behavior is further studied through overnight heat loss tests, which assess the rate at which the stored energy dissipates when the system is not actively heated. The discharging phase is explored using a design of experiments (DoE) approach where variables such as initial discharge temperature, discharge flow rate, and the presence of external fins on the U-tube heat exchanger are systematically varied. The discharging experiments are conducted in multiple runs, and the responses (including total solidification time and total effective energy discharged) are recorded and analyzed. Throughout the study, data acquisition is managed by an Arduino-based system that records temperature readings and controls water flow at regular intervals, while quality and repeatability are ensured by calibrating the instruments prior to experimentation as shown in Table 12 and Fig. 20 [23].

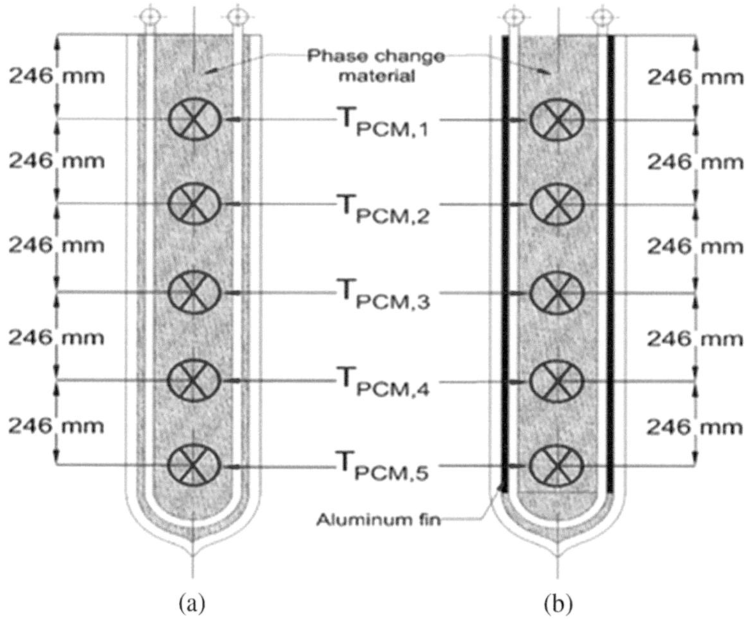

Fig. 19 The thermistors' location within the evacuated tube System A: fin-free, System B: fin-equipped [23]

Table 12 An overview of the measurement tools' random and systematic uncertainty [23]

Device	Location	Systematic uncertainty	Random uncertainty	Total uncertainty
Thermistor	TPCM	± 0.0038 °C	± 0.01 C	± 0.0038 °C
	Tin, Tout	± 0.009 °C	± 0.40 °C	± 0.40 °C
Pulse flow sensor	System A	± 0.006 L/min	± 0.12 L/min	± 0.12 L/min
	System B	± 0.007 L/min	± 0.16 L/min	± 0.16 L/min
Rotameter	System A	± 0.013 L/min	± 0.031 L/min	± 0.031 L/min
	System B	± 0.014 L/min	± 0.038 L/min	± 0.038 L/min

5.1.3 Result and Discussion

This study's results and discussion reveal a detailed evaluation of the thermal performance of two configurations of a direct flow evacuated tube solar water heater integrated with a phase change material (PCM). During the charging phase, the time-history of the PCM temperature and solar irradiance illustrated in Figs. 21 and 22 demonstrates how natural convection governs the heat transfer process. In the unfinned configuration (System A), the upper layers of the PCM reach higher temperatures, producing a higher daily system efficiency by approximately 14% compared to the finned configuration (System B) [23].

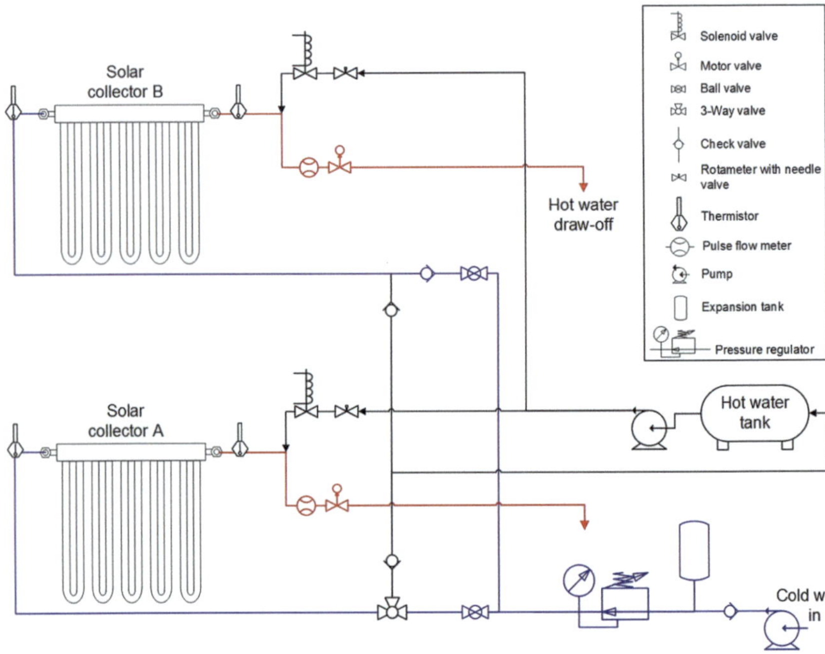

Fig. 20 Diagrammatic representation of the experimental setup [23]

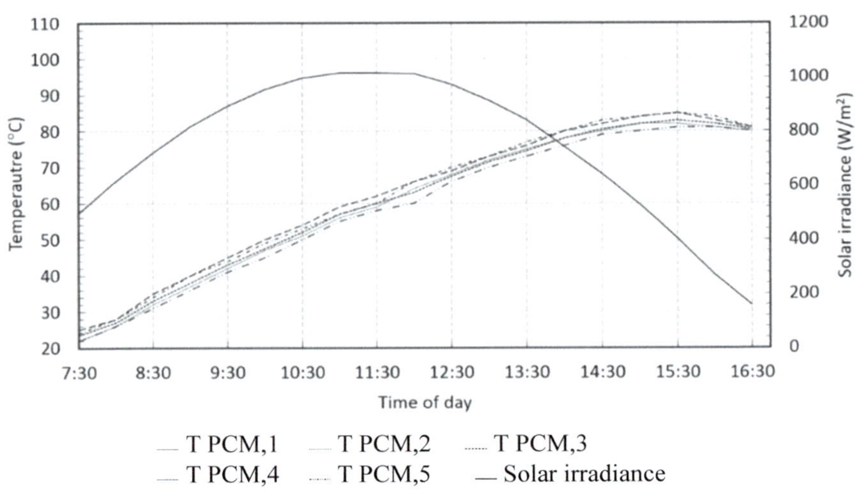

Fig. 21 PCM temperatures and solar irradiation during system B's charging phase [23]

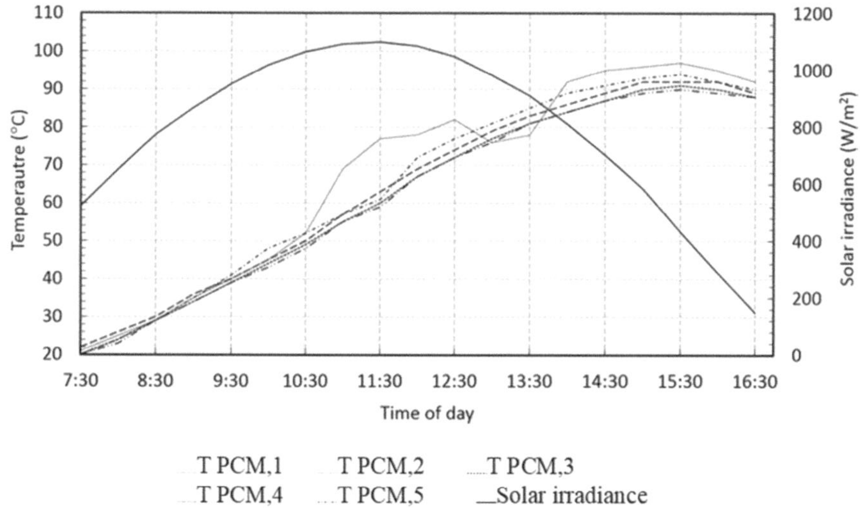

Fig. 22 Solar radiation and PCM temperatures during system B's charging phase [23]

However, the addition of external fins in System B promotes a more uniform temperature distribution across the PCM layers, reducing the maximum temperature difference and resulting in a lower heat loss coefficient. These observations are quantified in the daily efficiency plots and summarized in Fig. 23 along with related tabulated data [23].

The discharging phase experiments further highlight the contrasting performance between the two configurations. When evaluating water outlet temperature profiles and the solidification process at various discharge flow rates, the presence of fins is seen to effectively enhance the heat transfer rate at low discharge rates. For instance, under a discharge flow rate of 1 L/min at a high initial PCM temperature (85 °C), the fined system exhibits a faster discharge process with a total effective energy discharge (Q_D) improvement of about 35% relative to the un-finned system. In contrast, at higher discharge flow rates, the rapid formation of an insulating solidified

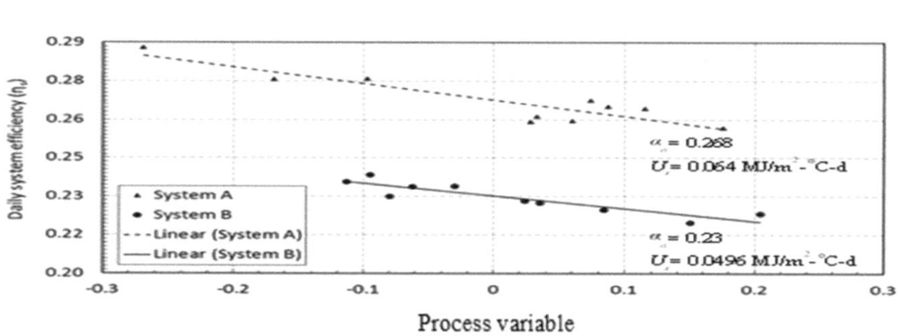

Fig. 23 Systems A and B's daily system efficiency [23]

layer diminishes the advantage provided by fins. The complete discharge behavior, presented over time in Figs. 24, 25, 26 and 27 confirms that both the total solidification time (t_D) and Q_D are predominantly sensitive to variations in the initial PCM temperature and the discharge flow rate [23].

Detailed statistical analysis via ANOVA indicates that the initial PCM temperature alone accounts for more than 50% of the variation in t_D and over 90% of the variation in Q_D, while the fin effect and discharge flow rate. Furthermore, regression models, including a linear model for t_D and a logarithmic model for Q_D, provide reliable predictive capabilities, which are confirmed by the repeatability tests shown in Fig. 28. Overall, the results underscore that optimum discharging performance is achieved at a discharge flow rate of 1 L/min and an initial PCM temperature of

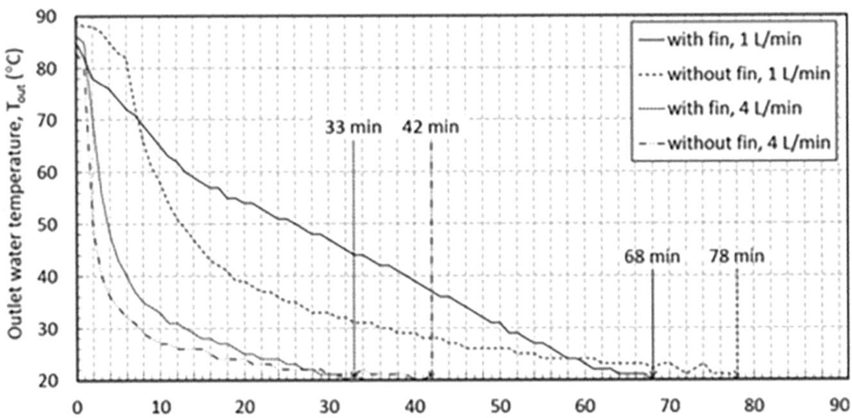

Fig. 24 Tout at $T_{PCMD} = 85$ C, V D 1, and 4 L/min during the discharging phase [23]

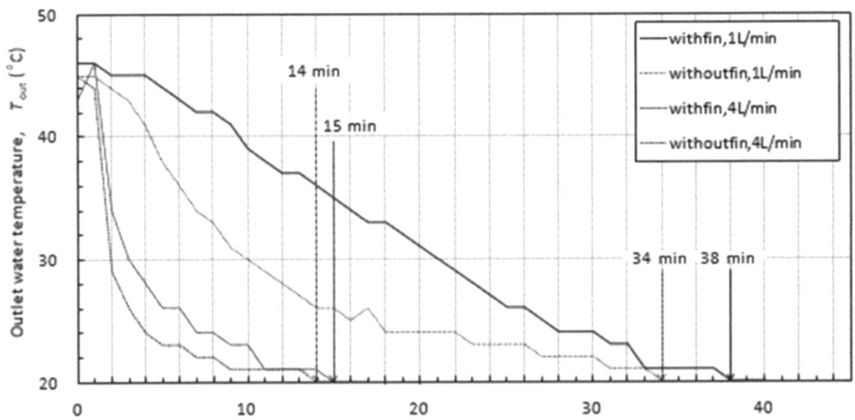

Fig. 25 Tout at T_{PCM} $D = 45$ C, V D 1, and 4 L/min during the discharging phase. [23]

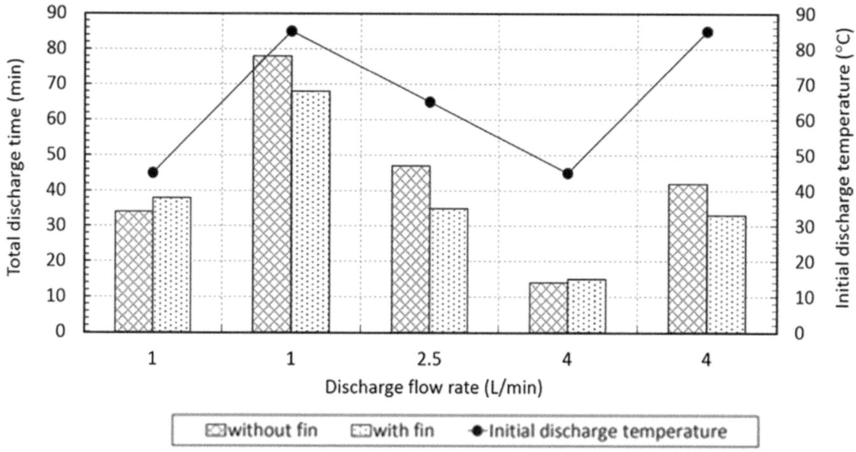

Fig. 26 Duration of total solidification at varying discharge flow rates and T_{PCM} [23]

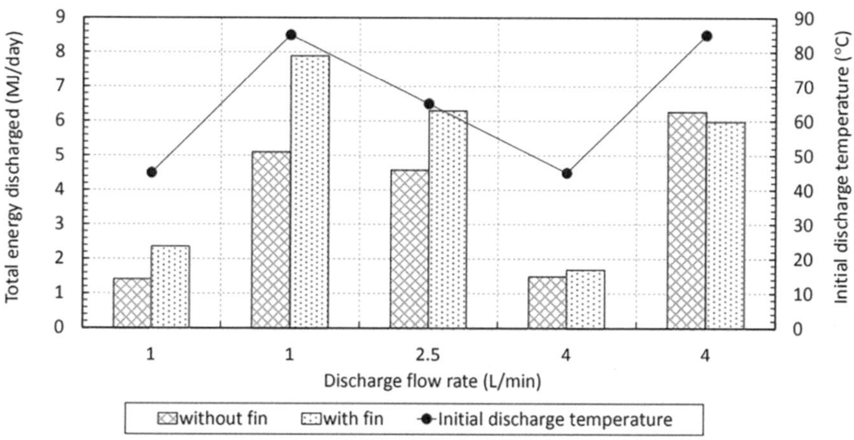

Fig. 27 Total effective energy released at varying $T_{PCM\ D}$ and discharge flow rates [23]

85 °C in the finned configuration, where experimental predictions closely match the observed outcomes with minimal error [23].

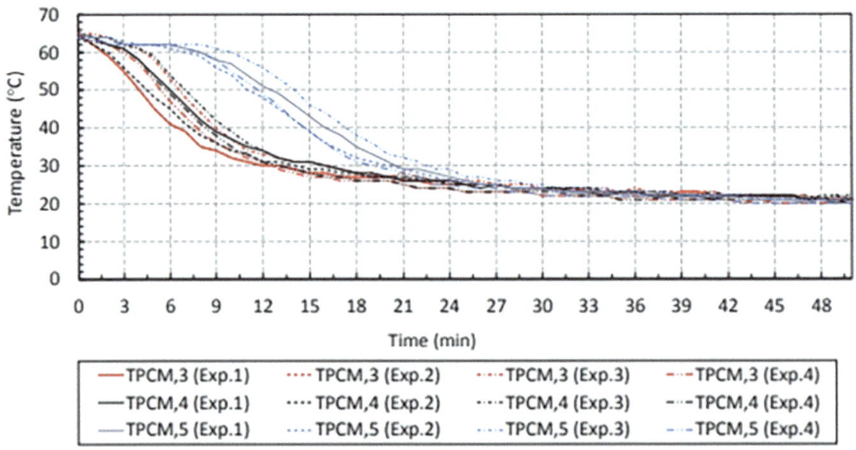

Fig. 28 PCM temperature (TPCM,3,4,5) for system A in four similar experiments [23]

6 Summary

This study concludes that integrating phase change material within a direct flow evacuated tube solar water heater significantly enhances thermal energy storage and discharge performance. The investigation demonstrated that although an unfinned configuration achieves a higher daily system efficiency due to higher PCM temperatures during the charging phase, the introduction of external fins facilitates a more uniform temperature distribution and improved energy retention during the discharging phase. The experiments revealed that the initial PCM temperature and the discharge flow rate are critical factors affecting both the total solidification time and the amount of energy effectively discharged. Statistical analyses and regression models confirmed the sensitivity of the system's performance to these parameters, with optimum performance achieved at a discharge flow rate of 1 L/min and an initial PCM temperature of 85 °C in the fined configuration. Overall, the study provides a comprehensive evaluation of the system's behavior under varied conditions, offering valuable insights into design and operational strategies for enhanced solar water heating systems [23].

6.1 Summary

This chapter provides a detailed description of the experimental methodologies and system configurations used to evaluate advanced solar water heating systems that integrate phase change materials (PCMs) for enhanced thermal energy storage. The chapter outlines three distinct case studies, each presenting a unique approach: one study investigates a thermal battery design where conventional evacuated tube heat pipe collectors are coupled with a PCM and porous media to improve heat distribution and storage efficiency; another examines a direct-absorption evacuated tube solar collector that incorporates a specially synthesized shape-stabilized PCM to eliminate bulky storage tanks and optimize charging/discharging cycles at various water flow rates; and the third evaluates a system that integrates PCM in both finned and un-finned configurations to assess the impact on system efficiency and heat recovery. Across all case studies, the chapter emphasizes the selection of appropriate materials such as paraffin wax with well-defined melting points, enhanced porous media to mitigate low thermal conductivity, and robust instrumentation for monitoring temperature, flow, and solar irradiance as well as the experimental setups designed to simulate real-world operating conditions, all aimed at optimizing solar thermal performance for sustainable domestic water heating applications are all summarized in Table 13.

7 Results and Discussion

7.1 Comparative Analysis of Case Studies

All three case studies demonstrate that integrating PCMs into solar collectors markedly improves thermal performance, though each system exhibits unique advantages and trade-offs. Case Study I—a solar water heater with a PCM "thermal battery" and porous media—achieved an average thermal efficiency of around 50% (approximately 44% on low-sun days and 56% on high-sun days). This system could heat over 120 L of water per day to at least ~ 38 °C under sunny conditions, effectively extending hot-water availability for typical domestic needs. Its design eliminated the need for a separate storage tank or pump by storing heat directly in the collector's manifold, thereby reducing complexity and preventing thermal stratification in the water. In contrast, Case Study II—which used a shape-stabilized PCM inside a direct-absorption evacuated tube collector—focused on maximizing energy capture within the collector itself. This configuration yielded a significantly higher stagnation thermal efficiency, improving from ~ 66% (with water-only) to ~ 82% when the PCM was integrated. The PCM in Case II effectively absorbed solar heat during "charging" (no flow) and then released it to water during "discharging"; tests showed that increasing the water flow rate from 10 to 27 LPH had minimal effect on heat-extraction efficiency, whereas a high flow of 40 LPH caused a noticeable drop in discharge efficiency. Case Study III—a direct-flow evacuated tube collector with PCM filling the tubes around a U-tube heat exchanger (comparing un-finned vs. finned tubes)—highlighted the impact of design modifications on performance. Notably, the *un-finned* configuration attained a higher daily efficiency, attributed to PCM reaching higher temperatures during charging (thus delivering higher-temperature output). The finned configuration, however, facilitated more uniform heat transfer into and out of the PCM, which prolonged the heat-release period and improved the total energy discharged under certain conditions.

Table 13 Case studies comparison table

Case study	Case study I	Case study II	Case study III
Method	Experimental	Experimental	Experimental
Type of solar collector	Evacuated tube heat pipe solar collector integrated with a thermal battery	Direct absorption evacuated tube solar collector	Direct-flow evacuated tube solar collector with a U-tube heat exchanger
Type of PCM	Organic	Eutectic	Organic
Name of PCM	Paraffin wax	Synthesized SSPCM based on sodium acetate trihydrate (SAT) with additives: sodium phosphate monohydrate (SPM), CMC, and expandable graphite	paraffin wax
PCM heat capacity	2.76 kJ/kg K (solid) and 2.48 kJ/kg K (liquid)	Not provided	2.3 kJ/kg·K
PCM melting temperature	53–57 °C (average ≈55 °C)	Not provided	58–62 °C (average ≈60 °C)
PCM density (solid/liquid)	990 kg/m^3 (solid at 15 °C)/ 916 kg/m^3 (liquid at 70 °C)	Not provided	0.920 kg/m^3 (solid at 20 °C)/0.795 kg/m^3 (liquid at 70 °C)
Outlet water temperature	In the discharging mode, the system delivers water warmed to about 38 °C	Not provided	Achieved 90 °C
Water flow rate	Exceeding 120 L/day,	Discharge tests performed at 10, 27, and 40 L/h (LPH)	Charging loop operated at 3 L/min
Surface area	Not provided	Not provided	Collector aperture area = 2.13 m^2
Solar intensity	Low 13.9–12.3 MJ/m^2 High 18.6–21.4 MJ/m^2	1 kW/m^2	7 MJ/m^2

(continued)

Table 13 (continued)

Case study	Case study I	Case study II	Case study III
System efficiency	Reported overall efficiency of the thermal battery is 50% ± 9.3%	Thermal efficiency improved from 66 to 82% with the integrated collector–storage system	While the study compared two configurations (un-finned vs. finned), only relative improvement was noted (≈ + 14% for un-finned)
Thermal conductivity	For the chosen paraffin wax 0.349 W/m K (solid) and 0.167 W/m K (liquid)	Not provided	0.21 W/m K (for the paraffin wax in both solid and liquid phases)

In summary, Case II achieved the highest peak efficiency (~ 82%) by virtue of direct PCM absorption, whereas Cases I and III prioritized system integration and heat-release management. All three clearly show that latent-heat integration stabilizes and enhances solar thermal performance, increasing efficiency extending the duration of useful heat delivery, and reducing thermal losses.

7.2 *Economic Implications and Energy Savings Potential*

Although PCM materials cost more per kg than water, their high latent-heat storage yields substantial energy savings, often offsetting the material investment. Common salt-hydrate PCMs can cost as little as US $0.10–2.50/kg yet store 0.2–0.3 kWh/kg of heat [36], delivering stored energy for under US$1.00/kWh. Specialized organic PCMs (paraffins, fatty alcohols) range higher (up to tens of US$ per kg) but are also steadily becoming more affordable as production scales up [37]. Importantly, field studies report payback periods often under 5 years due to reduced auxiliary-heating needs. For example, Case II's shape-stabilized PCM collector had an estimated payback of ~ 6 years thanks to its 16-pp efficiency boost and reduced backup heating. On a global scale, if PCM storage raises solar-fraction in existing solar-thermal installations by just 10–20%, the aggregate energy savings could be enormous, displacing large amounts of fossil-fuel heating [38].

7.3 *Environmental Benefits and Sustainability*

By extending solar-heat availability into evenings and cloudy periods, PCM storage maximizes renewable heating use and displaces grid electricity or fossil-fuel boilers, significantly cutting CO_2 emissions. Case II alone estimated annual CO_2 savings of

~ 5.4 tons per installation. Over typical lifetimes (15–20 years), cumulative reductions can exceed 50–100 tons of CO_2 per system. Moreover, most PCMs (paraffins, salt hydrates) are non-toxic and recyclable, minimizing end-of-life environmental impacts [39]. Enhanced efficiency also reduces the collector area needed for a given heat output, lowering material and embodied-energy footprints.

7.4 Future Work and Recommendations

While PCM-enhanced solar collectors have demonstrated clear benefits, further research is essential to address remaining challenges and optimize system performance. First, standardized benchmarking across different PCM-integrated designs such as heat-pipe versus direct-flow collectors under identical test conditions would provide objective metrics for efficiency, storage capacity, and output profiles [40]. Second, long-term durability studies are needed to assess PCM stability over thousands of melts–freeze cycles, focusing on phase separation, supercooling, and material fatigue; improved encapsulation and anti-leakage methods should be explored to maintain performance over decades [41]. Third, strategies to enhance PCM thermal conductivity, whether via metal foams, nanoparticle additives, or optimized fin geometries—require economic and technical evaluation to balance cost, manufacturability, and heat-transfer gains [42]. Fourth, expanding field trials to diverse climates (e.g., high-latitude winters, humid tropics) will inform adaptive PCM selection and system designs suited to variable solar resources. Finally, integrated techno-economic analyses and smart control strategies potentially combining PCM storage with PV-thermal modules or heat pumps can maximize overall renewable utilization, automate optimal charging/discharging, and improve return on investment. Addressing these areas will accelerate the maturity of PCM solar thermal systems, ensuring they deliver reliable, high-efficiency renewable heating across a broad range of applications.

8 Conclusion

In conclusion, this thesis demonstrates that integrating phase change materials into solar collector systems significantly boosts thermal efficiency by up to 16 percentage points extends useful heat delivery beyond periods of direct sunlight and stabilizes output across varying flow rates. Economic assessments indicate payback within 5–6 years and substantial energy-cost savings when deployed widely, while environmental analyses show annual CO_2 reductions of several tons per household and a higher solar-fraction utilization with minimal material impact. To fully realize these benefits, future work should establish standardized performance benchmarks, verify PCM durability over long cycling, develop cost-effective conductivity enhancements, and validate designs under diverse climatic conditions. Overall, PCMs provide a compact, cost-effective, and sustainable approach for next-generation solar thermal storage.

References

1. Butti K, Perlin J (1980) A golden thread: 2500 years of solar architecture and technology. Cheshire Books
2. Kalogirou SA (2004) Environmental benefits of domestic solar energy systems. Energy Convers Manage 45 (18–19):3075–3092. https://doi.org/10.1016/j.enconman.2003.12.019
3. Duffie JA, Beckman WA, Blair N (2020) Solar engineering of thermal processes, photovoltaics and wind, 5th ed. Wiley. https://doi.org/10.1002/9781119540328
4. Kalogirou SA (2013) Solar energy engineering: processes and systems, 2nd ed. Academic. https://doi.org/10.1016/C2011-0-07038-2
5. Stryi-Hipp G (2021) Renewable heating and cooling: technologies and applications. Woodhead Publishing. https://doi.org/10.1016/C2018-0-02250-0
6. Tian Y, Zhao CY (2013) A review of solar collectors and thermal energy storage in solar thermal applications. Appl Energy 104:538–553. https://doi.org/10.1016/j.apenergy.2012.11.051
7. IEA-SHC (2021) Solar heat worldwide: global market development and trends in 2020. International Energy Agency Solar Heating and Cooling Programme
8. Serale G, Goia F, Perino M (2016) Numerical model and experimental validation of a PCM-based thermal storage system for the improvement of the operation of CHP plants. Energy Build 126:447–461. https://doi.org/10.1016/j.enbuild.2016.05.052
9. de Gracia A, Cabeza LF (2015) Phase change materials and thermal energy storage for buildings. Energy Build 103:414–419. https://doi.org/10.1016/j.enbuild.2015.06.007
10. Fan L, Khodadadi JM (2011) Thermal conductivity enhancement of phase change materials for thermal energy storage: a review. Renew Sustain Energy Rev 15(1):24–46. https://doi.org/10.1016/j.rser.2010.08.007
11. International Renewable Energy Agency (2022) Renewable energy statistics 2022. IRENA. Retrieved from https://www.irena.org/publications/2022/Mar/Renewable-Energy-Statistics-2022
12. Al-Badi AH, Al-Hinai MH, Al-Maawali A (2018) Performance evaluation of a PCM-integrated solar thermal system for domestic water heating in Oman. Energy Convers Manage 174:550–561. https://doi.org/10.1016/j.enconman.2018.03.014
13. Farid MM, Khudhair AM, Razack SAK, Al-Hallaj S (2004) A review on phase change energy storage: materials, heat transfer analysis and applications. Energy Convers Manage 45(9–10):1597–1615. https://doi.org/10.1016/j.enconman.2003.09.015
14. Noro M, Suzuki T, Yamamoto K (2014) Thermal stability and cycling performance of phase change materials under prolonged high-temperature conditions. Sol Energy Mater Sol Cells 128:15–22. https://doi.org/10.1016/j.solmat.2014.05.003
15. Al-Waeli A, Al-Marri K, Al-Belushi N (2020) Engineering challenges in integrating phase change materials with solar thermal collectors in arid climates. Appl Therm Eng 172:115149. https://doi.org/10.1016/j.applthermaleng.2020.115149
16. Gastli I, Charabi Y (2016) Economic feasibility of integrating phase change materials in solar thermal energy systems in Oman. Energy Econ 58:303–311. https://doi.org/10.1016/j.eneco.2016.05.005
17. Kazem HM, Al-Farsi M, Al-Sharif A (2019) Impact of dust accumulation on the efficiency of solar thermal collectors in desert climates. Renew Energy 139:177–185. https://doi.org/10.1016/j.renene.2019.01.045
18. Naghavi Sanjani MS, Silakhori M, Ang BC, Metselaar HSC, Mousavi Gazafroudi SM, Noorollahi Y (2023) Experimental investigation on solar water heater integrated with thermal battery

using phase change material and porous media. Sustainability 15(8):6439. https://doi.org/10.3390/su15086439
19. Senobar H, Aramesh M, Shabani B (2022) Evacuated tube solar thermal collector with enhanced phase change material thermal storage: an experimental study. J Energy Storage 46:103838. https://doi.org/10.1016/j.est.2021.103838
20. Sadeghi G, Mehrali M, Shahi M, Brem G, Mahmoudi A (2022) Experimental analysis of shape-stabilized PCM applied to a direct-absorption evacuated tube solar collector exploiting sodium acetate trihydrate and graphite. Energy Convers Manage 269:116176. https://doi.org/10.1016/j.enconman.2022.116176
21. Sharma AK, Gupta B (2024) Performance enhancement of flat plate solar collector system with eutectic form phase change material. J Energy Storage 97:112770. https://doi.org/10.1016/j.est.2024.112770
22. Rizal TA, Amin M, Widodo SB, Abdul Rachman N, Amir F, Pane N, Mahlia TMI (2022) Integration of phase change material in the design of solar concentrator-based water heating system. Entropy 24(1):57. https://doi.org/10.3390/e24010057
23. Abokersh M, El-Morsi K, Sharaf M, Abdelrahman Y (2017) An experimental evaluation of a direct flow evacuated tube solar collector integrated with phase change material. Renew Energy 112:241–250. https://doi.org/10.1016/j.renene.2017.03.015
24. Shareef AS, Al-Mousawi FN, Sachit HS (2020) Experimental study of a PCM storage system integrated with a thermal solar collector. IOP Conf Ser Mater Sci Eng 671:012018. https://doi.org/10.1088/1757-899X/671/1/012018
25. Natesamurthi C et al (2022) Thermal performance enhancement of compound parabolic concentrator solar collectors using latent heat thermal energy storage. Environ Prog Sustain Energy 41(3):e13765. https://doi.org/10.1002/ep.13765
26. Almari RM, Abdelrahman MA, Moawad MA (2022) An experimental investigation of the evacuated tube solar collector in conjunction with a latent heat storage tank. Eng Res J (Shoubra) 51(1):20–28. https://doi.org/10.21608/erjsh.2022.224245
27. Algarni S, Mellouli S, Alqahtani T, Almutairi K, Khan A, Anqi A (2020) Experimental investigation of an evacuated tube solar collector incorporating nano-enhanced PCM as a thermal booster. Appl Therm Eng 180:115831. https://doi.org/10.1016/j.applthermaleng.2020.115831
28. Zhao J, Gao J, Liao J, Zhou B, Bai Y, Qiang T (2022) An experimental study on the heat storage and discharge performance (and economic analysis) of a flat plate PCM storage tank. Energies 15(11):4023. https://doi.org/10.3390/en15114023
29. Pathak SK, Tyagi VV, Chopra K, Pandey AK, Sari A, Abdulateef AM (2023) Energetic, exergetic, and heat transfer assessment of PCM-integrated heat-pipe-based ETSC for clear and cloudy weather conditions. Sustainability 15(12):9780. https://doi.org/10.3390/su15129780
30. Alshukri A et al (2022) Performance of a heat-pipe evacuated tube solar water heater with phase change material in collector and tank (experimental study). Sol Energy 233:271–281. https://doi.org/10.1016/j.solener.2022.01.082
31. Pawar S, Sobhansarbandi S (2022) Augmenting evacuated tube solar collectors using porous metal-integrated phase change materials. Sol Energy 229:103–113. https://doi.org/10.1016/j.solener.2021.09.080
32. Shafieian A et al (2021) Experimental evaluation of a heat-pipe ETC solar water heater with integrated PCM for cold-season performance. Sol Energy 216:194–204. https://doi.org/10.1016/j.solener.2020.12.080
33. Ahmed MA, Rofaiel IY, Essa MA (2020) Experimental study of an integrated solar collector water heater based on helical-finned heat pipes using PCM. Egypt J Eng Sci Technol 30:22–38
34. Li C, Zhang W, Huang Y (2021) Experimental study of a multi-cell phase change thermal storage tank for solar water heaters. Renewable Energy 171:736–746. https://doi.org/10.1016/j.renene.2021.02.014
35. Bonyadi R, Fazilati MA, Alemrajabi AA (2022) Phase change material for enhancing solar water heater: an experimental approach. Energy Convers Manage 171:138–145. https://doi.org/10.1016/j.enconman.2013.03.034

36. Hirschey J, Kumar N, Turnaoglu T, Gluesenkamp KR, Graham S (2021) Review of low-cost organic and inorganic phase change materials with phase change temperature between 0–65 °C. In: 6th International high performance buildings conference, Purdue University
37. Fan L, Khodadadi JM (2011) Thermal conductivity enhancement of phase change materials for thermal energy storage: a review. Renew Sustain Energy Rev 15(1):24–46
38. IEA Solar Heating and Cooling Programme (2021) Solar Heat Worldwide 2020: global market development and trends
39. Cabeza LF et al (2011) Environmental sustainability of latent heat thermal energy storage systems. Energy 36(9):4895–4901
40. Kalogirou SA (2004) Environmental benefits of domestic solar energy systems. Energy Convers Manage 45:3075–3092
41. Tiari S, Mahdavi M (2015) Nano-enhanced phase change materials for thermal energy storage in buildings—a critical review. Sol Energy Mater Sol Cells 141:70–85
42. Chen Z, Zhou G, Wang J, Liu H (2019) Heat transfer enhancement in phase change thermal energy storage: a review of fin structures and porous media. Renew Sustain Energy Rev 101:111–122

Design of an Energy Storage System Using PCM for a Solar Hot Water System

Ammar Al Reesi and Seyed Mojtaba Sadrameli

Abstract This research investigates the application of phase change material (PCM) in a solar water heating system (SWHS) to improve the efficiency of thermal energy storage (TES) and provide an extended hot water supply for domestic consumption. The study focuses on paraffin wax latent heat storage (LHS) and evaluates three experimental case studies. A compact flat plate SWHS (Case 1), a thermosyphon system with top mounted PCM (Case 2), a conventional system enhanced using RT42 paraffin under various inlet conditions (Case 3). The results showed that the placement of PCM, melting point selection, and operational parameters significantly impact system performance. The thermosyphon system demonstrated the most balanced improvement with enhanced efficiency of 16% and a prolonged discharging period of approximately 7 h. After analyzing the findings of these case studies an experimental design was proposed for implementation at the German university of Technology (GUtech) ECO-HOUSE in Oman. However, due to financial circumstances, the experiment was not conducted. The experimental design took the local climatic conditions into consideration, and it aimed to validate the theoretical results in future work. Moreover, utilizing PCM in SWHS under optimized thermal conditions has been proven to improve overall system performance.

Keywords Phase change material (PCM) · Solar water heating system (SWHS) · Thermal energy storage (TES) · Latent heat storage (LHS) · Renewable energy · Heat recovery · Paraffin wax · Experimental design

---

A. Al Reesi · S.M. Sadrameli (✉)
Department of Engineering, German University of Technology in Oman, Muscat, Oman
e-mail: seyed.sadrameli@gutech.edu.om

A. Azizi (ed.), *Mechatronic Innovations in Sustainable Energy*, Emerging Trends in Mechatronics, https://doi.org/10.1007/978-981-95-5863-6_7

1 Literature Review

This research provides a comprehensive overview of solar water heating system (SWHS) integration with phase change material (PCM)-based storage. It includes three Chapters: a combined literature review of SWHS and PCM technologies and previous studies in Chapter one, Chapter two focuses on case studies related to Oman's climate and the modeling of an SWHS with PCM-based storage for an ECO-HOUSE, and Chapter three discusses the results and provides a detailed analysis.

In recent years, the global energy transition has been increasingly prioritized to achieve sustainability and climate goals [1]. As the adoption of alternative energy sources is raising, solar energy is being utilized widely, due to its sustainability, green energy, and eco-friendly nature [2]. However, the issue with this energy is that its availability is limited only for a certain duration. Also, it is hard to control the supply of solar energy as it depends on natural occurrences related to weather conditions [3]. However, the efficient use of solar energy can be achieved through thermal energy storage (TES) [4]. On this basis, the latent energy storage (LHS), particularly in the area of PCMs has impressive properties, featuring high density storage at small temperature intervals and cost-effective storage media [5]. This will contribute to make future of Oman more sustainable and will decrease greenhouse emissions. By 2030 the government is willing to enlarge its electricity-generation with plans to generate at least 30% of electricity from renewable energy sources during that period [6]. As mentioned that PCMs are effective in terms of enhancing TES, this may fulfil Oman's goals of reaching net zero emissions in 2050 [7].

1.1 Overview

The study focuses on addressing one of the issues of renewable energy, especially in the field of solar energy as photovoltaic panels system could not offer electricity 24 h per day and only produce it at certain periods of the day since it is based on natural phenomena [8]. Therefore, this challenge could be mitigated by using energy storage systems synchronously [9]. With Oman categorized to be one of the Gulf Cooperation Council (GCC) countries that have highest solar generation capability due to their unique geographical location [10], SWHSs are considered a viable solution in terms of energy sustainable use, especially in regions that have significant average of solar irradiance annually [11]. Despite embracing SWHSs, they have some disadvantages such as lack of solar energy at night and seasonal changes [12]. By storing solar thermal energy these drawbacks can be fixed, in both forms of sensible heat or latent heat the thermal energy can be stored, in this context the PCMs may play the pivotal rule, when referring to latent thermal heat, because when the temperature decreases these materials release energy and return to their original physical state [13]. This may lead to hot water at non-sunny hours and enhance both efficiency and reliability of the system.

1.2 Problem Statement

One of the difficulties in SWHS is to achieve efficient TES. Noting that the potential setback of the conventional water storage tanks is that they experience a heat loss over time consequently that decreases the system efficiency [14]. Despite that the PCMs are considered a promised solution for energy applications [15], their real-world performance under extreme climatic conditions in Oman still needs thorough investigation. Therefore, current studies primarily focus on PCMs applications in temperate climate, emphasizing studying how these materials will act when exposed to high temperatures. Oman's energy sector is encouraged by implementing optimal PCMs for its environmental conditions to enhance the reliability and efficiency of SWHSs in the local region. This study, according to the region's climatic conditions, aims to design and model an optimized energy storage system using PCMs.

1.3 Research Objectives

The purpose of this research is to design an efficient energy storage system for SWHS using PCM in ECO-HOUSE of German University of Technology (GUtech) in Oman. This research will explore the behavior of distinct PCMs in enhancing thermal efficiency, performance, capabilities, and long-term stability under Oman's extreme climatic conditions. In addition, an analysis will be conducted with both traditional water storage tanks and PCM-based storage in water tanks to determine energy savings and thermal efficiency enhancements. By targeting these objectives, this will significantly support sustainable energy storage development solutions in Oman. Moreover, Fig. 1 shows the SWHS on top of ECO-HOUSE.

The key objectives of this study are:

- Designing and modeling a PCM-based TES system for SWHS of ECO-HOUSE existing study in GUtech, Oman.
- Evaluate the thermal performance of different PCMs under Oman's climatic conditions.
- Compare the efficiency of PCM-based storage with conventional water storage methods.

1.4 Literature Review of TES and PCM Applications

Energy systems that have a high level of irregularity such as wind and solar require an effective energy storage solution that can balance between demand and supply, during duration of significant energy production and modest demand

Fig. 1 SWHS in GUtech ECO-HOUSE [16]

excess energy is stored and released during high demand or low generation, noting that energy storage system will decrease dependence on fossil fuels, and will contribute to increase overall efficiency and enhance grid stability [17]. Studies show that energy storage plays a crucial role in increasing efficiency of renewable energy and reducing environmental impacts by utilizing excess energy that had been stored instead of being waste leading to lowering carbon dioxide (CO_2) emissions [18]. Moreover, a study's results show that the multi-energy system is the best solution in terms of cost-effectiveness, utilizing poly-generation technologies and energy storage in forms such as electrical, thermal, and chemical through the power-to-gas process [19].

While SWHSs offer a sustainable energy solution, they still face an obvious problem, as the main limitation is the absence of solar energy in the nighttime hours [20]. Therefore, implementation of TES can solve this issue and improve energy efficiency. As TES is categorized into three categories, sensible heat storage (SHS), LHS, and thermochemical heat storage (THS). LHS takes the lead in terms of higher energy storage density in contrast with SHS. Study results show that the PCM-based latent heat thermal storage energy system can successfully supply hot water for residential needs and enhance proportion of solar energy in building energy consumption [21]. The conventional water tank in SWHS suffers from heat loss over time due to the SHS that relies on temperature fluctuations, solving this issue by adopting PCMs that can act as an efficient heat storage medium, providing higher efficiency than the traditional water tanks. By implementing paraffin wax-based PCMs this will maintain the consistency of hot water temperature for longer duration, ensuring hot water availability when solar radiation levels are low [22].

PCM-based TES with a conventional SHS system is essential, to keep the system stable, to elevate the system performance, and to ensure the proper supply of hot water throughout the day and on cloudy days. As evidenced by the study results, it was revealed that the SWHS without the PCM storage system has experienced fluctuating efficiency over time, while the PCM-based system has shown consistent efficiency. Therefore, according to the findings, PCM tanks overcome conventional systems in all the following: cost-effective, space efficiency, steadily high temperature profile during heat accumulation. These advantages make the PCM energy storage system a remarkable choice in terms of sustainable solar thermal energy utilization [23].

PCM is the most efficient solution that bridges the gap between supply and demand maintains thermal management and has the potential to store long-term heat energy over an extended period of time. It faces challenges such as low thermal conductivity, production cost, and stability during initial and working stages in the SWHS. These challenges could be solved by developing research to address solutions for these drawbacks and overcome any obstacles that hinder PCMs usage [24].

The mechanism of SWHS is to harness the power of solar radiation to produce thermal energy that will heat water for domestic purposes. In addition, this system is environmentally friendly which is an alternative choice for conventional water heating methods, significantly reducing CO_2 emissions. It has simple operating principles, starting with absorption of solar radiation using a solar collector that uses a circulating fluid to transfer the heat to the water to heat it. However, SWHS face notable limitations, mainly the nature of solar radiation as it is not available throughout the day, which will affect its efficiency. Addressing these issues, TES technology can play a pivotal role in making sure of hot water availability [24].

According to Herez et al. [25], the most frequent PCMs that have been studied for SWHS applications are those having melting temperatures ranging between 40 and 80 °C. PCMs include paraffins, fatty acids, salt hydrates, and alcohols. They have found that integrating PCMs in SWHS can significantly enhance overall performance, increase thermal efficiency, minimize fluctuations in temperature, and stabilize operational conditions. It has been found that using PCMs as TES in SWHS will reduce temperature variations, through the absorption and release of heat during phase transition for this will provide a more consistent water temperature. This leads to reduction in the temperature fluctuations and improving overall efficiency and reliability of the SWHS through utilizing PCMs [25].

Considering the economic and environmental insights by Herez et al. [25], they conducted economic analysis of SWHS through case studies in Lebanon. It has been stated that the consumption of hot water depends on users, as when the demand for hot water is higher the payback period will be faster, given an example under moderate scenarios, families that have installed SWHS took the longest to recover their investment approximately 14.2 years, because the demand of daily hot water is low, instead in hotels, the demand of daily hot water is higher, and they successfully achieved to recover their initial investment faster averaging approximately 4.4 years. Moreover, regarding the environmental aspects of integrating SWHS, they found a relationship between utilizing SWHS and reduction of CO_2 emissions [25].

A study about encapsulated PCMs performance in a SWHS has been done by Khayyaminejad et al. [26]; they conducted the study on a residential building in Thessaloniki, Greece, by using software of MATLAB and TRNSYS wherin the system was modeled. They selected a paraffin wax PCM with melting temperature of approximately 60 °C. The setup of encapsulated PCM was in spherical capsules inside a storage tank, and the heat transfer function depended on conduction and convection mechanisms for the circulating water to ensure efficient progression. They noted that the modeling was for transient daily variations within realistic weather conditions. The outcome of the study shows that the hybrid PCM-integrated system successfully managed to supply hot water at the desired temperature throughout sunny hours and nighttime [26].

The simulation results by Khayyaminejad et al. [26] showed that the system successfully achieved during peak solar radiation hours to reach a cumulative storage capability of approximately 56.44 MJ. This maintains a consistent hot water temperature without relying on additional electric heating sources thus preventing any temperature drops during non-sunny hours. Also, it improves the efficiency of heat storage and prevents overheating of water storage tanks using the integration of the PCM system. Moreover, in case of reliability and accuracy, the model validation was compared with experimental data and showed mean relative errors below 3.79% for TRNSYS model and 0.13% for MATLAB model; these results confirm that the approach is reliable.

The schematic diagram of the proposed combination system of solar and encapsulated-PCM water heating system, has been simulated in TRNSYS, as shown in Fig. 2 [26].

Mellouli et al. [27], have done a study on a SWHS integrated with PCMs to evaluate its potential for thermal load shifting by using a numerical parametric analysis on the system. The analysis reveals that the thermophysical properties of PCM impact significantly on performance and system efficiency highlighting that the melting temperature, density, and latent heat are the main key properties that had affect the performance. The results demonstrate that increasing the density of PCM from 1400 kg/m^3 to 3200 kg/m^3 leads to multiple system improvements such as significantly enhancing thermal storage capacity, prolonging the effective heat collection time to approximately 21.1%. Furthermore, the PCM latent heat capacities increase positively to system stability, slowing down the water temperature fluctuations during both charging and discharging phases. The optimal PCM characteristics for optimizing the system performance were identified to be melting temperature of 40 °C, 3200 kg/m^3, and latent heat around 520 kJ/kg. Under these

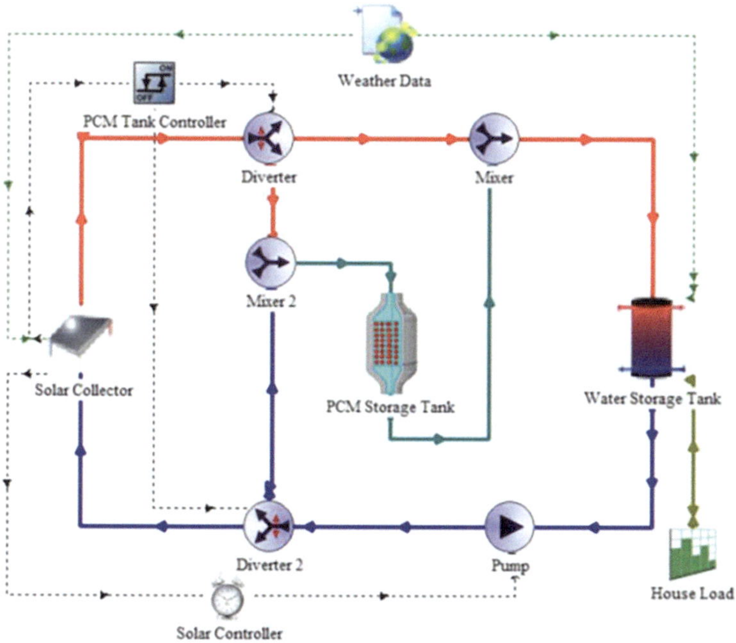

Fig. 2 Hybrid PCM-integrated system [26]

conditions, the system shifted about 20.2% of thermal loads from peak daytime to off peak nighttime, mentioning that the water temperatures remained stable in the night period [27].

For the selection criteria of PCM in various energy storage applications, a study has been conducted by Rao et al. [28], they stated that it is difficult to choose an optimal PCM for a given application, as it has many requirements that should be sufficient, such as thermal properties, physical properties, kinetic properties, chemical properties, cost, and factors taken into account by management. Moreover, None of the PCMs possess all favorable properties, for example, abundant thermal conductivity for better storage operation, both high specific heat and latent heat for large storage capability, a desired melting point for such an application, and economical. In addition, they reported that researchers have built up a methodology known as multi-attribute decision-making (MADM), it is characterized by the accuracy for selecting the most optimal PCM for such an application. However, the MADM methods have their advantages and disadvantages, for this reason, a new MADM has been developed by Rao et al. [28] and is titled 'Best Holistic Adaptable Ranking of Attributes Technique (BHARAT)'. It used to identify and choose the best PCM selection for distinct energy storage applications. Also, to approve this method, Rao et al. [28] have used BHARAT in three different case studies: choosing the best PCM for solar cooker box, selecting optimal PCM for a ground source heat pump, and the last case study was about identifying the best PCM that is suitable for thermal comfort in a vehicle. After comparing BHARAT results with the most used MADM methods, it has been concluded that the BHARAT method has promising potential for future PCM selection, being verified by valid results of the three discussed case studies. As well as that the BHARAT method is easy to implement, avoids fuzzy logic while offering a clear approach to attribute the characteristic properties and is adaptable for PCM selection applications in various scenarios [28].

Based on a paper of Caramitu and Lungu [29], they have stated that PCMs are useful in terms of conservation of heat and solar energy. They have demonstrated how the PCMs behave according to LHS technology, given examples for these materials whose phase changes from liquid to solid, including organic (paraffins, fatty acids, esters, alcohols, and glycols (PEG stands for polyethylene glycol), inorganic materials (salt hydrates, nitrate salts, metals, etc.), eutectics which are a mixture of organic-organic, inorganic-inorganic, and organic and inorganic. Moreover, they mentioned that SHS faces several disadvantages, for example, the need for large storage volumes for low energy density, implementation of an effective insulation to reduce heat losses, etc. Furthermore, LHS has advantages such as providing a high energy density by utilizing both sensible and latent heat together and it is considered as a proper choice for low temperature applications. However, there are some disadvantages, for instance, salt and metal PCMs can chemically interact with surroundings resulting in system components corrosion. While THS is more efficient than LHS and SHS, although it is more complex and costly. Caramitu and Lungu [29], provided in Tables 1 and 2, are discussing various PCMs, focusing on their melting point and latent heat energy [29].

Research by Mahmood et al. [30], describe the function of paraffin PCM in convective-vacuum drying systems, providing in-depth details of thermal energy accumulators based on paraffin wax and methods for improving their efficiency and techniques to enhance their thermal conductivity. In addition, demonstrating that integrating heat accumulators with PCMs will increase the overall energy efficiency of the drying systems, this has been confirmed by researchers results from experimental studies. They stated that during a phase change process the substance accrues most of its heat which is directly connected to its latent heat. Moreover, the technology relies on reliable LHS through which PCM absorbs and releases heat during the charging and discharging phases of the phase change process due to its high heat of fusion, as illustrated by Mahmood et al. [30] in Fig. 3, the PCM TES cycle involves charging, storing, and discharging phases [30].

Furthermore, Mahmood et al. [30], concluded that designing thermal energy accumulators using a PCM is a challenging task when integrated into a system. They focused on using paraffin wax as the main PCM for the convective-vacuum impulsive drying unit in their research, due to the features this PCM has, such as availability, cost-effective, no supercooling process, non-toxic, chemically stabilized, high latent heat of fusion, non-corrosive and safe handling. They presented paraffins thermal properties in both the liquid and solid phase and given the chemical formula, as shown in Table 3. Also, visual appearance of paraffin wax is displayed in Fig. 4 [30].

Table 1 Thermophysical properties of organic PCMs for thermal energy applications [29]

Organic PCMs for thermal storage applications	Melting temperature (°C)	Latent heat of fusion (kJ/kg)
Paraffins class		
n-Heptadecane ($C_{17}H_{36}$)	22.6	214
n-Octadecane ($C_{18}H_{38}$)	28.4	244
PRS® paraffin wax (RepSol-YPF)	40–45	202
Fatty acids class		
Lauric acid ($C_{12}H_{24}O_2$)	42	171
Myristic acid ($C_{14}H_{28}O_2$)	54	190
Palmitic acid ($C_{16}H_{32}O_2$)	64	185.4
Stearic acid ($C_{18}H_{36}O_2$)	69	209
Esters class		
Ethylene glycol distearate ($C_{38}H_{74}O_4$)	63.2	215.8
Alcohols class		
1-Dodecanol ($C_{12}H_{26}O$)	23.3	184
1-Tetradecanol ($C_{14}H_{30}O$)	39.3	221
Glycols class		
PEG400 ($C_{10}H_{22}O_6$)	4.2	117.6
PEG600 ($C_4H_{10}O_3$)	12.5	129.1
Paraffins class		
PEG1000 ($C_2H_6O_2$)	40	168.6
PEG10000 ($C_{34}H_{66}O_6$)	65.9	171.6
PEG35000 ($C_4H_{10}O_3$)	68.7	166.9
Organic eutectics class		
Lauric acid-stearic acid (75.5/24.5)	37	182.7
Myristic acid-palmitic acid (58/42)	42.6	169.7
Myristic acid-stearic acid (65.7/34.3)	52	162

Table 2 Thermophysical properties of inorganic PCMs for thermal energy applications [29]

Inorganic PCMs for thermal storage applications	Melting temperature (°C)	Latent heat of fusion (kJ/kg)
Salt hydrates class		
$Na_2S_2O_3 \cdot 5H_2O$	48	200
$NaCH_3COO \cdot 3H_2O$	58	240
$Ba(OH)_2 \cdot 8H_2O$	78	265.7
Nitrate salts class		
Sodium nitrate ($NaNO_3$)	306	172
Potassium nitrate (KNO_3)	335	266
Metals class		
Zinc (Zn)	419	113
Aluminum (Al)	660	397
Copper (Cu)	1084	208

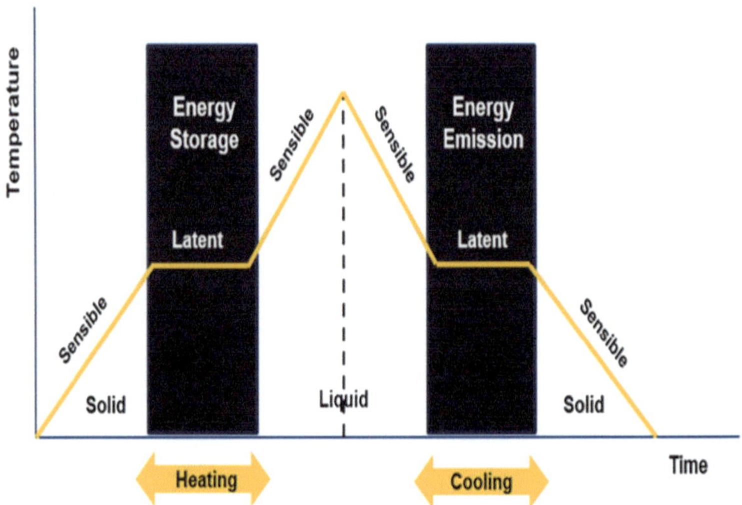

Fig. 3 PCM TES cycle [30]

Table 3 General thermal properties of paraffins [30]

Property	Liquid phase	Solid Phase
Boiling point	>370 °C	
Melting temperature	57 °C	57°C
Thermal conductivity	0.25 W/m °C	0.23 W/m °C
Specific heat capacity	1.6 kJ/kg °C	2 kJ/kg °C
Latent heat of fusion	2.1 kJ/kg °C	
Chemical formula	C_nH_2n+2	

While being paraffin wax is the most suitable choice to be used as heat accumulator in such an application. However, according to Mahmood et al. [30], that this PCM is having a drawback, namely, low thermal conductivity, as its slowdowns the process of melting and solidification and impacting the thermal performance of the chosen system. To overcome this, thermal conductivity enhancement techniques are proposed, as a solution to increase the thermal conductivity for paraffin wax PCM. Mahmood et al. [30] have presented several effective methods. Starting with the use of nanoparticles (NP) with paraffin wax, studies have shown that integrating NP with paraffin wax accelerates both charging and discharging time, resulting in increasing the thermal conductivity. The second method is increasing the thermal conductivity of paraffin wax by using expanded graphite as a supplement material. Given third method, integration of metallic foams as a solution to increase thermal conductivity. The latest proposed method and most cost-effective technique as Mahmood et al. [30] mentioned is encapsulating the PCM in finned containers, gives examples for the internal in external fins in Fig. 5 [30].

Fig. 4 Visual appearance of paraffin wax [30]

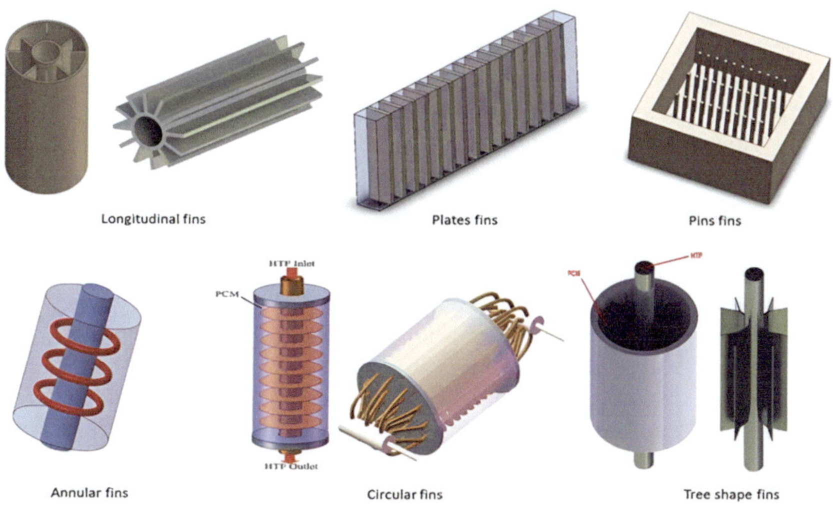

Fig. 5 Internal/external fins for encapsulation of PCM in TES systems [30]

A study was conducted by Albalushi et al. [31] on evacuated tube solar water heater performance in Al Seeb, Oman. The system's performance was monitored on different inclination angles and weather conditions, specifically, 45°, 56°, and 65°. Based on experimental results, it was revealed that at an inclination angle of 45°, the system performed the best, achieving the highest temperature of 90° among tested angles. They reported that at 45° angle, the heat absorption reaches its max at 5:30 pm. Also, they stated that dust, dirt, sulfate compounds, nitrates, and organic carbon have an impact on the evacuated tube solar water heater performance. According to Albalushi et al. [31], ambient air temperature and sun power significantly affect solar water heaters, in terms of achieving maximum solar energy [31].

Another study done by Ihnayyish et al. [32], focused on simulation, to investigate the thermal performance of U-shaped copper tube within an evacuated glass tube solar collector using PCMs. Three cases were simulated in this experiment, each with a different geometry of inner diameter of copper tube with fixed settings of paraffin wax PCM at a temperature range of 54°–57°, and a glass tube with inner diameter of 3.7 cm. The three cases respectively were smooth copper pipe, two lobes with copper tube, and four lobes with copper tube. According to the results of this study, it shows that case 1 outstands as the maximum temperature among the simulated cases, with a temperature of 45°, the remaining cases respectively were 41° and 42°. Ihnayyish et al. [32] proved by numerical study that using geometry of smooth pipe for inner diameter of copper tube embedded in evacuated glass tube with paraffin wax is the optimal case, in terms of overall efficiency [32].

Jegadheeswaran and Sundaramahalingam [33] have done a study on enhancing thermal conductivity of paraffin wax by embedding it with copper NP. The study successfully proved the enhancements of both PCM phases liquid and solid. Also, Jegadheeswaran and Sundaramahalingam [33] concluded from the experiments and results that using more copper NP can achieve the purpose of increasing the thermal conductivity and delays thermal degradation, meaning better thermal stability. However, by more addition of copper NP the latent heat will slightly decrease due to the reduction of PCM volume and particle interference. According to the authors, they stated that the optimal weight of copper NP for paraffin wax is 2% for best performance, beyond this it may cause agglomeration [33].

An experiment study has been done by Al-Abdali and Ammari [34] on evacuated tube solar collector for domestic water in Mutah University, Jordan. The paper includes two experiments, one with paraffin wax PCM (Receiver A), one without PCM (Receiver B), as shown in Fig. 6 is about the setup configuration of the two receivers, and Fig. 7 is about the schematic diagram of both models A and B. It is reported that the experiment analysis has been done in two stages. For the first stage they used circulated water flow rate of 7 L/h. The second stage is without using flow rate. After analyzing the data, the results indicate that the receiver which contained PCM was more efficient than the other receiver in both stages, where in the first stage the model that used PCM resulted in an efficiency of 47.7 and 41.6% in the second stage. On the other hand, the second receiver, which did not have integrated PCM in it, yielded 40.6% in the first stage, and 35.2% in the second stage. Moreover, they proved that integrating paraffin wax PCM in solar evacuated tube water-in-glass thermal collectors improved TES and increased overall efficiency by 8.6%, and 7% respectively. Also, Al-Abdali and Ammari [34] addressed the financial benefits of Integrating SWHS with PCM-based storage, by

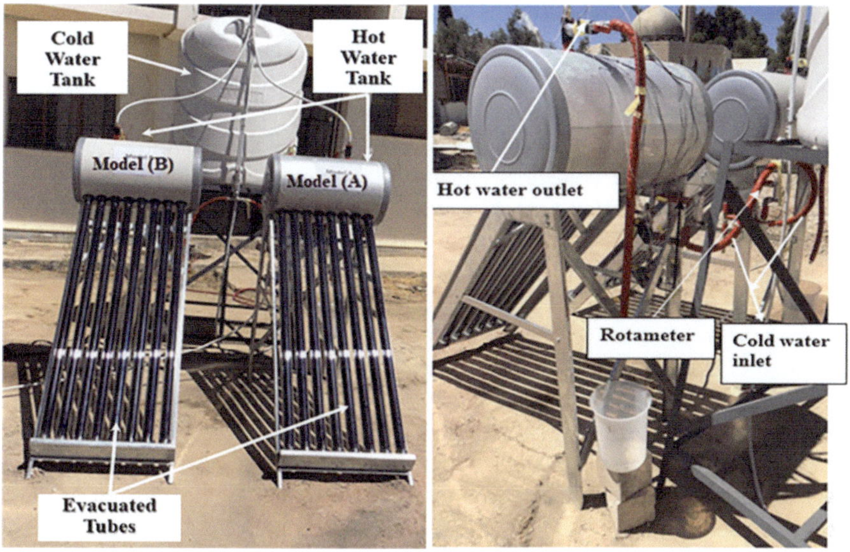

Fig. 6 Setup of SWHS models A and B [34]

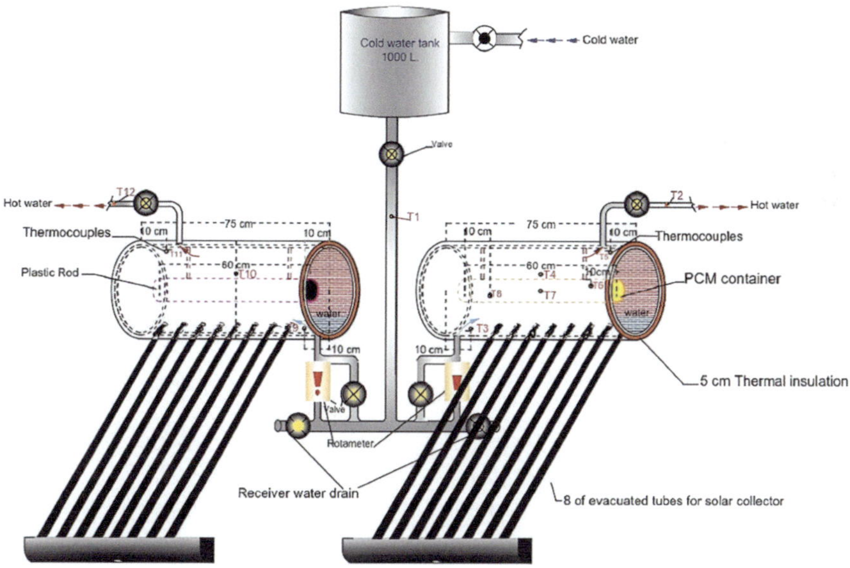

Fig. 7 Schematic diagram of models A and B [34]

comparing the annual costs of solar collectors with PCM and the system that uses electrical geyser showed that the first receiver payback period was 5.35 years [34].

Summarizing the review literature, the studies showed that using PCM in SWHS improves energy efficiency and thermal storage capacity. Al-Abdali and Ammari [34] proved by experimental study that integrating paraffin wax PCM in evacuated tube solar collectors enhanced the overall efficiency of SWHS and reduced the payback period [34]. Mahmood et al. [30] presented various enhancement methods to increase thermal conductivity of thermal energy accumulators, specifically focusing on paraffin PCM, and highlighted techniques such as integrating PCM with NP, using expanded graphite as a supplement material to PCM, and the most cost-effective one is encapsulating PCM in a finned container [30]. Ihnayyish et al. [32] demonstrated a numerical study to improve the performance of PCM in an evacuated glass tube solar collector, wherein they experimented various geometries of the internal diameter of U-shaped copper tubes embedded in an evacuated glass tube with paraffin wax. They concluded that the optimal inner diameter geometry for obtaining the highest outlet temperature of 45 °C is using a smooth pipe [32]. Another study has been mentioned in the literature by Jegadheeswaran and Sundaramahalingam [33] to overcome the disadvantage of paraffin wax by embedding it to copper NP, reporting that the optimal copper NP by weight is 2%, beyond this will cause obstacles [33]. For decision-making criteria, Rao et al. [28] proposed the BHARAT decision-making method, which is an approach that identifies suitable change PCM through neglecting fuzzy logic [28]. Overview of TES, classifications, and key performance criteria of PCM were introduced by Caramitu and Lungu [29]. Despite PCM being a solution for storing thermal energy for the future, there are gaps that should be bridged, there is a lack of experimental research of enhanced-PCMs in real world application, feasible study in the project of life cycle, and in-depth optimization and placement of the enhanced-PCM, and areas of controlling the integration of PCM remain unexplored. In the future, studies should focus on the practical deployment of nano-enhanced PCMs, extending thermal cycling tests, and developing smart and cost-effective systems fitting different climates and usage profiles especially the climate of Oman although there is a lack of studies on integrating PCM in SWHS in Oman.

2 Case Study and Modeling

2.1 Overview

This chapter will present a detailed case study analysis for three selected experimental studies focused on the integration of PCM in SWHS. The reviewed cases studies are compact solar water heaters with paraffin storage, investigations into PCM-enhanced thermosyphon systems, and thermal improvements using encapsulated PCMs. Each study delivers important information about PCM behavior, energy storage capabilities, and operational effects of PCM integration across different working conditions. This chapter will also outline the modeling of a PCM-based SWHS for ECO-HOUSE project at GUtech. Proposing strategies for optimizing thermal performance are tailored to the climatic conditions of Oman.

2.2 *Case Study One*

Title: Performance Analysis of a Compact Solar Water Heater with Paraffin Wax Storage Unit

Authors: Dr. Aouf Abdulrahman Al-Tabbakh, Amir Fadhil Noori, Raid Salman Fahad [35]

2.2.1 Materials Used in the Experiment

This experiment studied the thermal performance of a compact solar water heater in (Baghdad, Iraq), that contain a solar collector and PCM-based thermal energy unit. The tank unit is made of galvanized steel and split into two identical sections. The upper section that receives solar radiation is filled with water, and the lower section was filled with paraffin wax with a melting point of 58 °C. Lateral sides and back face were thermally insulated by using glass wool with thickness of 5 cm. Upper face is covered using a glass sheet over 4 cm air gap. A replacement tank is installed in the compact solar water heater to prevent any drainage losses from daily use. Also, the whole unit was put into an aluminum casing tilted at 45° relative to horizon. A K-type thermocouple device has been used to measure the temperatures of both water and PCM, and a thermometer instrument is used to measure the ambient temperature. In addition, a solar digital meter is used to record solar radiation intensity. The following Table 4 represents the specifications and dimensions of the materials used in this experiment.

2.2.2 Methods Used in the Experiment

In the system, the 5 nodes are evenly distributed in each section along the long side. Average the 5 readings to a single reading to represent T_w and T_{PCM}. At the beginning of the experiment, the upper and below section of the tank was filled with water and paraffin wax respectively. Also, a caution procedure is taken when filling the PCM tank to avoid filling it completely. As the process of melting PCM causes an expansion of its volume, a gap of 10% should be considered to prevent facing inefficiencies. The system is positioned to face south with an inclination of 45 relative to the horizon. For receiving temperatures data of the system, the 10 thermocouple temperature probes were installed to each section distributed as

Table 4 Materials specifications and dimensions [35]

No	Item	Dimension	Units
1	Collector aperture area	1.05	m^2
2	Water tank volume	0.052	m^3
3	Water tank dimensions	1.5 * 0.7 * 0.05	m
4	PCM volume	0.052	m^3
5	PCM melting point (paraffin wax)	58	°C
6	PCM latent heat of fusion	140	kJ/kg
7	Air gap depth	4	cm
8	Water tank depth	5	cm
9	PCM tank depth	5	cm

5 for water tank and 5 for PCM tank and connected to a digital instrument. Solar radiation intensity was measured using a sensor connected to solar radiation meter, the sensor is placed along the collector glass cover, measured in W/m^2. Moreover, a digital thermometer has been used to record the ambient temperature. The experiment's duration was 8 h starting from 8 am to 4 pm for the two chosen days (15 and 16 of March), the data was recorded every 30 min, at the end of each interval. The mean temperatures of water and PCM were calculated by averaging the three readings. Moreover, Fig. 8 represents the schematic diagram of compact solar heat.

In this experiment, the overall efficiency of the system is to be defined as the useful amount of heat stored in PCM and Water tanks to the amount of solar radiation that the system receives during the intervals of 30 min, all units in kJ. The following Eq. (1) is used to calculate the overall efficiency of the system at the end of the specified time interval.

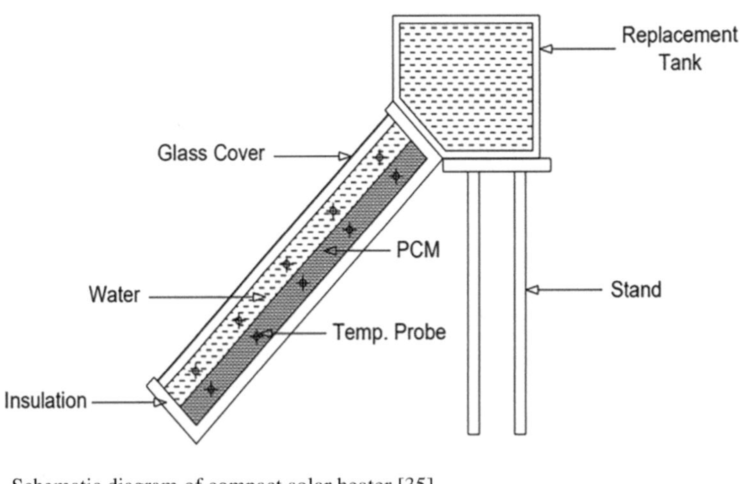

Fig. 8 Schematic diagram of compact solar heater [35]

$$\eta_0 = \frac{\{Q_{\{SW\}} + Q_{\{SPCM\}} + Q_{\{LPCM\}} - Q_{\{Loss\}}\}}{\{H_T\}} \quad (1)$$

In this study, the sensible heat terms are $Q_{\{SW\}}$ for water and $Q_{\{SPCM\}}$ for PCM, latent heat of PCM that is stored over a duration of 30 min and the heat loss to the surroundings are presented by $Q_{\{LPCM\}}$ and $Q_{\{LOSS\}}$ respectively. The total amount of solar energy that the system received in 30 min is represented by $H_{\{T\}}$. The below Eqs. (2)–(5) are to calculate all of $Q_{\{SW\}}$, $Q_{\{SPCM\}}$, $Q_{\{LPCM\}}$, and $Q_{\{LOSS\}}$, respectively.

$$Q_{\{SW\}} = M_W C_{\{PW\}} (T_{start} - T_{end})_w \quad (2)$$

$$Q_{\{SPCM\}} = M_{PCM} C_{PPCM} (T_{start} - T_{end})_{PCM} \quad (3)$$

$$Q_{\{LPCM\}} = \alpha M_{PCM} h_{fu} \quad (4)$$

$$Q_{\{LOSS\}} = U_L A_c (T_w - T_{ambient}) \quad (5)$$

The melting ratio and latent heat of fusion of the paraffin wax are α and *hfu*, respectively. In addition, each reading corresponds to 1/5 of the total volume. As the 5 temperature readings of PCM tank registered to be above or equal to the melting temperature then the melting ratio is taken 1, but if only one of the 5 probes registered a temperature equal to or above the melting point of the PCM then the melting ratio is taken 0.2.

2.2.3 Results

In the experiment, the authors changed intervals of recording data from 30 to 15 min. The experiment was undertaken from 8 am to 4 pm and is conducted over a span of two days (15th and 16th of March). In both experimental days, the weather was clear, so variation of solar radiation illustrated a symmetrical shape as shown in Fig. 9 for 15th of March and Fig. 10 for 16th of March. After observing the data, it appears that the water temperature was the highest along PCM and the ambience, due to that water tank absorbs solar radiation directly and delivers the excess to PCM tank. Figures 11 and 12 show through the test period the changes in the average temperatures of the water, PCM, and ambient temperature. The low thermal conductivity of PCM in addition to its indirect exposure to solar radiation created a temperature difference between water and PCM. Only two probes have been recorded to reach the melting range of (55–60 °C) during late afternoon, indicating the occurrence of melting. After monitoring PCM tank temperature, it has been observed that the melting ratio (liquid volume/total volume) yielded no more than 25% over the course of the two days. To overcome the issue of PCM not reaching the melting range, authors suggested in future studies, to either choose PCM with low melting point, or minimizing PCM volume, ensuring complete utilization of the PCM.

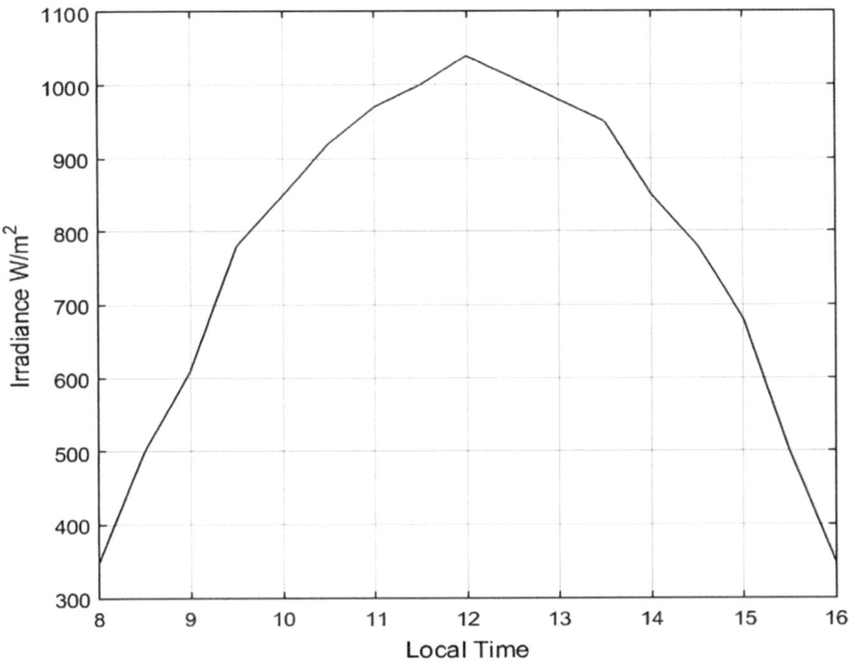

Fig. 9 Solar Radiation Intensity curve on March 15th [35]

While recording experiment data, the system indicates high performance between 10am and 1pm resulting in overall efficiency remaining above 70% between 10am and 1pm. After 1pm the system experienced noticeable drop until the end of the test. Comparing this result to the trend of conventional flat-plate solar collectors, it can be observed that using PCM in a SWHS does not only act as TES but also serves as insulation to increase overall system efficiency. If PCM is not installed in the SWHS, the system will experience a higher temperature that will be rejected to the surroundings, ending up with higher thermal losses. PCM keeps the system with high efficiency with lower relatively temperature, noting that the calculation of the overall system efficiency included with the presence of PCM, results in fetching high values.

Variation of overall system efficiency (as defined by Eq. (1)) over the course of both testing days are presented in Figs. 13, and 14.

Designing this compact solar water heat via installing PCM (paraffin wax) with a melting range of (55–60 °C). The experiment proved that this compact system, led to high overall efficiency and reduced thermal energy losses. However, this system lacks the full utilization of PCM melting, this can be solved by either using PCM with low melting point or decreasing the amount of PCM used. Also, authors suggest that future studies should focus on choosing the optimum PCM/water ratio.

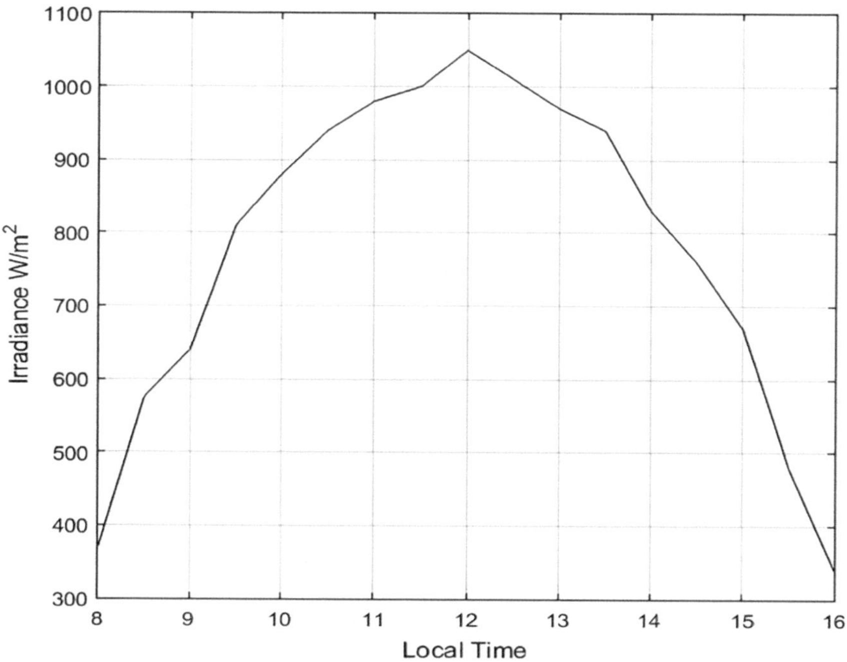

Fig. 10 Solar Radiation Intensity curve on March 16th [35]

2.3 Case Study Two

Title: An Experimental Study of PCM-Incorporated Thermosyphon Solar Water Heating System

Authors: G. Murali, K. Mayilsamy, and T. V. Arjunan [36]

2.3.1 Materials Used in the Experiment

Figure 15 shows the visual representation of the experimental apparatus. The system consists of, an insulated cylindrical tank material that is made from stainless steel, a flat plate is the type of solar collector, and its area is 2 m^2, an aluminum cylindrical container for PCM, a HEATCON 8003 temperature data logging device has been used since it has the feature of measuring with precision of 1 °C, six thermocouples for temperature measurements, a TES 1333 solar power meter with precision of $\pm$ 10 W/m^2 used to measure the solar radiation, and insulated pipes for both the inlet and outlet. Both coconut coir and thermocol are used as insulation for the 78-L stainless-steel TES tank. PCM cylindrical containers are 20 mm thick, installed inside the stainless-steel tank to ensure heat transfer between

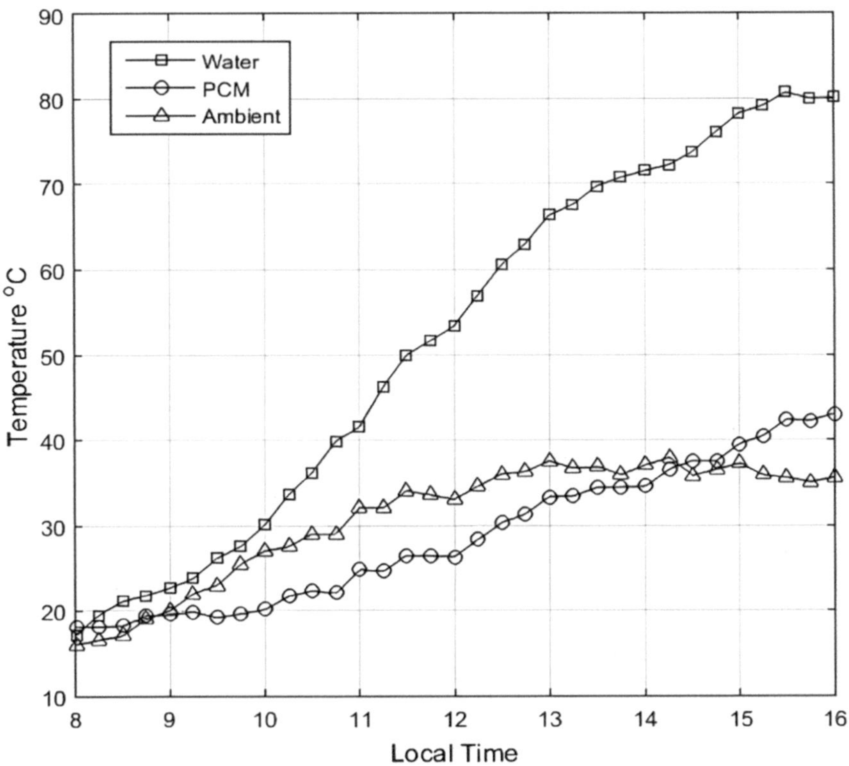

Fig. 11 Changes in water, PCM, and ambient temperatures throughout the testing period on March 15th [35]

the heat transfer fluid and PCM. The abundance of coconut coir from agriculture waste makes it a favorable option to be used as an insulator for the system, since it has low thermal conductivity ranging between 0.054 and 0.143 W/m K. PCM that has been used in this experiment is paraffin wax with melting point of 56 °C, the 2.5 kg by weight PCM is inserted into a cylindrical aluminum structure with a 121mm inner diameter, a 300 mm vertical length, and a wall measuring 3 mm in thickness. Tables 5 and 6 provide the specifications of the flat plate collector used in the solar water heater and thermal properties of paraffin wax, respectively.

2.3.2 Methods Used in the Experiment

This experiment was conducted in Hosur in the South of India in February, where the average temperature in summer is between 25 and 35 °C, and in winter it is between 13 and 25 °C. The authors used water as the heat transfer fluid to transfer heat from tubes to the storage tank. Thermosyphon principle has been applied

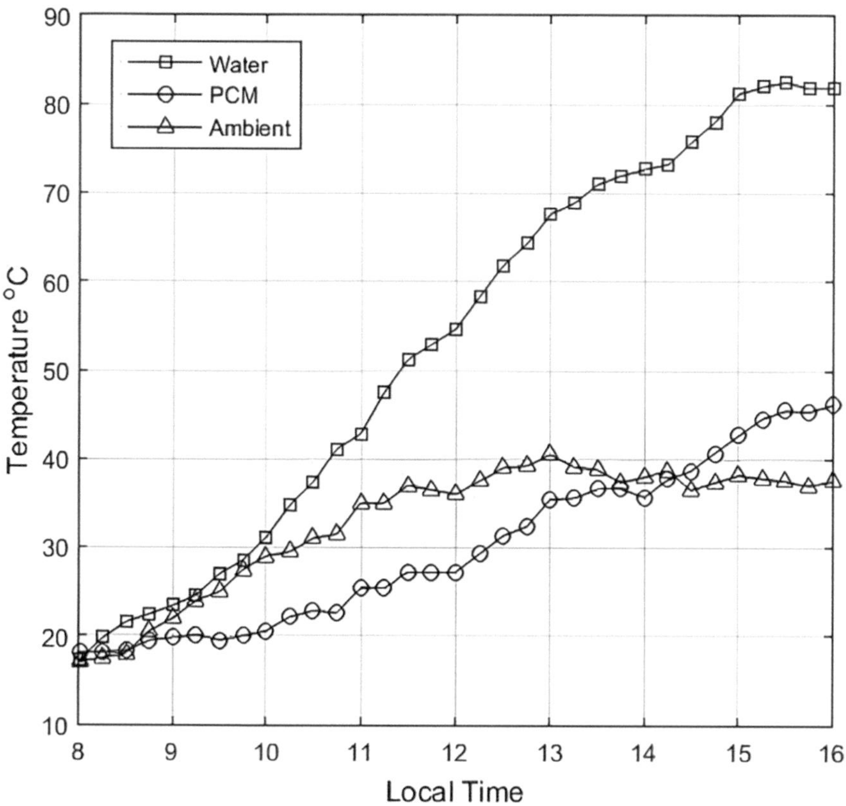

Fig. 12 Changes in water, PCM, and ambient temperatures throughout the testing period on March 16th [35]

to this system, as when the system heats the water, consequently it will elevate the water through the pipe directly to the insulated storage tank, and to compensate for the displaced water the cold water is supplied to the bottom inlet of the flat plate collector. To examine the stratification process of the solar system, the authors sealed the insulated storage tank by closing both hot water outlet and the cold water inlet. Before starting the experiment, the authors prevented PCM expansion volume issues by melting it above its melting point, after that they put it in the aluminum cylinder. Temperature readings and flow rate of the heat transfer fluid are taken every 30 min. Moreover, in the storage tank, six thermocouples were installed and distributed with equal distance to record data of each layer temperature. Away from the insulated storage tank central axis, the thermocouples are installed with a distance of 170 mm, with an 80 mm gap separating each one. Also, there are two thermocouples installed, one to monitor the outlet temperature of hot water leaving the insulated storage tank, and the other to measure PCM

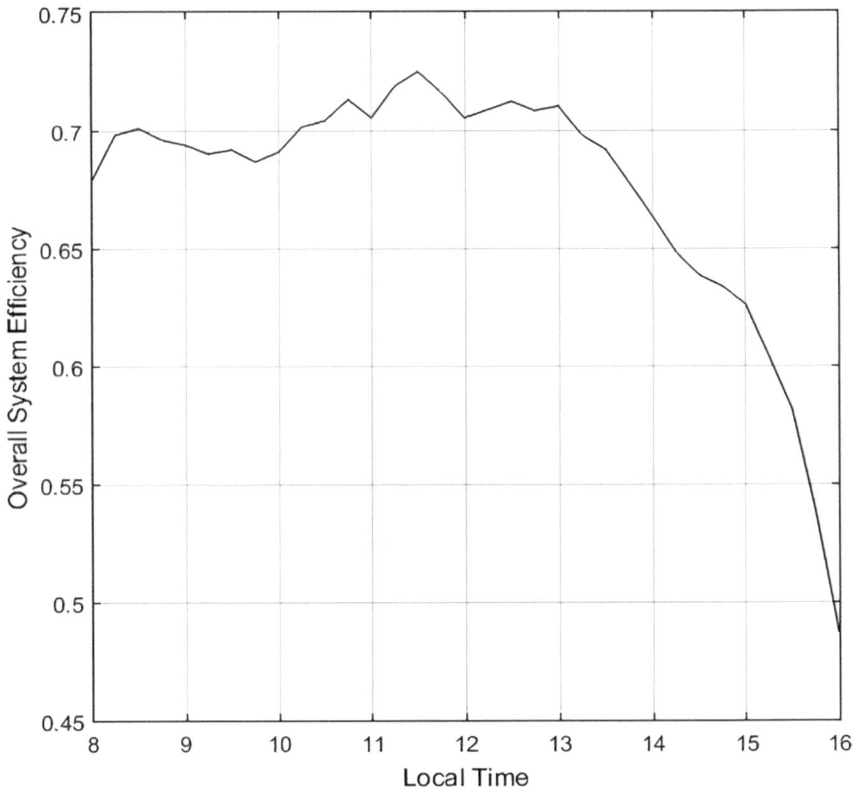

Fig. 13 Trend in overall system efficiency during test periods on March 15th [35]

temperature inside the aluminum cylinder. Figure 16 represents the schematic view of the experimental setup.

In the charging process, inside both water storage tank and collector plate the heat transfer fluid is circulated in a loop, driven solely by natural circulation. As the heat transfer fluid remains at room temperature, it starts to absorb solar energy from the radiation of the sun and transfers the heat to PCM. Initially PCM undergoes sensible heating, as the temperature increases it reaches the melting point. By isothermal melting process at 56 °C the paraffin wax achieved the stage of being an energy storage, raising the temperature above that makes the PCM enter the superheat region which will result in repetition of undergoing sensible heating. After reaching an average temperature of 65 °C for both insulated storage tank and PCM, the study fulfilled its purposes, as the charging process is done with and without PCM, the performance of TES tank can subsequently be evaluated.

While in the discharging process, it starts naturally when the heat source is no longer present, the heat transfer fluid exchanges the heat with the surroundings, as a result of the temperature difference between the heat transfer fluid and ambient

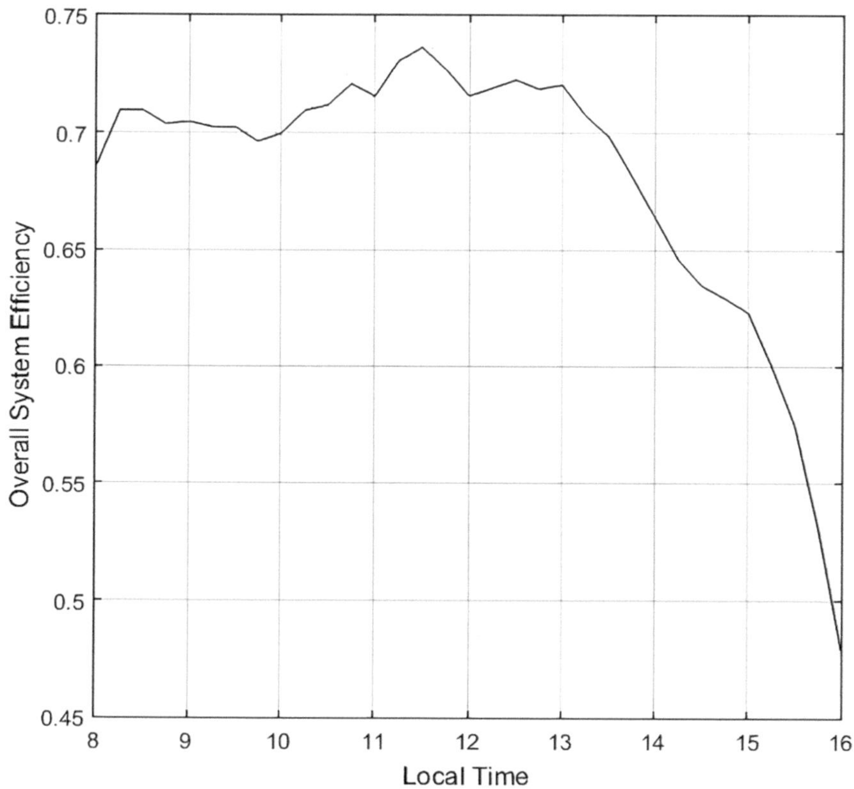

Fig. 14 Trend in overall system efficiency during test periods on March 16th [35]

environment. Just before the discharging process starts, the system must be isolated, by using a thick covering to cover the flat plat collector, providing insulation, and by closing the gate valve that controls hot water inlet that entering the insulated storage tank. As both PCM and water average temperature reach below 45 °C, the water cannot be used for domestic daily uses, and the study of the stratification number can be evaluated. To assess enhancements in the SWHS, the duration of heat discharge both with and without PCM is evaluated. In addition, we need to determine the time needed for the stratification process, which separates hot and cold water within the same tank.

"The charging energy efficiency Eq. (6) is driven by the first law of thermodynamics. It is defined as the ratio of the heat absorbed by the water to the heat supplied to it."

$$\eta_{ch} = \frac{T_{avg(t)} - T_{ini}}{T_{inlet} - T_{ini}} \tag{6}$$

Fig. 15 Captured image of the system setup [36]

Table 5 Flat plate collector specifications [36]

Parameter	Specification/value
Area of the collector	2 m^2 (2 m × 1 m)
Cover area of the collector	Single tempered glass, 3-mm thickness
Emissivity/absorptivity of the glass cover	0.85
Glass refractive index relative to air	1.5
Absorber plate length	1.95 m
Absorber plate width	0.95 m
Absorber plate material	Copper
Plate material thermal conductivity	386 W/m K
Plate material density	8954 kg/m^3
Thickness of the plate	0.16 mm
Tube diameter	6.35 mm
Tube spacing (center-to-center)	100 mm
Total tubes	9
Header pipes diameter	12.7 mm
Collector tray placement	Hosur
Angle of the collector	30°

Table 6 Thermal properties of paraffin wax [36]

Properties	Solid phase	Liquid phase
Melting point (°C)	56	56
Latent heat (kJ/kg)	142.7	142.7 ß
Thermal conductivity (W/m °C)	0.4	0.2
Density (kg/m^3)	670	640
Specific heat (kJ/kg °C)	2.4	1.6

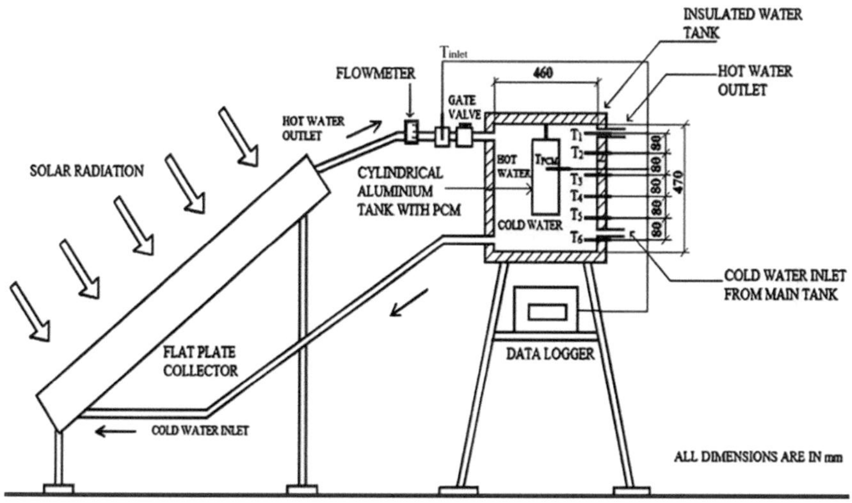

Fig. 16 Diagrammatic representation of the experimental setup [36]

"For the stratification number Eq. (7) and vertical temperature gradient (8), as proposed by, evaluates the thermal stratification of water within a storage tank. It is defined as the ratio of the average temperature gradient at each time interval to the gradient at the initial time."

$$\mathrm{Str} = \frac{\overline{\left(\frac{\partial T}{\partial Z}\right)_t}}{\left(\frac{\partial T}{\partial Z}\right)_{t=0}} \tag{7}$$

$$\overline{\frac{\partial T}{\partial Z}} = \frac{1}{j-1}\left[\sum_{j=0}^{j-1}\left(\frac{T_{j+1} - T_j}{\Delta z}\right)\right] \tag{8}$$

"The thermal efficiency Eq. (9) measures the storage tank's performance and is expressed as the ratio of the useful heat captured by the solar water heater to the solar energy received, with a flat plate collector efficiency of 0.55."

$$\eta_{\mathrm{th}} = \frac{m_{\mathrm{cp}}(T_{\mathrm{f}} - T_{\mathrm{ini}})}{A c_{\mathrm{G}}} \tag{9}$$

2.3.3 Results

After investigating the influence of utilizing a PCM cylinder within a thermosyphon solar water tank featuring stratified storage, the authors made sure that they chose days where global radiation is similar for both PCM and without PCM, as presented in Fig. 17. This approach ensured an easier comparison of charging experiments performed on separate days. Charging and discharging experiments were performed using the solar water heater without a PCM cylinder on February 11 and 12, 2013 and then repeated with the PCM cylinder on February 27 and 28, 2013. Figures 18 and 19 presented the temperature variations across different layers of the storage tank during charging experiments conducted with and without PCM, respectively. The 30-min intervals for temperature measurements were relatively spaced apart which reduces visibility of PCM energy absorption in the plotted graphs. More frequent readings would have revealed the slope reduction which happens when PCM melts. The storage tank demonstrates proper thermal stratification throughout the midday period because the six temperature layers maintain uniform distribution throughout the charging phase. The PCM addition resulted in an evident temperature disparity between the upper and lower tank layers from start to finish during charging. This indicates the advantage of using PCM inside the stratified storage tank. Through the experiments the authors found that the discharging process without PCM resulted in a significant reduction in temperature variation of different layers in a storage tank, as shown in Fig. 20. In contrast, the discharging process using PCM maintained higher temperature variations, as

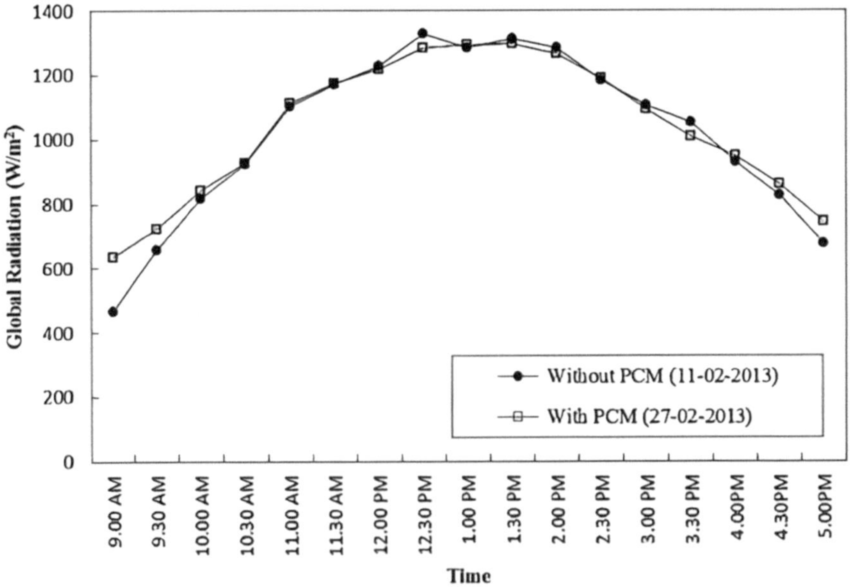

Fig. 17 Global radiation comparison recorded during the charging experiments [36]

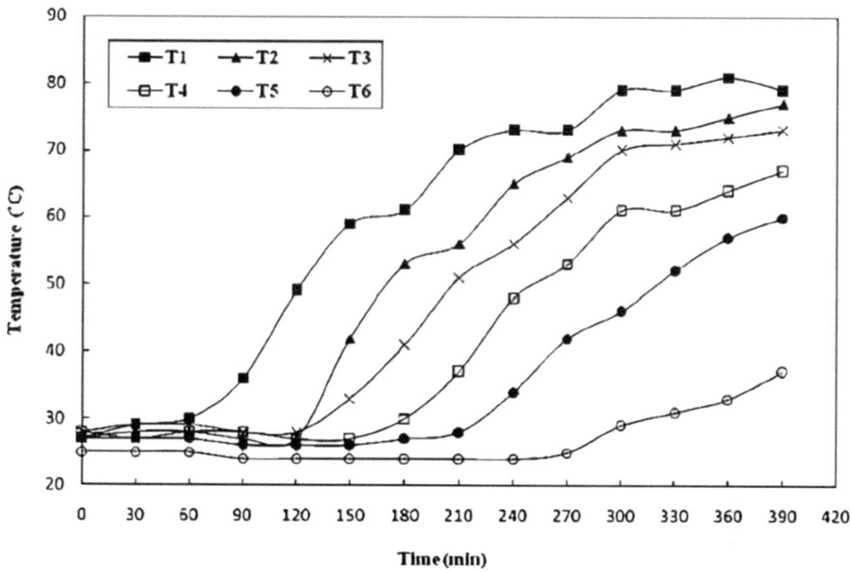

Fig. 18 Temperature variations during the charging process without PCM [35]

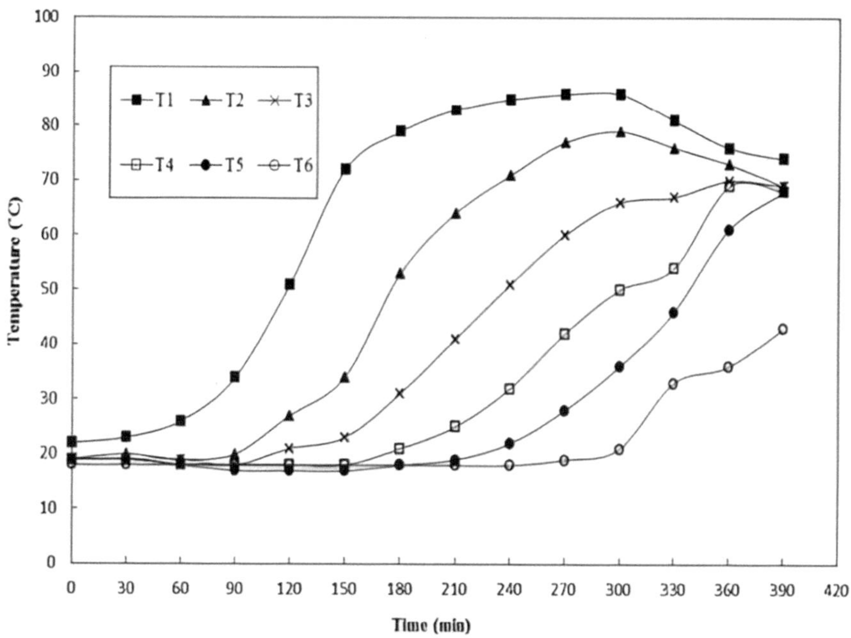

Fig. 19 Temperature variations during the charging process with PCM [36]

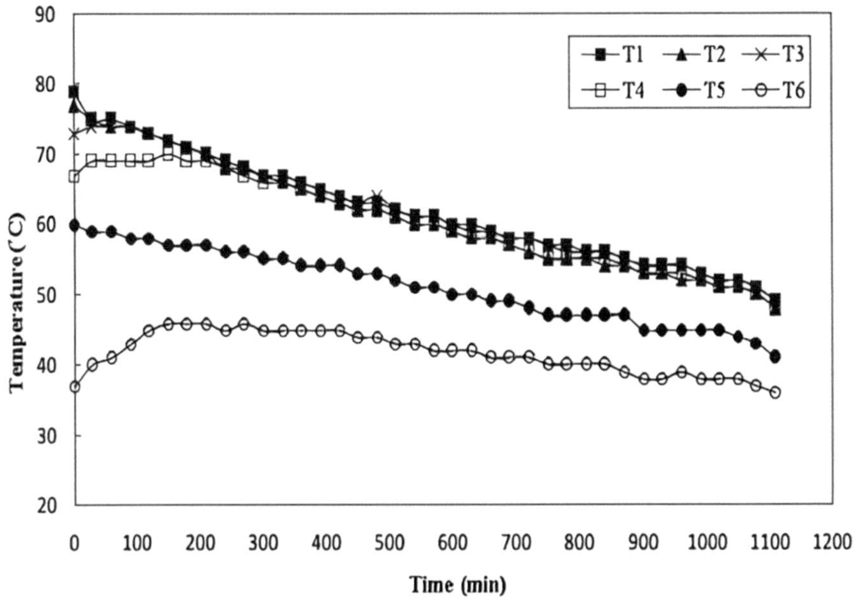

Fig. 20 Discharging process temperature variations without PCM [36]

presented in Fig. 21. In the charging process of the SWHS with and without PCM, the water heated to approximately 65 °C and took 390 min. On the other hand, for the discharging process both with and without PCM, the temperature dropped from 65 to 45 °C. However, the discharging durations for the SWHS were 1530 min with PCM and 1110 min without PCM, respectively. This reflects the benefit of utilizing PCM in SWHS as it extends the time of cooling the water inside the storage stank. Figure 22 provides an average temperature during discharging process experiments of both SWHS with and without PCM. From the data, it is evident that the SWHS without PCM reached 45 °C 440 min earlier than that with PCM, due to the low heat transfer rate of PCM with its surroundings. These results proved that utilizing PCM in SWHS improves the performance of the system and enhances the stratification.

Figure 23 shows comparison of the stratification number during the charging process of SWHS with and without PCM. More stratification number indicates more thermal performance of SWHS and more overall efficiency. The authors observed from discharging experiments' graphs that the stratification number with using PCM is always 10% higher than without PCM indicating that PCM contributes in enhancing overall performance by bettering stratification number. Stratification efficiency of the SWHS tank is studied by evaluating its charging efficiency. Figure 24 illustrates the comparison of the charging efficiency of water during the charging process of SWHS with and without PCM. At the beginning, as a portion of the heat is utilized to melt the PCM, the charging efficiency is less

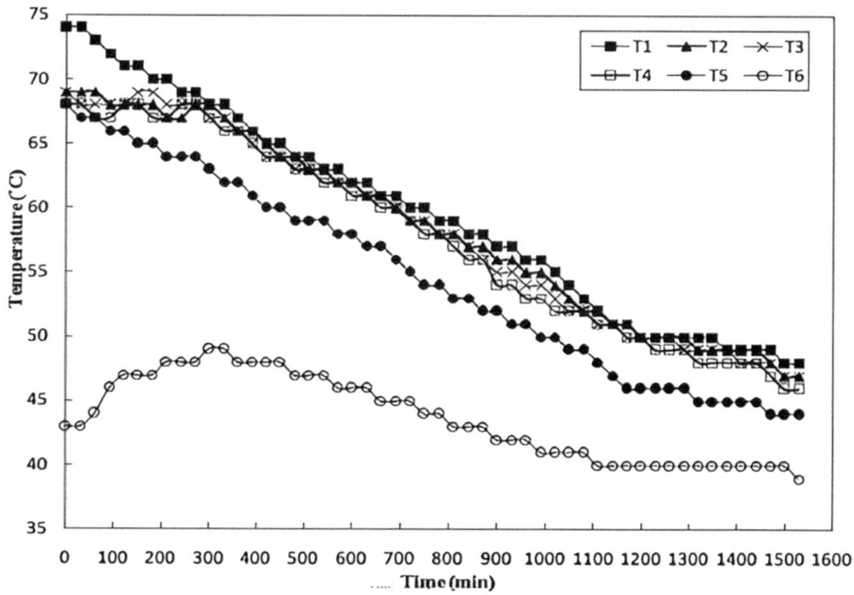

Fig. 21 Discharging process temperature variations with PCM [36]

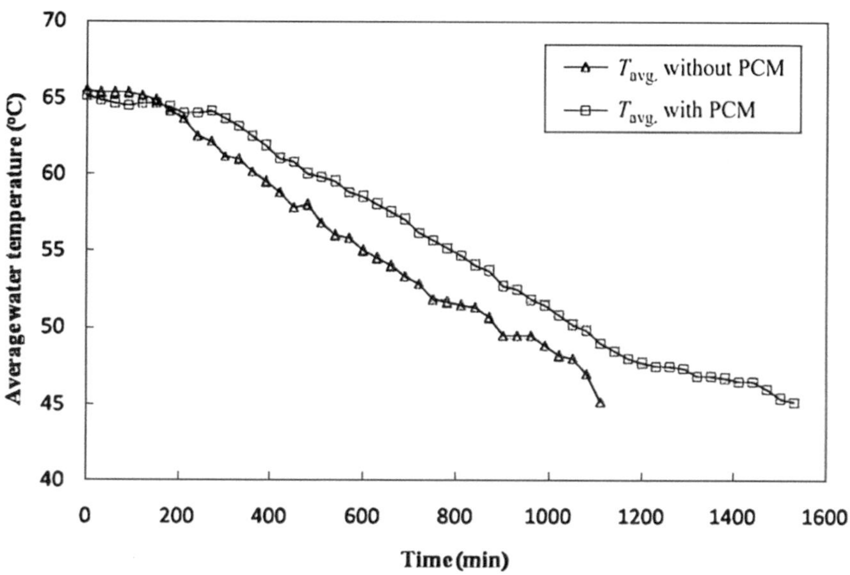

Fig. 22 Comparison of T_{avg} during discharging process for both without and with PCM [36]

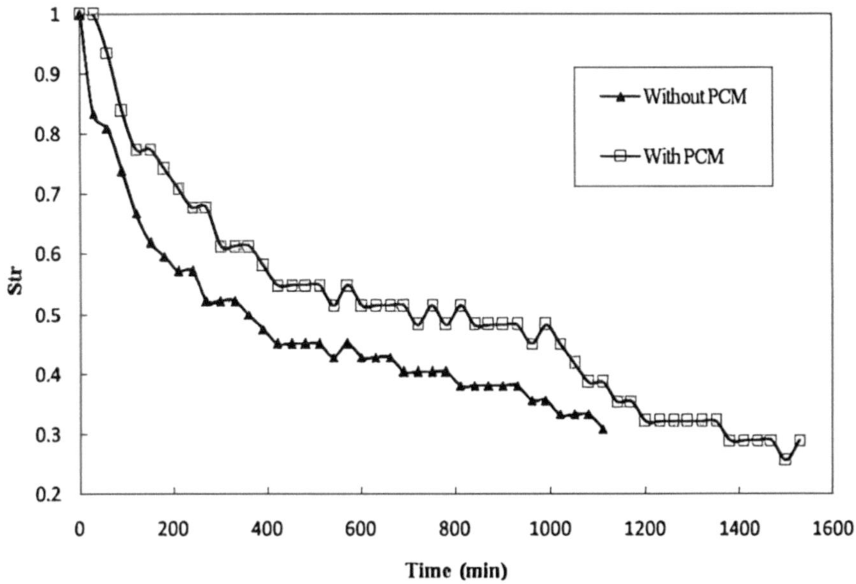

Fig. 23 Stratification number comparison during the charging process with and without PCM [36]

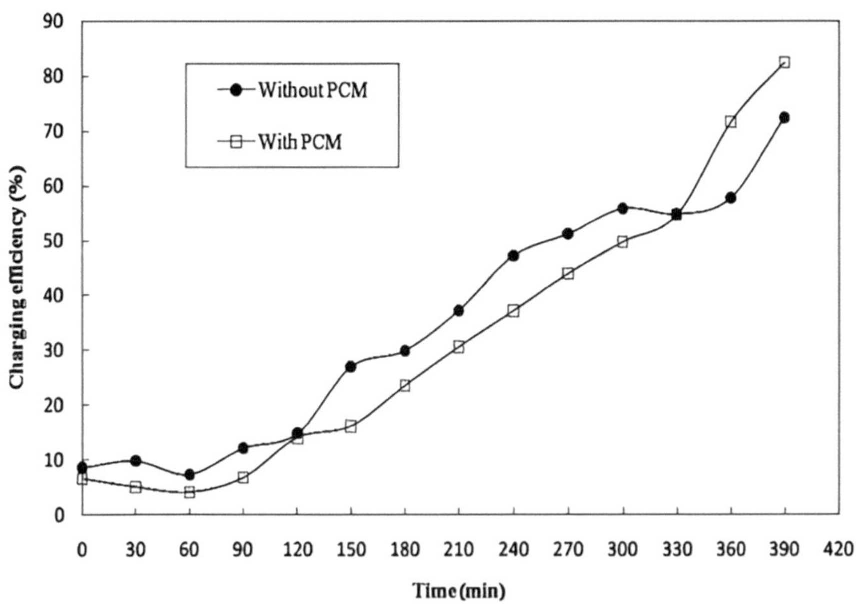

Fig. 24 Charging efficiency comparison during the charging process with and without PCM [36]

compared to the process without PCM. Just when the melting process was completed, the trend of PCM curve showed significant rise, which indicates that the charging efficiency of PCM surpassed that of the system without PCM. SWHS during discharging experiments was used to study thermal efficiency with and without PCM. During discharging process, SWHS was analyzed to evaluate thermal efficiency both with and without PCM.

The comparison of thermal efficiency with and without PCM is shown in Fig. 25. From the graph it can be observed that there is a noticeable difference between thermal efficiency of SWHS tank with and without PCM. Throughout the discharging process, it is evident that the thermal efficiency with PCM is greater than that without PCM. At the beginning of the experiments, the difference in thermal efficiency was 10%, by the time it increased to 16%, concluding that the thermal efficiency of the storage tank increased by 10–16% due to the latent heat stored in 2.5 kg of paraffin wax PCM. These results conclude that integrating PCM into the SWHS system increases stratification numbers, boosts charging efficiency, and enhances overall thermal efficiency. Moreover, the authors have done a simulation study focused on using software to model the discharging process with and without PCM, identifying the temperature distribution of the heat transfer fluid across various layers at the onset of discharging. Also, the simulation model reinforces the reliability of using PCM in SWHS and serves as basis for comparison with experimental results. Figure 26 presents the comparison of average temperature during the discharging process. The close similarity between the simulation results and the experimental results demonstrates a strong agreement.

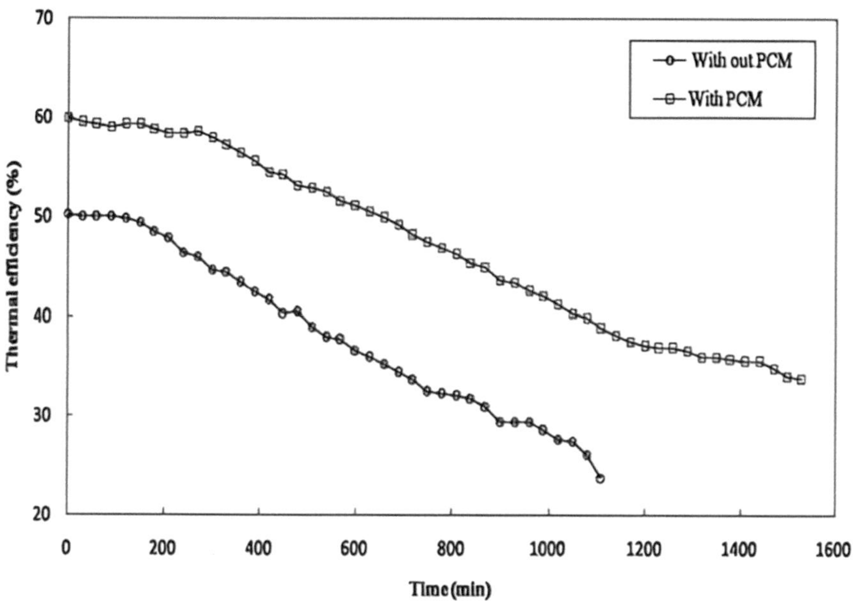

Fig. 25 Thermal efficiency comparison of the storage tank during the discharging process [36]

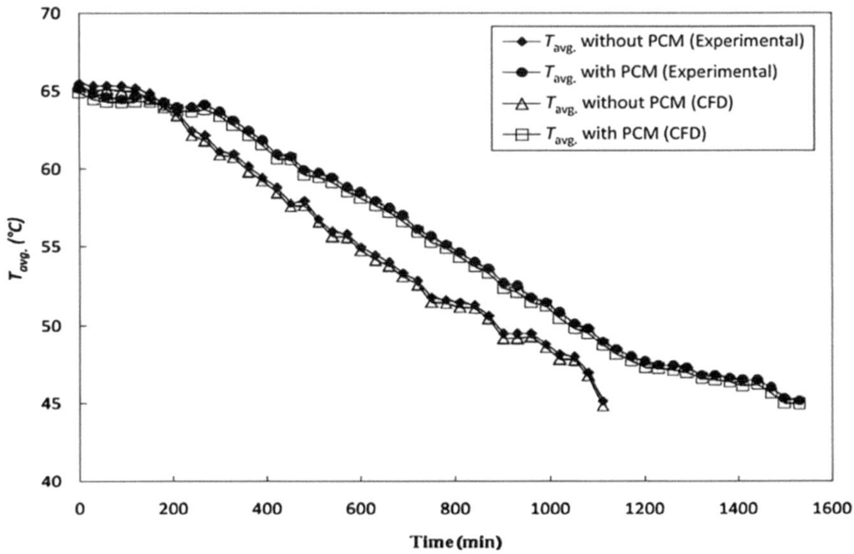

Fig. 26 Comparison of average temperatures both experimentally and through simulations for cases with and without PCM [36]

This experimental study showed that incorporating 2.5 kg of PCM improved system performance by enhancing temperature stratification and thermal efficiency. The charging efficiency yielded 82.44% at 2 pm. Stratification and thermal efficiency during discharging were consistently 10–16% higher with PCM. The simulation study results were too close to the experimental results which confirms that integrating the PCM will boost SWHS performance.

2.4 *Case Study Three*

Title: Thermal Enhancement of Solar Energy Storage Using Phase Change Materials

Authors: Bahzad Darwesh Darwesh, Idres Azzat Hamakhan, Banipal Nanno Yaqob [37].

2.4.1 Materials Used in the Experiment

This experiment was conducted in Eribil, a city in Kurdistan. The system consists of flat plate solar collectors, pumps for circulation purposes inside the water storage tank, a large water storage tank (1), PCM-based water storage tank (2), pipes, sensors to measure the temperature, pressure, and flow throughout the system. In addition, an expansion valve for heating purposes, and water softener. Moreover, all the data related to temperature, flow meter, and pressure are monitored using an application called DESIGO INSIGHT, a software application compatible with computer devices. Furthermore, PCM that has been used in this experiment is paraffin wax type RT42 with a melting point of 43 °C. Figure 27 illustrates the physical appearance of the PCM, while Table 7 outlines its thermal properties.

Fig. 27 Paraffin wax type RT42 [37]

Table 7 RT42 thermal properties [37]

Type of property	Numeric value	Temperature variation (°C)
Density (ρ)	880 kg/m^3 760 kg/m^3	15 80
Latent heat capacity (Δh)	165,000 J/kg	
T_m	42–43 °C	
T_s	38 °C for melting 37 °C for solidification	
C_p	2000 J/kg K	42

2.4.2 Methods Used in the Experiment

The experiment took place on the rooftop of the Erbil Research Center building. Ten flat plate solar collectors, each with a surface area of 2 m^2 were used, these collectors were arranged into two rows, with five in each row positioned parallel to one another. As shown in Fig. 28, the solar plates were tilted at a 60° angle from the horizontal. Figure 29 presents the 1 m^3 water storage tank, commercially known as WBO 1005 UNO/DUO. This tank is capable of storing heated water at temperatures of up to 130 °C and can withstand a maximum pressure of 16 bars.

The authors made configurations on the 0.5 m^2 PCM-based water storage tank (2) to enhance its thermal performance. The storage tank is filled with RT42 paraffin PCM and water, comprising 81% water with a total volume of 0.405 m3, and 17% paraffin PCM with a volume of 0.085 m^3. In addition, the tank is filled with 18 aluminum cylinders, each with an external diameter of 0.1 m, an internal diameter of 0.094 m, and a height of 0.66 m. Moreover, the base of the tank occupies 0.01 m^3 which represents 2% of the total tank volume. Figure 30 illustrates various aspects of the PCM-based water storage tank: (a) the insulated tank, (b) its dimensions, and (c) the 18 aluminum cylinders containing PCM. Figure 31 shows the installation of thermometer sensors on the 18-alumnium cylindrical PCM. The authors only mentioned the placement of three temperature sensors of the twelve temperature sensors, one at the top, one at the midpoint, and one at the bottom, respectively. These sensors are connected to a temperature monitoring device, the DS18B20 thermometer. Figure 32 presents a schematic diagram of the SWHS, the system involves sensors of temperature, pressure, flow meters, pump, two flat plate collectors, a water storage tank (1), and a PCM-based water storage tank (2). The study aimed to evaluate energy savings with and without the integration of paraffin PCM. Consequently, tank (2) was positioned before the load and after tank (1) to facilitate the comparison.

Fig. 28 Solar flat plate collectors in the experiment [37]

Fig. 29 The 1 m^3 water storage tank (1) [37]

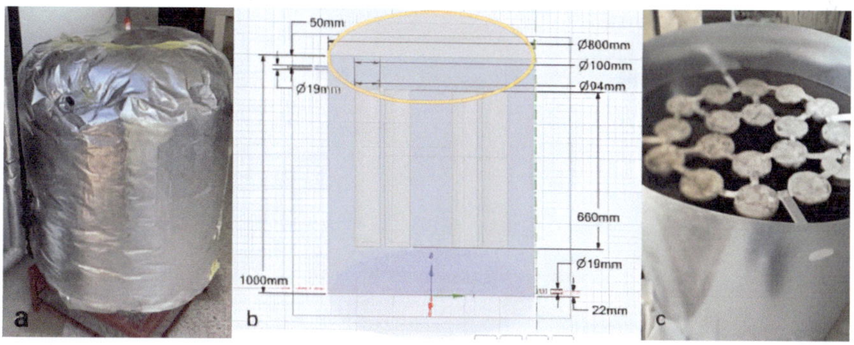

Fig. 30 PCM-water storage tank: insulation (**a**), its dimensions (**b**), 18-alumunium cylindrical PCM (**c**) [37]

Calculating the basis energy accumulation using Eq. (10), as Δh denotes latent heat capacity, the energy accumulation is influenced by both the heat capacity and the percentage volume of the PCM. Equation (11) defines x as the melting fraction

Fig. 31 Placements of thermocouples in PCM [37]

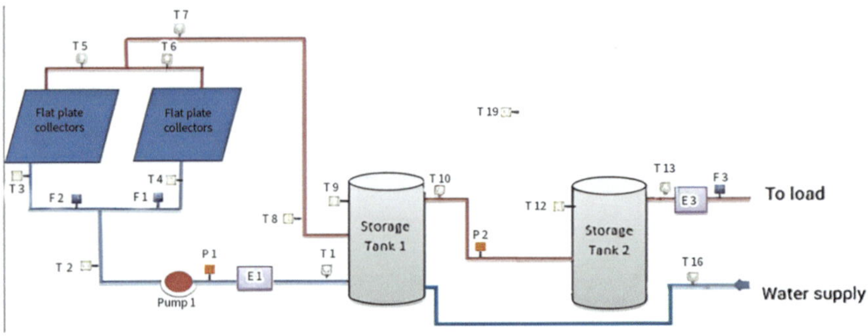

Fig. 32 SWHS experimental setup schematic diagram [37]

of the PCM. Moreover, T_s and T represent solidus temperature of PCM and mid-point temperature of PCM, respectively. "The Eq. (12) illustrates the energy accumulation or heat absorbed (stored) during the charging process."

$$Q_{st} = M\left[c_{ps}(T_m - T_i) + x\Delta h + c_{pl}(T_f - T_m)\right] \tag{10}$$

(11)

(12)

During the discharging process, the rejected heat (Q_{rej}) can be calculated as illustrated in Eq. (13). The energy recovery ratio (ERR) in Eq. (14) is defined as the recovered heat value after the discharging process ends. "Major factors leading to high ERR, for example to the minimum difference value between discharge and charging process, or the short duration of discharging process." In addition, if (t) is considered in both the charging and discharging processes, then the ERR is calculated as demonstrated in Eq. (15).

(13)

(14)

(15)

The potential of choosing PCM with high energy storage will provide high thermal capacity for the system, allowing it to maintain water at elevated temperatures for extended periods. Equations (16) and (17) illustrate the total accumulated energy (E_{tot}) without and with PCM, respectively. Furthermore, Eq. (18) represents the calculation of thermal storage gain (Q_{ths}). To calculate the system efficiency with and without PCM, Eqs. (19) and (20) represent the system efficiency with and without PCM, respectively.

(16)

(17)

(18)

(19)

(20)

The authors used Taylor series approach to estimate the propagation effect and also to identify uncertainty of the experimental work using Eq. (21). Table 8 provides the average uncertainties of the main parameters.

Table 8 The main parameters average uncertainty [37]

Parameter	Uncertainty (Ur)
Temperature	±0.5 °C
Mass flow rate	≤2%
Solar radiation	≤1.8%
PCM mass	±2%

(21)

2.4.3 Results

The cylindrical design of the aluminum significantly increased the surface area of the PCM, which in turn enhanced the rate of heat transfer. Nine cases were selected for the charging process of RT42, as shown in Table 9. The relationship between a shorter charging period and the amount of heat stored is direct. A shorter charging period indicates the system can store the maximum amount of energy in the least time, resulting in higher system efficiency. The shortest charging duration for RT42 was accounted, when the system operated under the highest inlet water temperature (T_{in}) of (60 °C) and with a mass flow rate of (0.1 kg/s), these values made the system achieve peak efficiency. However, the longest charging time was observed at the lowest inlet temperature of (50 °C) and a mass flow rate of (0.06 kg/s). When both parameters were set at intermediate values, the charging duration also fell within a moderate range.

Figure 33 illustrates charging process time of RT42 as inlet temperature and mass flow rate change. From the graph, it can be observed that as T_{in} increases, the process duration decreases. Similarly, a higher mass flow rate results in a shorter time. Nevertheless, the maximum time required for the nine cases that are mentioned in this experiment is clearly shown in Table 10. The highest T_{in} (33 °C) combined with the lowest mass flow rate (0.06 kg/s) results in the maximum discharging time. While the lowest T_{in} (24 °C) and the highest mass flow rate (0.1 kg/s) led to the minimum discharging time. Also, the medium conditions for discharging were observed at moderate values of Tin and mass flow rate.

Table 9 Duration of the charging period (hours) for each case [37]

Inlet temperature (°C)	Charging time at 0.06 kg/s	Charging time at 0.08 kg/s	Charging time at 0.10 kg/s
50	2.93	2.35	2.00
56	2.49	2.02	1.80
60	1.88	1.75	1.53

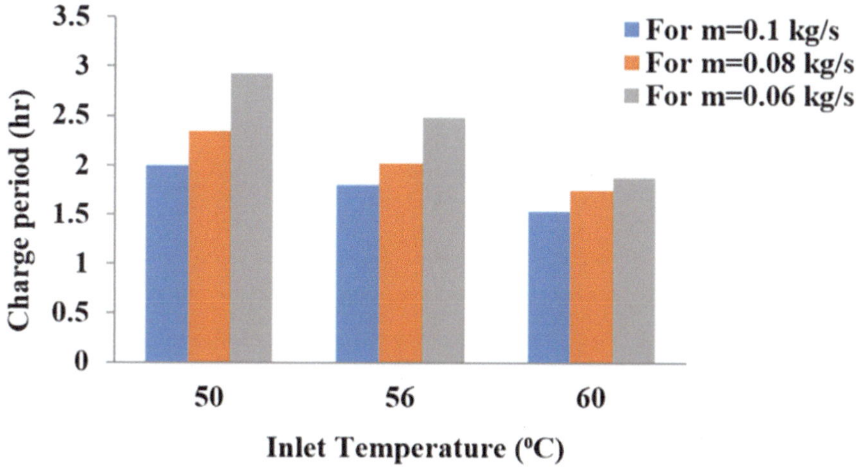

Fig. 33 Variation of temperature and water flow rate during the charging process of RT42 [37]

Table 10 Discharging period in hours for each case [37]

Inlet temperature (°C)	Charging time at 0.06 kg/s	Charging time at 0.08 kg/s	Charging time at 0.10 kg/s
33	3.2	3.05	2.48
28	2.68	2.15	2
24	2.12	1.98	1.65

Moreover, in Fig. 34 the variation in the discharge period of RT42 with changes in inlet temperature and mass flow rate are shown. An analysis of the graph reveals that the charging period shortens as the T_{in} decreases. On the other hand, the discharging time decreases as the mass flow rate increases since the colder fluid at higher flow rates enables more efficient heat transport. Figure 35 represents an example of the charging and discharging processes, showcasing PCM temperatures; during charging, the initial and final temperatures are 18 and 56 °C, while for discharging, they are 56 and 18 °C. The scenario includes a mass flow rate of 0.1 kg/s. with T_{in} set to 60 °C for charging and 33 °C for discharging.

Figure 36 illustrates the comparison of system efficiency with and without PCM. The findings indicate that integrating the PCM enhances energy storage, by increasing system efficiency under same boundary conditions across all cases by 12–25%, thus achieving improved thermal performance.

Figure 37 depicts for the nine cases the increase of efficiency in the system by integrating PCM in the system. It can be observed that the system efficiency significantly increased by 25% when utilizing the highest inlet temperature (60 °C) combined with a mass flow rate of 0.06 kg/s. The least improvement in system efficiency was measured at 12% and was observed under the lowest T_{in} and mass flow rate conditions, specifically at (50 °C) and (0.06 kg/s). In addition, increasing T_{in} will enhance system efficiency. However, increasing the mass flow rate at a

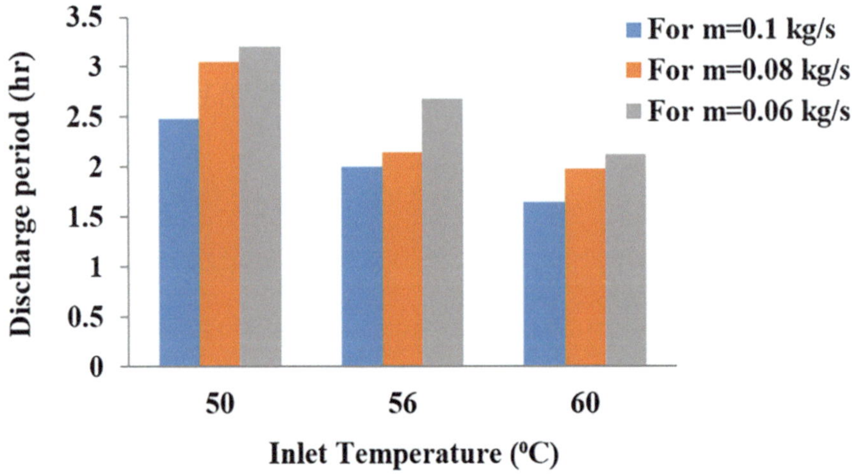

Fig. 34 Variation of temperature and water flow rate during the discharging process of RT42 [37]

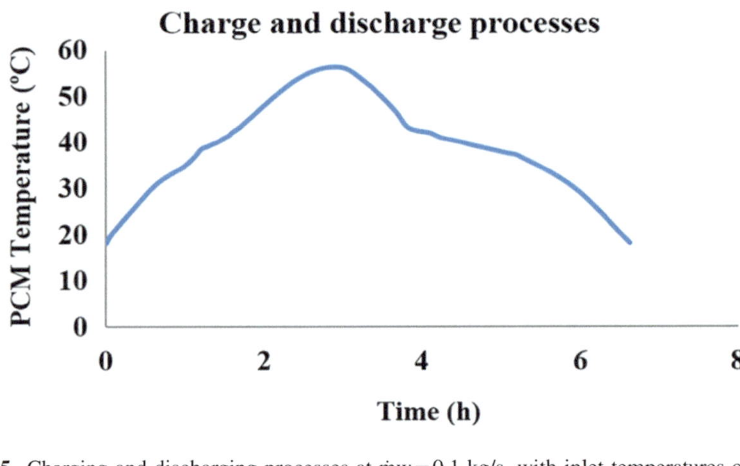

Fig. 35 Charging and discharging processes at ṁw = 0.1 kg/s, with inlet temperatures of 60 °C for charging and 33 °C for discharging [37]

constant T_{in} could either raise or lower system efficiency, depending on the energy storage in the storage tank (2).

The amount of energy stored inside the PCM depends on both its initial and final temperatures throughout the charging process. Greater energy storage is achieved when the PCM starts at a lower initial temperature and reaches a higher final temperature. Figure 38 shows the accumulated energy stored by RT42 across all nine test cases. Moreover, the conditions of T_{in} at (60 °C) combined with mass flow rate of (0.1 kg/s) resulted in the maximum energy storage reaching 2.12 kWh. Conversely, the system stored the lowest amount of energy (1.11 kWh)

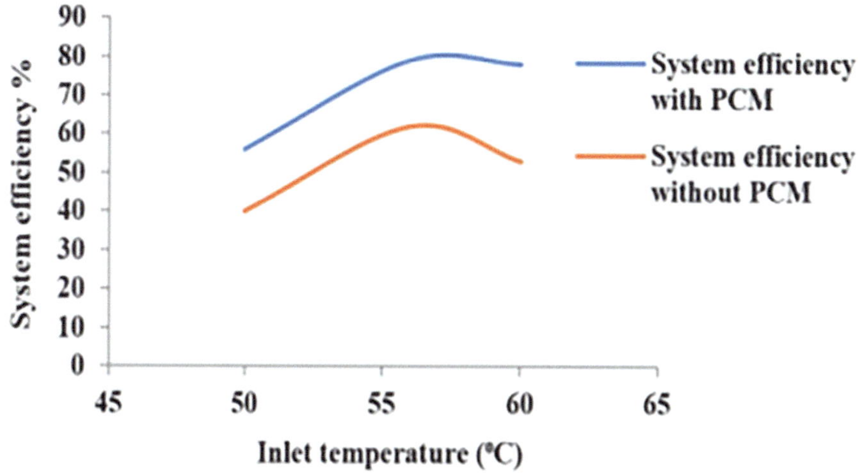

Fig. 36 Comparison between the system efficiency with and without PCM [37]

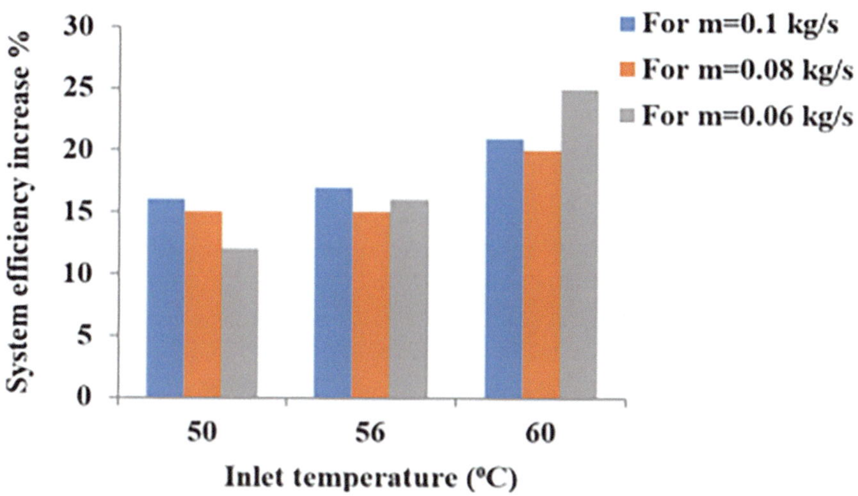

Fig. 37 By incorporating PCM, system efficiency improves across the nine cases [37]

when operated at T_{in} (50 °C) with mass flow rate of (0.06 kg/s). Nevertheless, Fig. 39 presents the total energy accumulated by the system across different days of the year. The comparison indicates that using the same boundary conditions, the PCM-based system consistently stored more energy than the system without PCM; this improvement is due to the additional sensible heat stored within the water-PCM tank. The data indicates that the maximum energy stored with PCM reaches 10.9 kWh on December 15 and its minimum of 7.1 kWh on November 29. These variations are mainly due to fluctuations in solar radiation between days, with a lesser influence from the operating boundary conditions.

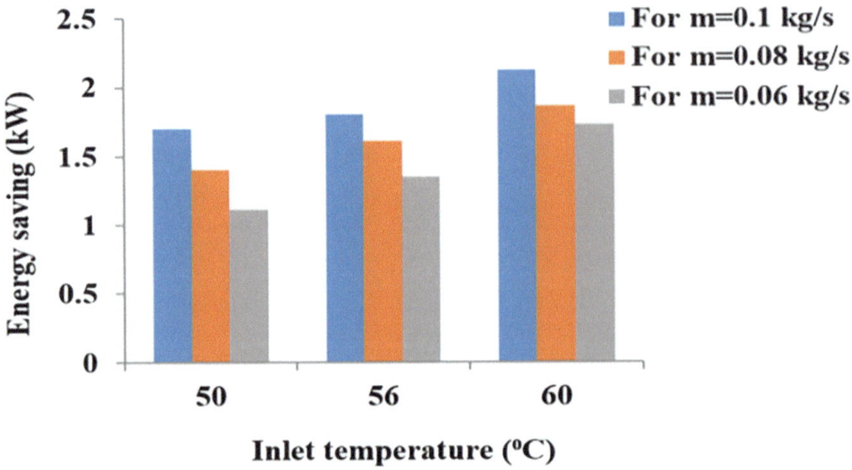

Fig. 38 Energy saving for RT42 for different T_{in} and $\dot{m}_w$ [37]

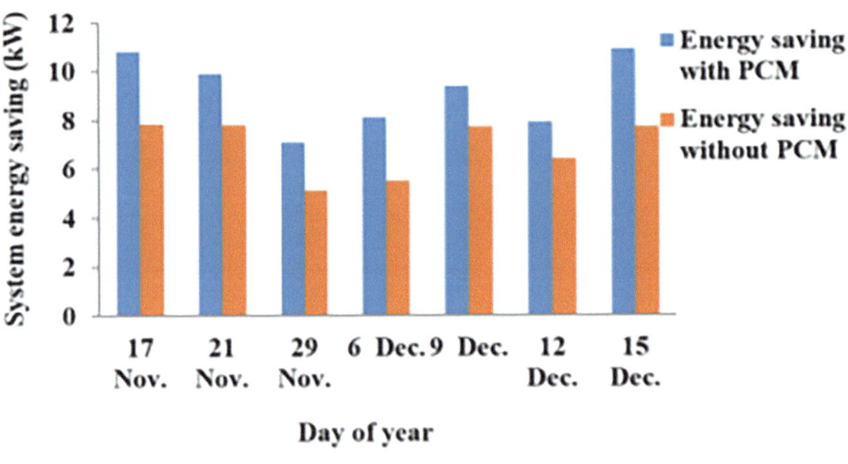

Fig. 39 The total energy savings of the system, with and without paraffin wax RT42, across various days in November and December 2021 [37]

Thermal storage gain (Q_{ths}), defined as the ratio of the total energy stored with PCM to the total energy stored without PCM, is a parameter to evaluate the effectiveness of PCM integration in SWHS. The thermal enhancement capacity of RT42 material increases when Q_{ths} values rise, as shown in Fig. 40. Moreover, the system efficiency reached its maximum level and Q_{ths} values obtained their highest values when operating conditions included under T_{in} at (60 °C) combined with mass flow rate of (0.06 kg/s). The highest recorded value of Q_{ths} was 1.48, while the lowest value 1.2, and was observed at T_{in} at (56 °C) and a flow rate of 0.1

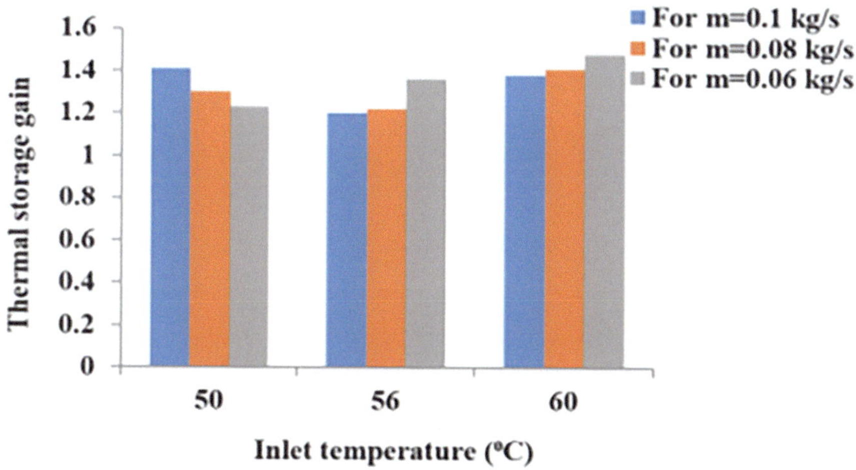

Fig. 40 Changes in Q_{ths} due to variations in inlet temperature and mass flowrate [37]

kg/s. In addition, the system demonstrates a high ERR due to the absence of hysteresis and sub-cooling errors, as shown in Fig. 41. Furthermore, the energy savings in kWh can be calculated by multiplying the energy savings in kW by the discharge duration in hours. Figure 42 clearly shows the maximum stored that is 4.75 kWh and was obtained at the lowest T_{in} at (50 °C) combined with the highest mass flowrate (0.1 kg/s). On the other hand, the minimum stored energy observed is 3.2 kWh, was at the highest T_{in} at (60 °C) with the same mass flow rate of (0.1 kg/s). Also, the system achieved in storing the highest energy of 30.2 kWh

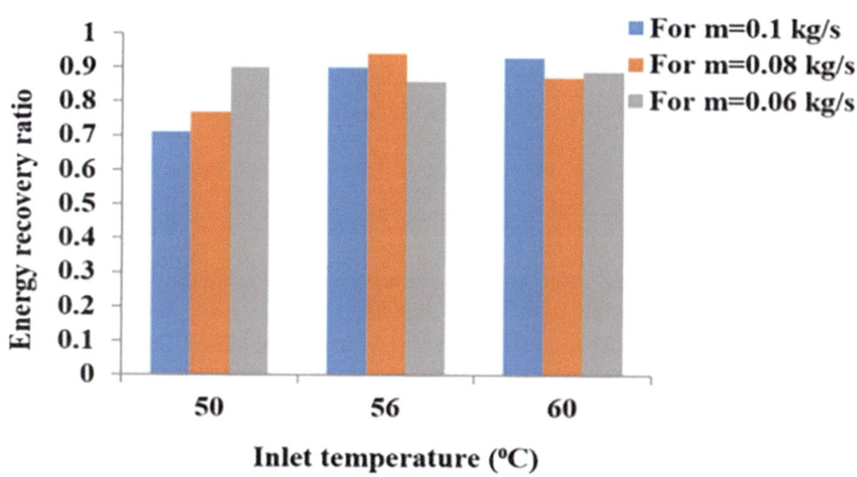

Fig. 41 Changes in ERR resulting from variations in inlet temperature and mass flow rate [37]

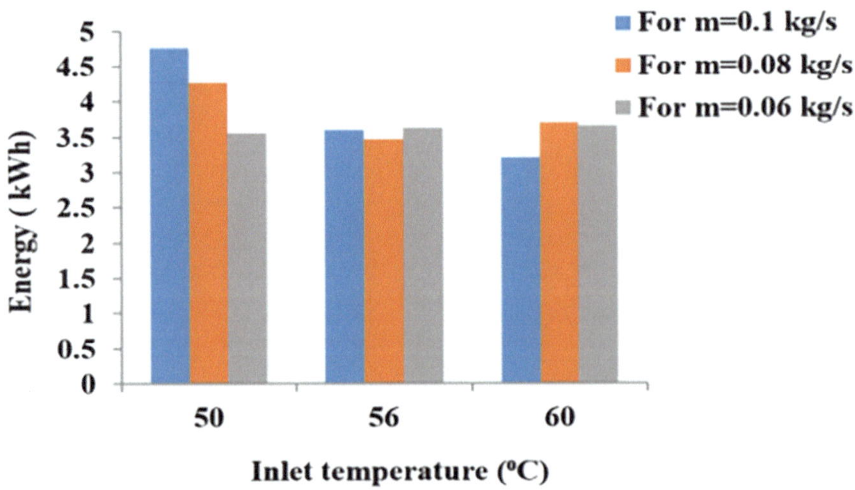

Fig. 42 Changes in energy savings with RT42 due to variations in inlet temperature and mass flow rate [37]

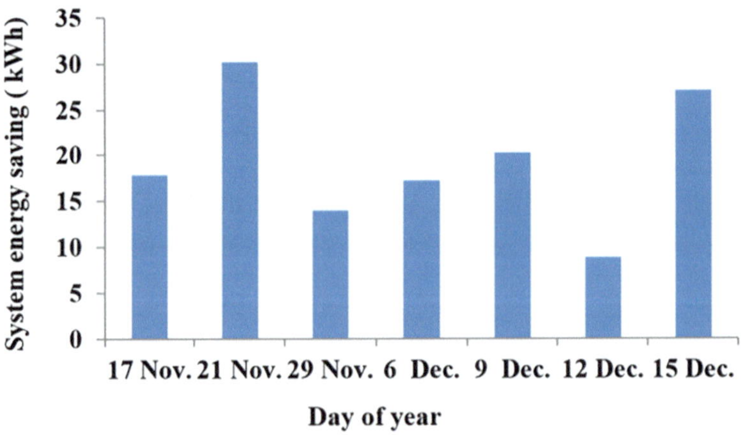

Fig. 43 System energy savings in kWh measured on different days throughout the year [37]

on November 21, while the lowest energy stored was 8.8 kWh on December 12. Additionally, the system gathered on average 19.3 kWh of energy storage daily throughout the study period, as shown in Fig. 43.

The experimental investigation confirmed that PCM-based TES represents a dependable solution to reduce fossil fuel usage while increasing SWHS thermal efficiency. The aluminum cylindrical PCM proved to be the best module for integrating PCM into SWHS applications. In addition, excess heat that the PCM absorbed during daytime can be used to heat for domestic applications during nighttime. The authors recommend future studies focus on investigating other PCMs performances in SWHS applications. Also, estimate implementation costs to evaluate potential savings.

3 Summary and Comparison of Experimental Case Studies

Three experimental case studies were selected to investigate the performance of PCM integration on SWHS. The first study is about a compact solar water heater that contains paraffin wax PCM layer in the system. The second study discussed the thermosyphon system, as PCM is directly integrated to the storage tank to evaluate the performance of the system during the discharging process. The third study evaluates the thermal performance of using aluminum cylinders filled with PCM type RT42, evaluating its behavior when varying the inlet temperatures and mass flow rates. Table 11 presents a comparison table of the three case studies.

3.1 Experimental Design and Scenario Proposal

The experimental study aims to design and implement an experimental small-scale energy storage simulation system for SWHS in GUtech ECO-HOUSE, with a tank which incorporates PCM technology. The system consists of two distinct water containers where (Tank 1) serves water heating and (Tank 2) stores thermal energy through paraffin wax cylinders. This experiment will demonstrate that PCM improves the thermal storage capacity and overall efficiency compared to conventional water storage systems. Table 12 presents the thermal properties of paraffin wax which will be used in this experiment.

The technical specifications of paraffin wax including melting point, oil content, penetration at 25 °C, and color are reported in the manufacturer datasheet. However, key thermal properties such as latent heat of fusion, specific heat capacity for both liquid and solid phase, and density are missing from the datasheet provided by the manufacturer. The values were obtained from peer-reviewed literature, since testing them and determining their thermal characteristics require expensive equipment and technical expertise. The values were estimated from the scientific paper of Ukrainczyk et al. [39] who performed a thermophysical analysis of different commercial paraffin waxes. The values derived from literature were scientifically proven approximations that suit experimental needs of this experiment.

Table 11 Comparison table of the three case studies

Aspect	Case study 1 (compact SWHS)	Case Study 2 (thermosyphon system)	Case Study 3 (thermal enhancement with PCM)
System type	Compact SWHS (tilted flat plate)	Thermosyphon SWHS	Forced circulation SWHS (pumped system)
PCM used	Paraffin wax (58 °C)	Paraffin wax (56 °C)	RT42 paraffin wax (42–43 °C)
PCM form	Layer behind water tank	Cylindrical capsules	Aluminum-encapsulated cylinders
Objective	Increase midday efficiency, reduce midday decline in system	Enhance discharging, maintain temperature	Improve charging/discharging, analyze energy savings
Testing method	Ten thermocouples (five for water temp, five for PCM temp)	Six thermocouples	Twelve Thermocouples with varying temperatures and flow rates
T_{in}	Around 60 °C	Around 65 °C	50, 56, and 60 °C
T_{out}	Drops to around 45 °C by the end of day	Drops down to 45–50 °C during discharging	Varies depending on case; around 37–48 °C
Aspect	Case Study 1 (compact SWHS)	Case Study 2 (thermosyphon System)	Case Study 3 (thermal enhancement with PCM)
Key results	PCM delayed efficiency drop, improved efficiency	PCM delayed cooling, better temperature stability	Higher thermal storage, better energy recovery ratio
Limitations	Performance highly dependent on PCM melting and solidification timing, heat losses through glass and air gap	Low thermal conductivity of PCM slows heat transfer	Experimental setup requires precise control
Location	Iraq	India	Kurdistan region of Iraq

Table 12 Thermal properties of paraffin wax used in the experimental study [38, 39]

Property	Test method	Value
Melting point	ASTM D127	58–62 °C
Oil content	ASTM D721	3–5%
Penetration at 25 °C (hardness)	ASTM D1321	1.5–3.5 mm
Color	Visual	White to pale yellow
Latent heat of fusion	Estimated	206 kJ/kg
Specific heat (liquid)	Estimated	2.49 kJ/kg °C
Specific heat (solid)	Estimated	2.1 kJ/kg °C
Density (solid)	Estimated	893 kg/m^3

3.1.1 Experimental Materials

In this experiment, the materials involved in the study include the two tanks made of either stainless steel or aluminum, braided flexible hoses for inlets and outlets, temperature sensors, 3 kg of transformer paraffin wax with a melting point of 58–62 °C, encapsulated cylinders, an immersion rod water heater to heat water, aluminum bubble wrap for insulation purposes, a small pump that can handle hot temperatures, manual ball valves to trap water, and a mechanical flow meter.

3.1.2 Experimental Methodology

The experiment involves two vertical cylindrical tanks as the main components, hot water tank (Tank 1) and PCM tank (Tank 2). Tank 1 dimensions are 40 cm inner diameter and 55 cm height. The top of the tank will have a removable lid, allowing water to fill and permitting installation of immersion rod heater and temperature sensors. Also, the lid will be drilled for wiring purposes related to the immersion rod heater and temperature sensors, the holes will be slightly larger than the diameter of the wire, acting as air gaps preventing buildup of water vapor and water expansion during the heating process. Furthermore, the outlet hole of Tank 1 is approximately 5–10 cm above the tank bottom. Tank 2 is similar to Tank 1, but the difference is that its inner diameter is 30 cm. Additionally, the inlet and outlet holes of Tank 2 are near the top of the tank and 5–10 cm above the tank floor, respectively. The encapsulated paraffin wax will be filled in it, each with dimension of 5 cm diameter and 10 cm height. The manual ball valves are installed at the outlet of Tank 1, and inlet and outlet of Tank 2. A small pump is used to transfer hot water from Tank 1 to Tank 2 via stainless steel braided flexible hoses, passing by a mechanical flow meter to monitor and control the flowrate of the hot water. Moreover, both Tank 2 and pipes are insulated using aluminum bubble wrap to minimize heat losses. The temperature sensors are installed in Tank 1, inlet and outlet of Tank 2, and inside Tank 2. Figure 44 illustrates the schematic diagram of the experimental setup for the PCM-based TES system. TR represents temperature recorder as it functions as a temperature sensor, and FC stands for flow controller which is flow meter in this scenario.

3.1.3 Experimental Scenario

The key objective of this experiment is to develop and test an operational PCM-based energy storage system for SWHS, while validating theoretical benefits of LHS through real experimental data. Focusing on enhancing the discharging period of a thermal storage tank by incorporating PCM. The experiment consists of two main scenarios, charging and discharging processes. At the beginning of the experiment, Tank 1 is filled with water and heated using an immersion rod heater until the water reaches 70 °C, after that the manual ball valve at its outlet is set to

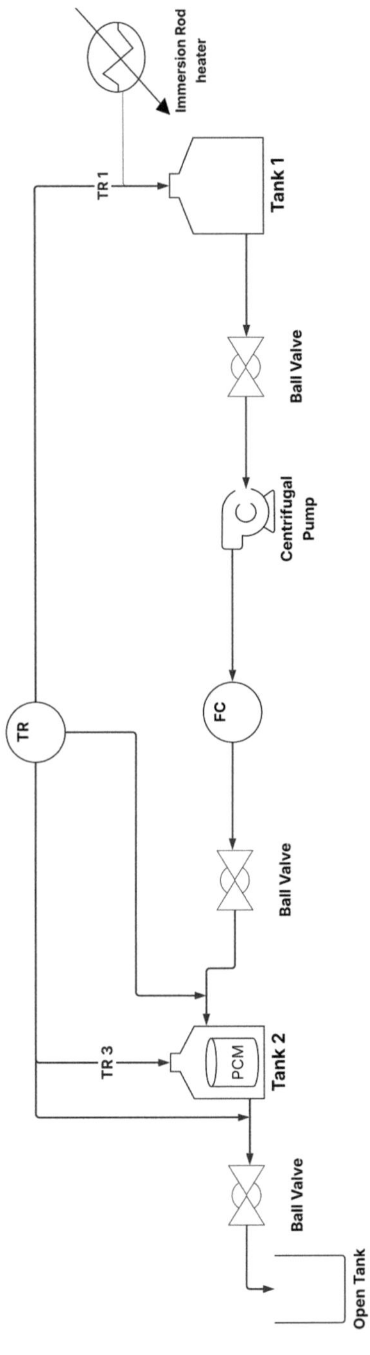

Fig. 44 Schematic diagram of the experimental setup, designed by using Lucidchart

allow the hot water to flow into the pipe, subsequently the pump increases the flow pressure, pushing the hot water through a mechanical flowmeter into Tank 2. Once Tank 2 reached the desired temperature, the charging process ends. The discharging process starts when Tank 2 reaches 70 °C and ends as its temperature drops to reach ambient temperature. Furthermore, temperature readings will be recorded every 5 min throughout the experiment, in both charging and discharging period. Flowrate of the water will be monitored using a mechanical flowmeter during the charging process. In addition, the collected data will be transferred to software for analysis and graph plotting. The resulting graphs will represent the effect of integrating PCM in TES tank during the discharging period. Also, a graph shows the discharging performance of the system with PCM versus without PCM.

4 Results and Discussion

4.1 Overview

This chapter will discuss the findings from the cases studied in Chapter "Biofuel Production from Corn Stover by Pyrolysis". A compact flat-plate solar water heater with an integrated paraffin wax storage (Case 1), a thermosyphon flat-plate collector with a top mounted PCM tank (Case 2), and a thermal enhancement of the solar collector using RT42 PCM (Case 3). From the discussed studies, the planned experimental modeling of PCM integration into SWHS at GUtech ECO-HOUSE in Oman will validate and extend these findings by testing them under local climatic conditions. Finally, this chapter concludes by summarizing critical insights drawn from all case studies, highlighting the potential of PCM to enhance system performance, improve hot water accessibility, and prolong TES solutions in SWHSs of Oman and similar regions.

In Case 1, the compact system reached a high temperature of 82 °C by 4 PM. Also, the integrated PCM in compartment remained significantly cooler around 47 °C at the same time. This trend is illustrated in Figs. 11 and 12, showing that the water reaches higher temperatures than the PCM. According to the authors, paraffin wax never fully melted during the test period, since its temperature stayed below its 55–60 °C melting point range. The authors identified the need for a reduced amount of PCM or a lower melting point PCM as potential solutions to address this limitation. Regardless of this limitation, the compact system experienced a peak of 82 °C water temperature and attained maximum efficiency of about 74% at 11:30 AM. This efficiency is considered significant for a combined storage unit, but it suggests that the collected heat had some unutilized portion in the PCM. Next, Case 2 examines the thermosyphon system with a cylindrical PCM module in storage tank, which showed effectiveness of PCM in terms of LHS, as it provided a longer discharge period and improved thermal stratification. Figure 22 showed that the useful hot water remained for about 25.5 h when incorporated with PCM, retaining heat for a longer duration compared to the system without PCM, which lasted only 18.5 h, indicating that the addition of PCM extended the duration by 7 h. As this extended discharge is relatable to PCM, the process of releasing the stored energy by PCM occurs after sunset, leading to maintaining the water temperature. The experiment demonstrated that the tank temperature decreased from 65 to 45 °C more slowly with PCM than without, confirming that the PCM provides a thermal backup. Moreover, the incorporation of PCM showed more thermal stratification number than the one without, meaning that PCM keeps the hottest water available at the outlet for a longer time, as shown in Fig. 23. During the discharge process, the stratification number in the storage tank incorporated with PCM was about 10% higher than in the no PCM storage tank, indicating that the water stayed hotter at the top and cooler at the bottom due to the PCM absorbing heat at the top and releasing it gradually. In terms of efficiency, Case 2 also outperformed the compact Case 1 system. The thermosyphon system's charging efficiency reached a peak of about 82.4% around 2 PM, with PCM storage. Figure 24 highlights that the PCM initially decreased charging efficiency since some incoming heat was used to melt the paraffin wax, but once the PCM melted completely, the system with PCM reached higher efficiency than the system without PCM during same time. As this is attributed to PCM absorbing excess heat and reducing losses, it leads to efficient storage of solar energy. At the end of the charging process, the thermal energy stored in the 2.5 kg paraffin wax has increased the tank storage thermal efficiency by roughly 10–16% when compared to the one without PCM. In other words, PCM started with a 10% efficiency advantage during discharging, which increased to 16% at the end of the process, reflecting the continued release of latent heat as the water cooled. This supports the conclusion that integrating PCM in SWHS improves overall system performance by maintaining a high water temperature for longer periods of time. The thermosyphon Case 2 outperformed the compact system Case 1, in terms of achieving a high peak efficiency. In addition, the maximum output temperature of Case 2 is around 60-70C°, which is within the range of typical flat plate heaters. Figure 22 illustrates the discharging temperature in the presence of PCM compared to its absence. Moreover, Case 3 involved thermal enhancements of a conventional SWHS through PCM integration, studying in depth the impact of PCM on system efficiency and thermal storage. The study investigated the performance of RT42 paraffin wax PCM equipped in SWHS, focused on variations in the inlet temperatures (50 °C, 56°, 60°) and flowrates (0.06, 0.08, 0.1 kg/s). From Case 3, the authors showed the relationship between increased inlet temperatures and improved system efficiency. The incorporation of PCM increased the system's efficiency by approximately 25% when operated at the highest inlet temperature of 60 °C and with the lowest flowrate of 0.06 kg/s. However, the system only increased approximately 12% when operated at the lowest inlet temperature and the lowest flowrate of 0.06 kg/s. The operating conditions significantly affect PCM performance during the charging process. RT42 achieved its highest TES of 2.12 kWh during operation at the highest inlet temperature of 60 °C combined with the lowest flowrate of 0.06 kg/s. The TES was reduced by half when operated at the lowest inlet temperature of 50 °C and the lowest flowrate of 0.06 kg/s, resulting in 1.11 kWh. The successful outcome of this case study demonstrates that PCM performance reaches its peak when specific thermal and operational parameters are carefully selected and managed effectively. The incorporation of PCM in SWHS proves to enhance system performance, providing tangible benefits in both thermal storage duration and efficiency improvements. Each case demonstrated essential aspects of PCM optimization. In summary, Case 1 explained the critical selection of PCM melting points with achievable collector temperatures to fully exploit LHS. Case 2 demonstrated the strategic placement of PCM to allow for better thermal stratification performance which resulted in extended hot water availability and increased efficiency. Case 3 demonstrated crucial factors of operational parameters such as temperature and flowrates to maximize PCM charging efficiency and subsequent energy recovery, showcasing how specific adjustments can drastically impact PCM performance.

5 Conclusion

In conclusion, three case studies discussed paraffin wax phase change material (PCM) enhancements in overall efficiency and thermal energy storage (TES) capacity for the solar water heating system (SWHS). A compact solar water heater (Case 1) demonstrated the importance of choosing PCM melting points that match solar collector temperatures to fully utilize latent heat storage (LHS). The system showed high efficiency, although the PCM did not melt completely, showing the importance of PCM sizing. The thermosyphon system (Case 2) produced successful outcomes that simultaneously improved thermal stratification and extended hot water availability duration. The experimental results showed that the system achieved 10–16% higher efficiency and extended the discharge period approximately 7 h. The thermal enhancement experiment (Case 3) demonstrated the effectiveness of operational parameters particularly inlet temperature and water flowrate, which influenced thermal storage performance, efficiency, and heat recovery, showing the need for strategic operational optimization for maximum PCM utilization. Although budget constraints prevented the experimental setup of energy storage system from being conducted, the system design and methodology were carefully developed, laying a strong foundation for potential future testing in the GUtech ECO-HOUSE under local climatic conditions. Advancing the findings of these studies, the planned energy storage design of PCM integrated system of SWHS at GUtech ECO-HOUSE in Oman aims to validate and expand these insights under local climatic conditions. This model will consider the environmental and operational aspects of ECO-HOUSE to strategically integrate PCM. Moreover, integrating PCM in SWHS significantly boosts thermal energy storage and extends hot water availability compared to the conventional SWHSs. Optimizing PCM advantages requires detailed consideration of PCM characteristics alongside suitable system design and strategic placement and the optimization of operational conditions.

References

1. International Renewable Energy Agency (IRENA) (2022) World energy transitions outlook 2022. IRENA. https://www.irena.org/Digital-Report/World-Energy-Transitions-Outlook-2022. Accessed: 24 Feb 2025
2. Maka AOM, Alabid JM (2022) Solar energy technology and its roles in sustainable development. Clean Energy 6(3):476–483. https://doi.org/10.1093/ce/zkac023
3. Mofijur M, et al (2019) Phase change materials (PCM) for solar energy usages and storage: an overview. MDPI AG. https://doi.org/10.3390/en12163167
4. Lanahan M, Tabares-Velasco PC (2017) Seasonal thermal-energy storage: a critical review on BTES systems, modeling, and system design for higher system efficiency. MDPI AG. https://doi.org/10.3390/en10060743
5. Silakhori M, Fauzi H, Mahmoudian MR, Metselaar HSC, Mahlia TMI, Khanlou HM. Preparation and thermal properties of form-stable phase change materials composed of palmitic acid/polypyrrole/graphene nanoplatelets. Energy Build 99:189–195. https://doi.org/10.1016/j.enbuild.2015.04.042
6. International Trade Administration. Oman—Renewable energy. U.S. Department of Commerce. https://www.trade.gov/country-commercial-guides/oman-renewable-energy. Accessed 2025
7. Oman Ministry of Energy and Minerals. Oman announces 2050 Net Zero commitment and unveils ambitious green hydrogen strategy. Ministry of Energy and Minerals, Oman.
8. Akhlaghi MM, Alavijeh AS, Hosseinalizadeh R, Nasri S, Ghazinoory S (2025) Systematic failures in the development of photovoltaic systems: The case study of Iran's solar energy. Energy Strategy Rev 57. https://doi.org/10.1016/j.esr.2024.101637
9. Dimitriadis CN, Tsimopoulos EG, Georgiadis MC (2023) Optimization-based economic analysis of energy storage technologies in a coupled electricity and natural gas market. J Energy Storage 58. https://doi.org/10.1016/j.est.2022.106332
10. Al-Awadhi T, Al Ramimi W, Al Jabri K, Abulibdeh A (2025) Solar farms suitability analysis using GIS-based analytic hierarchy process (AHP) multi-criteria: a case study of Al Duqm—Oman. Renew Energy 241. https://doi.org/10.1016/j.renene.2024.122295
11. Eze F, Egbo M, Anuta UJ, Ntiriwaa O-BR, Ogola J, Mwabora J (2024) A review on solar water heating technology: Impacts of parameters and techno-economic studies. Bull Natl Res Cent 48(1). https://doi.org/10.1186/s42269-024-01187-1
12. Rawat P, Mehrotra S, Debbarma M, Sudhakar K (2014) Performance of a solar panel with water immersion cooling technique. www.ijset.net
13. Rawat P, Kishor Purohit J (2022) Analysis of solar water heater with aerogel-based thermal energy storage medium. www.jetir.org
14. Patel S (2023) Solar water heater systems: a comprehensive review of design and performance. www.ijcspub.org
15. Hamzat AK, et al (2025) Phase change materials in solar energy storage: Recent progress, environmental impact, challenges, and perspectives. Elsevier. https://doi.org/10.1016/j.est.2025.115762
16. Oman Eco House (2025) Exterior view of Oman Eco House. https://www.flickr.com/photos/omanecohouse/14914215383/. Accessed 08 May 2025
17. Kayode Olajiga O, Chigozie Ani E, Queen Sikhakane Z, Michael Olatunde T, Author C (2024) Assessing the potential of energy storage solutions for grid efficiency: a review. Eng Sci Technol J 5(3):1112–1124. https://doi.org/10.51594/estj/v5i3.974
18. Arbabzadeh M, Sioshansi R, Johnson JX, Keoleian GA (2019) The role of energy storage in deep decarbonization of electricity production. Nat Commun 10(1). https://doi.org/10.1038/s41467-019-11161-5
19. Bartolini A, Carducci F, Muñoz CB, Comodi G (2020) Energy storage and multi energy systems in local energy communities with high renewable energy penetration. Renew Energy 159:595–609. https://doi.org/10.1016/j.renene.2020.05.131

20. Nanda IR, Pambudi NA, Aziz M (2023) Review on the progress of solar water heaters and their future perspectives. Wiley. https://doi.org/10.1002/ente.202300191
21. Modi N, Wang X, Negnevitsky M (2023) Solar hot water systems using latent heat thermal energy storage: perspectives and challenges. Energies (Basel) 16(4). https://doi.org/10.3390/en16041969
22. Fadzlin WA, Hasanuzzaman M, Rahim NA, Amin N, Said Z (2022) Global challenges of current building-integrated solar water heating technologies and its prospects: a comprehensive review. MDPI. https://doi.org/10.3390/en15145125.
23. Morkane N, Sakhare S, Joshi L, Ingale B (2024) Enhancing solar thermal efficiency: PCM integration for compact energy storage systems. https://irjaeh.com
24. Mourad A, Aissa A, Said Z, Younis O, Iqbal M, Alazzam A (2022) Recent advances on the applications of phase change materials for solar collectors, practical limitations, and challenges: a critical review. Elsevier Ltd. https://doi.org/10.1016/j.est.2022.104186.
25. Herez A, et al (2023) Solar water heating: comprehensive review, critical analysis and case study. Int J Thermofluids 20. https://doi.org/10.1016/j.ijft.2023.100503
26. Pour SS, Momeni M, Khayyaminejad A, Fartaj A (2024) Integration of a solar water heating system with encapsulated phase change material storage for enhanced energy efficiency in domestic hot water systems. In: 2024 ASHRAE winter conference, ASHRAE. https://doi.org/10.63044/w24pou114
27. Mellouli S, Alqahtani T, Algarni S (2022) Parametric analysis of a solar water heater integrated with PCM for load shifting. Energies (Basel) 15(22). https://doi.org/10.3390/en15228741
28. Rao RV, Pavel T, Ravipudi JL (2024) Selection of phase change materials for energy storage applications using BHARAT decision-making methodology. Jordan J Mech Ind Eng 18(4):691–704. https://doi.org/10.59038/jjmie/180405
29. Caramitu AR, Lungu MV (2024) An overview of thermal energy storage (TES) materials and systems for storage applications. EEA Electrotehnica Electronica Automatica (EEA) 72(4):28–42. https://doi.org/10.46904/eea.24.72.4.1108003
30. Mahmood MAS, Viktorovich RY, Viktorovich SA (2023) Application of thermal energy accumulators based on paraffin phase change materials in convective-vacuum impulsive drying units: a brief-focused overview of characteristics and thermal conductivity enhancement techniques. J Renew Energy Environ 10(4):99–106. https://doi.org/10.30501/jree.2023.370842.1501
31. Albalushi Z, Chala GT (2023) Performance analysis of solar water heater-a case study in Al Seeb, Oman
32. Ihnayyish IL, Ahmed AQ, Mohammad ATh, Al-Syyab AKS () Numerical study to investigate the performance of U-shaped flat plate solar collector using phase change materials (PCMs). J Tech 5(2):74–80. https://doi.org/10.51173/jt.v5i2.1302
33. Jegadheeswaran S, Sundaramahalingam A (2023) Synresearch and characterization of copper nanoparticles-embedded paraffin wax for solar energy storage. J Therm Anal Calorim 148(9):3417–3425. https://doi.org/10.1007/s10973-023-11952-5
34. Al-Abdali AM, Ammari H (2022) Thermal energy storage using phase-change material in evacuated-tubes solar collector. AIMS Energy 10(3):486–505. https://doi.org/10.3934/energy.2022024
35. Abdulrahman Al-Tabbakh A, Fadhil Noori A, Salman Fahad R (2019) Performance analysis of a compact solar water heater with paraffin wax storage unit تحليل اداء سخان ماء شمسي مدمج مزود بوحدة خزن مه شمع البارافيه. J Eng Sustain Dev 23(01). https://doi.org/10.31272/jeasd.2019.1.10
36. Murali G, Mayilsamy K, Arjunan TV (2015) An experimental study of PCM-incorporated thermosyphon solar water heating system. Int J Green Energy 12(9):978–986. https://doi.org/10.1080/15435075.2014.888663
37. Darwesh BD, Hamakhan IA, Yaqob BN. Thermal enhancement of solar energy storage using phase change materials. Int J Heat Technol 40(3): 758–766. https://doi.org/10.18280/ijht.400314

38. RX Marine International. Paraffin wax for transformer insulation: technical bulletin RXSOL-37-6605-025
39. Ukrainczyk N, Kurajica S, Šipušiae J (2009) Thermophysical comparison of five commercial paraffin waxes as latent heat storage materials

MIX
Papier aus verantwortungsvollen Quellen
Paper from responsible sources
FSC® C105338

If you have any concerns about our products,
you can contact us on
ProductSafety@springernature.com

In case Publisher is established outside the EU,
the EU authorized representative is:
Springer Nature Customer Service Center GmbH
Europaplatz 3, 69115 Heidelberg, Germany

Printed by Libri Plureos GmbH
in Hamburg, Germany